Lecture Notes in Artificial Intelligence   11816

Subseries of Lecture Notes in Computer Science

Carlos Martín-Vide · Matthew Purver ·
Senja Pollak (Eds.)

Statistical Language and Speech Processing

7th International Conference, SLSP 2019
Ljubljana, Slovenia, October 14–16, 2019
Proceedings

 Springer

Editors
Carlos Martín-Vide 🆔
Rovira i Virgili University
Tarragona, Spain

Matthew Purver 🆔
Queen Mary University of London
London, UK

Senja Pollak 🆔
Jožef Stefan Institute
Ljubljana, Slovenia

ISSN 0302-9743 ISSN 1611-3349 (electronic)
Lecture Notes in Artificial Intelligence
ISBN 978-3-030-31371-5 ISBN 978-3-030-31372-2 (eBook)
https://doi.org/10.1007/978-3-030-31372-2

LNCS Sublibrary: SL7 – Artificial Intelligence

© Springer Nature Switzerland AG 2019
This work is subject to copyright. All rights are reserved by the Publisher, whether the whole or part of the material is concerned, specifically the rights of translation, reprinting, reuse of illustrations, recitation, broadcasting, reproduction on microfilms or in any other physical way, and transmission or information storage and retrieval, electronic adaptation, computer software, or by similar or dissimilar methodology now known or hereafter developed.
The use of general descriptive names, registered names, trademarks, service marks, etc. in this publication does not imply, even in the absence of a specific statement, that such names are exempt from the relevant protective laws and regulations and therefore free for general use.
The publisher, the authors and the editors are safe to assume that the advice and information in this book are believed to be true and accurate at the date of publication. Neither the publisher nor the authors or the editors give a warranty, expressed or implied, with respect to the material contained herein or for any errors or omissions that may have been made. The publisher remains neutral with regard to jurisdictional claims in published maps and institutional affiliations.

This Springer imprint is published by the registered company Springer Nature Switzerland AG
The registered company address is: Gewerbestrasse 11, 6330 Cham, Switzerland

Preface

These proceedings contain the papers presented at the 7th International Conference on Statistical Language and Speech Processing (SLSP 2019), held at the Jožef Stefan Institute, Ljubljana, Slovenia, during October 14–16, 2019.

The SLSP conference series deals with topics of theoretical and applied interest concerning the employment of statistical models (including machine learning) within language and speech processing. Specific areas covered in this and previous installments include, but are not limited to:

- Anaphora and coreference resolution
- Authorship identification, plagiarism, and spam filtering
- Corpora and resources for speech and language
- Data mining, term extraction, and semantic web
- Dialogue systems and spoken language understanding
- Information retrieval and information extraction
- Knowledge representation and ontologies
- Lexicons and dictionaries
- Machine translation and computer-aided translation
- Multimodal technologies
- Natural language understanding and generation
- Neural representation of speech and language
- Opinion mining and sentiment analysis
- Part-of-speech tagging, parsing, and semantic role labeling
- Question-answering systems for speech and text
- Speaker identification and verification
- Speech recognition, transcription, and synthesis
- Spelling correction
- Text categorization and summarization
- User modeling

SLSP 2019 received 48 submissions. Each submission was reviewed by three Program Committee (PC) members, with some external experts consulted. After a thorough and vivid discussion phase, the PC decided to accept 25 papers (which represents an acceptance rate of about 52%). The conference program also included three invited talks (one of which is included as a paper in this volume) and a number of poster presentations of work in progress.

The excellent facilities provided by the EasyChair conference management system allowed us to deal with the submissions successfully and handle the preparation of these proceedings in time.

We would like to thank all invited speakers and authors for their contributions, the PC and the external reviewers for their diligent cooperation, and Springer for its very professional publishing work.

July 2019

Carlos Martín-Vide
Senja Pollak
Matthew Purver

Organization

Program Committee

Jon Barker	University of Sheffield, UK
Roberto Basili	University of Rome Tor Vergata, Italy
Pushpak Bhattacharyya	Indian Institute of Technology Bombay, India
Fethi Bougares	University of Le Mans, France
Philipp Cimiano	Bielefeld University, Germany
Carol Espy-Wilson	University of Maryland, USA
Nikos Fakotakis	University of Patras, Greece
Robert Gaizauskas	University of Sheffield, UK
Julio Gonzalo	National Distance Education University, Spain
Reinhold Häb-Umbach	Paderborn University, Germany
John Hershey	Google, USA
Julia Hirschberg	Columbia University, USA
Jing Huang	JD AI Research, USA
Mei-Yuh Hwang	Mobvoi AI Lab, USA
Nancy Ide	Vassar College, USA
Martin Karafiát	Brno University of Technology, Czech Republic
Vangelis Karkaletsis	National Center for Scientific Research "Demokritos", Greece
Tomi Kinnunen	University of Eastern Finland, Finland
Sandra Kübler	Indiana University, USA
Carlos Martín-Vide (Chair)	Rovira i Virgili University, Spain
David Milne	University of Technology Sydney, Australia
Marie-Francine Moens	KU Leuven, Belgium
Preslav Nakov	Qatar Computing Research Institute, Qatar
Elmar Nöth	University of Erlangen-Nuremberg, Germany
Senja Pollak	Jožef Stefan Institute, Slovenia
Stephen Pulman	University of Oxford, UK
Matthew Purver	Queen Mary University of London, UK
Mats Rooth	Cornell University, USA
Tony Russell-Rose	UX Labs, UK
Horacio Saggion	Pompeu Fabra University, Spain
Tanja Schultz	University of Bremen, Germany
Efstathios Stamatatos	University of the Aegean, Greece
Erik Tjong Kim Sang	Netherlands eScience Center, The Netherlands
Isabel Trancoso	Instituto Superior Técnico, Portugal
Josef van Genabith	German Research Center for Artificial Intelligence, Germany
K. Vijay-Shanker	University of Delaware, USA

Atro Voutilainen	University of Helsinki, Finland
Hsin-Min Wang	Academia Sinica, Taiwan
Hua Xu	University of Texas, Houston, USA
Edmund S. Yu	Syracuse University, USA
François Yvon	CNRS - Limsi, France
Wlodek Zadrozny	University of North Carolina, Charlotte, USA

Additional Reviewers

Arvanitis, Gerasimos	Maier, Angelika
Baskar, Murali Karthick	Pittaras, Nikiforos
Grezl, Frantisek	Rentoumi, Vassiliki
Grimm, Frank	Sayyed, Zeeshan Ali
Hu, Hai	Schaede, Leah
Kaklis, Dimitrios	Sun, Sining
Kukurikos, Antonis	ter Horst, Hendrik

Contents

Text Analysis and Classification

Invited Talk

The Time-Course of Phoneme Category Adaptation in Deep Neural Networks

Junrui Ni[1], Mark Hasegawa-Johnson[1,2], and Odette Scharenborg[3(✉)]

[1] Department of Electrical and Computer Engineering,
University of Illinois at Urbana-Champaign, Champaign, IL, USA
[2] Beckman Institute, University of Illinois at Urbana-Champaign,
Champaign, IL, USA
[3] Multimedia Computing Group,
Delft University of Technology, Delft, The Netherlands
o.e.scharenborg@tudelft.nl

Abstract. Both human listeners and machines need to adapt their sound categories whenever a new speaker is encountered. This perceptual learning is driven by lexical information. In previous work, we have shown that deep neural network-based (DNN) ASR systems can learn to adapt their phoneme category boundaries from a few labeled examples after exposure (i.e., training) to ambiguous sounds, as humans have been found to do. Here, we investigate the time-course of phoneme category adaptation in a DNN in more detail, with the ultimate aim to investigate the DNN's ability to serve as a model of human perceptual learning. We do so by providing the DNN with an increasing number of ambiguous retraining tokens (in 10 bins of 4 ambiguous items), and comparing classification accuracy on the ambiguous items in a held-out test set for the different bins. Results showed that DNNs, similar to human listeners, show a step-like function: The DNNs show perceptual learning already after the first bin (only 4 tokens of the ambiguous phone), with little further adaptation for subsequent bins. In follow-up research, we plan to test specific predictions made by the DNN about human speech processing.

Keywords: Phoneme category adaptation · Human perceptual learning · Deep neural networks · Time-course

1 Introduction

Whenever a new speaker or listening situation is encountered, both human listeners and machines need to adapt their sound categories to account for the speaker's pronunciations. This process is called perceptual learning, and is defined as the temporary adaptation of sound categories after exposure to deviant speech, in a manner such that the deviant sounds are included into pre-existing sound categories, thereby improving intelligibility of the speech (e.g., [1–6]). A specific case of perceptual learning is lexically-guided perceptual learning [2], in which the adaptation process is driven by lexical information. Human lexically-guided perceptual learning has been shown to be fast, and requires only a few instances of the deviant sound [5, 6]. Automatic speech recognition (ASR) systems typically adapt to new speakers or new listening conditions

© Springer Nature Switzerland AG 2019
C. Martín-Vide et al. (Eds.): SLSP 2019, LNAI 11816, pp. 3–15, 2019.
https://doi.org/10.1007/978-3-030-31372-2_1

using both short-time adaptation algorithms (e.g., fMLLR [7]) and longer-term adaptation techniques (e.g., DNN weight training [8]). In previous work [9], we showed that Deep Neural Networks (DNNs) can adapt to ambiguous speech as rapidly as a human listener by training on only a few examples of an ambiguous sound. Here, we push this research further and ask the following questions: Are ambiguous sounds processed in the same way as natural sounds; and, how many examples of the ambiguous sound are needed before the DNN adapts?

In short, the aim of this paper is two-fold: (1) we investigate the time-course of phoneme category adaptation in a DNN in more detail focusing on the amount of deviant speech material and training needed for phoneme category retuning to occur in a DNN, (2) with the larger aim to investigate the DNN's ability to serve as a model of human perceptual learning. In order to do so, we base our research on the experimental set-up and use the stimuli of a human perceptual learning experiment (see for other examples, e.g., [9–11]).

In a typical human lexically-guided perceptual learning experiment, listeners are first exposed to deviant phonemic segments in lexical contexts that constrain their interpretation, after which listeners have to decide on the phoneme categories of several ambiguous sounds on a continuum between two phoneme categories (e.g., [1–6]). This way the influence of exposure to the deviant sound can be investigated on the phoneme categories in the human brain. In this paradigm, two groups of listeners are tested. Using the experiment from which we take our stimuli [4] as an example: one group of Dutch listeners was exposed to an ambiguous [l/ɹ] sound in [l]-final words such as *appel* (Eng: *apple*; *appel* is an existing Dutch word, *apper* is not). Another group of Dutch listeners was exposed to the exact same ambiguous [l/ɹ] sound, but in [ɹ]-final words, e.g., *wekker* (Eng: *alarm clock*; *wekker* is a Dutch word, *wekkel* is not). After exposure to words containing the [l/ɹ], both groups of listeners were tested on multiple steps from the same continuum of [l/ɹ] ambiguous sounds from more [l]-like sounds to more [ɹ]-like sounds. For each of these steps, they had to indicate whether the heard sound was an [l] or an [ɹ]. Percentage [ɹ] responses for the continuum of ambiguous sounds were measured and compared for the two groups of listeners. Lexically-guided perceptual learning shows itself as significantly more [ɹ] responses for the listeners who were exposed to the ambiguous sound in [ɹ]-final words compared those who were exposed to the ambiguous sound in [l]-final words. A difference between the groups is interpreted to mean that listeners have retuned their phoneme category boundaries to include the deviant sound into their pre-existing phone category of [ɹ] or [l], respectively.

We base our research on the time-course of adaptation found in human listeners in the experiment in [5]. Their question was similar to ours: Are words containing an ambiguous sound processed in the same way as "natural" words, and if so, how many examples of the ambiguous sound are needed before the listener can do this? Participants had to listen to nonsense words, natural words, and words containing an ambiguous sound, and were instructed to press the 'yes' button as soon as possible upon hearing an existing word and 'no' upon hearing a nonsense word. Yes/no responses and reaction times to the natural and "ambiguous" words were analyzed in bins of 5 ambiguous words. They found that words containing an ambiguous sound were accepted as words less often, and were processed slower than natural words, but this difference in acceptance disappeared after approximately 15 ambiguous items.

2   Methods

In our DNN experiment, we follow the set-up used in [9]. To mimic or create a Dutch listener, we first train a baseline DNN using read speech from the Spoken Dutch Corpus (CGN; [12]). The read speech part of the CGN consists of 551,624 words spoken by 324 unique speakers for a total duration of approximately 64 h of speech. A forced alignment of the speech material was obtained using a standard Kaldi [13] recipe found online [15]. The speech signal was parameterized using a 64-dimensional vector of log Mel spectral coefficients with a context window of 11 frames, each has a segment length of 25 ms with a 10 ms shift between frames. Per-utterance mean-variance normalization was applied. The CGN training data were split into a training (80% of the full data set), validation (10%), and test set (10%) with no overlap in speakers.

Because we aim to investigate the DNN's ability to serve as a model of human perceptual learning, we used the same acoustic stimuli as used in the human perception experiment [4] for retraining the DNN (also referred to as retuning). The retraining material consisted of 200 Dutch words produced by a female Dutch speaker in isolation: 40 words with final [ɹ], 40 words with final [l], and 120 'distractor' words with no [l] and [ɹ]. For the 40 [l]-final words and the 40 [ɹ]-final words, versions also existed in which the final [l] or [ɹ] was replaced by the ambiguous [l/ɹ] sound. Forced alignments were obtained using a forced aligner for Dutch from the Radboud University. For four words no forced alignment was obtained, leaving 196 words for the experiment.

2.1   Model Architecture

All experiments used a simple fully-connected, feed-forward network with five hidden layers, 1024 nodes per layer, with logistic sigmoid nonlinearities as well as batch-normalization and dropout after each layer activation. The output layer was a softmax layer of size 38, corresponding to the number of phonemes in our training labels. The model was trained on CGN for 10 epochs using an Adam optimizer with a learning rate of 0.001. After 10 epochs, we reached a training accuracy of 85% and a validation accuracy of 77% on CGN.

2.2   Retuning Conditions

To mimic the two listener groups from the human perceptual experiment, and to mimic a third group with no exposure to the ambiguous sound (i.e., a baseline group), we used three different configurations of the retuning set:

- Amb(iguous)L model: trained on the 118 distractor words, the 39 [ɹ]-final words, and the 39 [l]-final words in which the [l] was replaced by the ambiguous [l/ɹ].
- Amb(iguous)R model: trained on the 118 distractor words, the 39 [l]-final words, and the 39 [ɹ]-final words in which the [ɹ] was replaced by the ambiguous [l/ɹ].
- Baseline model: trained on all 196 natural words (no ambiguous sounds). This allows us to separate the effects of retuning with versus without the ambiguous sounds.

In order to investigate the time-course of phoneme category adaptation in the DNNs, we used the following procedure. First, the 196 words in the three retuning sets were split into 10 bins of 20 distinct words, except for the last two bins, which each contained only 18 words. In order to be able to compare between the different retuning conditions, the word-to-bin assignments were tied among the three retuning conditions. Each word appeared in only one bin. Each bin contained: 4 words with final [r] (last bin: 3 words) + 4 words with final [l] (penultimate bin: 3 words) + 12 'distractor' words with no [l] or [r] (last two bins: 11 words). The difference between the retuning conditions is:

- AmbL: the final [l] in the 4 [l]-final words was replaced by the ambiguous [l/ɹ] sound.
- AmbR: the final [ɹ] in the 4 [ɹ]-final words was replaced by the ambiguous [l/ɹ] sound.
- Baseline: only natural words.

The [l]-final, [ɹ]-final, and [l/ɹ]-final sounds of the words in bin t from all three retuning sets, combined, functioned as the test set to bin $t - 1$. As all the acoustic signals from the test bin were unseen during training at the current time step, we denote this as "open set evaluation". Figure 1 explains the incremental adaption. Note that the final bin was only used for testing; because at $t = 10$, there is no subsequent bin that could be used for testing.

Retuning was repeated five times, with five different random seeds for permutation of data within each bin, for each retuning condition/model. Each time, for every time step of incremental adaptation, we retrained the baseline CGN-only model using bin 0 up to bin $t - 1$ of the retraining data for 30 epochs using an Adam Optimizer with a learning rate of 0.0005. The re-tuning accuracy on the training set after 30 epochs always reached an accuracy of 97.5–99%.

```
for each retuning set from {Baseline, AmbL, AmbR}
    Test the CGN-only model using bin 0 from the test set

    for t in [1,9]:
        Retrain the CGN-only model using bin 0 up to bin t-1
        Test the retrained model from bins 0 through t-1 using test set bin t
```

Fig. 1. Incremental retuning procedure for the open set evaluation.

3   Classification Rates

In the first experiment, we investigated the amount of training material needed for perceptual learning in a DNN to occur. Classification accuracy was computed for all frames, but since we are primarily interested in the [l], [ɹ], and the ambiguous [l/ɹ] sound, we only report those. Figure 2 through 4 show the proportion of correct frame classifications as solid lines, i.e., [l] frames correctly classified as [l] and [ɹ] frames correctly classified as [ɹ], for each of the 10 bins ($0 \leq t \leq 9$). Dashed lines show, for example, the proportion of [l] frames incorrectly classified as [ɹ], and of [ɹ] frames incorrectly classified as [l]; the rate of substitutions by any other phone is equal to 1.0 minus the solid line minus the dashed line. The interesting case is the classification of the [l/ɹ] sound (see triangles), which is shown with a dashed line when classified as [l] and with a solid line when classified as [ɹ]. Note, in the legend, the capital letter denotes the correct response, lowercase denotes the classifier output, thus, e.g., L_r is the percentage of [l] tokens classified as [ɹ].

Figure 2 shows the results for the baseline model retrained with the natural stimuli. The baseline model shows high accuracy in the classification of [ɹ]. The [l] sound is classified with high accuracy at $t = 2$, then drops for increasing t, up to $t = 8$. The [ɹ] sound, on the other hand, is classified with very high accuracy after seeing a single bin of retuning data, with very little further improvement for subsequent bins. The [l/ɹ] sound (not part of the training data for this model) is classified as [ɹ] about 70% of the time, and as [l] about 10% of the time, with the remaining 20% of instances classified to some other phoneme.

Figure 3 shows the results for the model retrained with the ambiguous sound bearing the label of /l/. The AmbL model has a high accuracy in the classification of the [ɹ]; however, the accuracy of natural [l] is less than 50% after the first bin and continues to worsen as more training material is added. The lexical retuning dataset contains no labeled examples of a natural [l]; apparently, in this case, the model learns the retuning data so well that it forgets what a natural [l] sounds like.

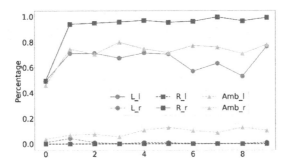

Fig. 2. Proportion of [l] and [ɹ] responses by the baseline model, retrained with natural stimuli, per bin.

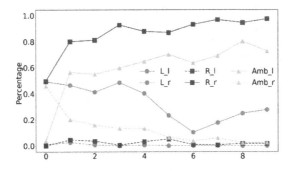

Fig. 3. Proportion of [l] and [ɹ] responses by the AmbL model, retrained with [l/ɹ] labeled as [l], per bin.

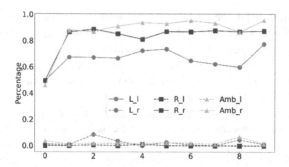

Fig. 4. Proportion of [l] and [ɹ] responses by the AmbR model, retrained with [l/ɹ] labeled as [ɹ], per bin.

Importantly, after the first bin, the network has correctly learned to label the [l/ɹ] sounds as [l], indicating 'perceptual learning' by the AmbL system. The classification of [l/ɹ] as [l] continues to rise slightly for subsequent bins. While the AmbL model already correctly recognizes most [l/ɹ] sounds as [l] after the first bin, recognition further improves for subsequent bins.

Figure 4 shows the results for the model retrained with the ambiguous sound labeled as /r/. The AmbR model has high accuracy for both the [l] and [ɹ] sounds. So, unlike the AmbL model, the AmbR model did not forgot what a natural [ɹ] sounds like. Moreover, after the first bin, this model has learned to classify [l/ɹ] as [ɹ] more than 85% of the time, which is a 10% increase over the model trained on the natural sounds at the same time step, thus showing perceptual learning. Unlike the AmbL model, additional [l/ɹ] training examples show little tendency to further increase the classification of [l/ɹ] as [ɹ], up to and including the last point.

Interestingly, all phones, including the natural [l] and [ɹ] as well as the ambiguous phone, show classification accuracy of around 50% prior to retraining. This rather low accuracy is most likely due to the differences in recording conditions and speaker between the CGN training set and the retraining sets. After retraining with the first bin, the classification accuracies make a jump in all models, with little further adaptation for subsequent bins, although the AmbL shows a small increase in adaptation for later bins, while this is not the case for the baseline and AmbR models. This adaptation suggests that the neural network treats the ambiguous [l/ɹ] exactly as it treats every other difference between the CGN and the adaptation data: In other words, exactly as it treats any other type of inter-speaker variability. In all three cases, the model learns to correctly classify test tokens after exposure to only one adaptation bin (only 4 examples, each, of the test speaker's productions of [l], [r], and/or the [l/ɹ] sound).

All three models show little tendency to misclassify [l] as [ɹ], or vice versa. This indicates that the retraining preserves the distinction between the [l] and [ɹ] phoneme categories.[1]

To investigate where the retuning takes place, we examined the effect of increasing amounts of adaptation material on the hidden layers of the models using the inter-category distance ratio proposed in [9]. This measure quantifies the degree to which lexical retuning has modified the feature representations at the hidden layers using a single number. First, the 1024-dimensional vector of hidden layer activations is re-normalized, so that each vector sums to one, and averaged across the frames of each segment. Second, the Euclidean distances between each [l/ɹ] sound and each [l] segment are computed, after which the distances are averaged over all [l/ɹ]-[l] token pairs, resulting in the average [l]-to-[l/ɹ] distance. Third, using the same procedure the average [ɹ]-to-[l/ɹ] distance is computed. The inter-category measure is then the ratio of these two distances, and is computed for each of the ten bins.

Figures 5 through 7 show the inter-category distance ratio ([l/ɹ]-to-[l] over [l/ɹ]-to-[ɹ]) for the baseline model, the AmbL model, and the AmbR model, respectively, for each of the 5 hidden layers, for each of the bins.

Figure 5 shows that for earlier bins in the baseline model, the distance between the ambiguous sounds and the natural [l] category and natural [ɹ] category is approximately the same for the different layers, with a slight bias towards [ɹ] (the ratio is >1); the lines for the five layers are close together and do not have a consistent ordering. From bin 5 onwards, and particularly for the last 3 bins, the distance between [l/ɹ] and the natural [l] category decreases from the first (triangles) to the last layer (diamonds), suggesting that [l/ɹ] is represented closer to the [l] category. However, this cannot be observed in the classification scores: Fig. 2 shows that [l/ɹ] is primarily classified as [ɹ]. The adaptation of [l/ɹ] towards natural [l] for the later bins suggests that adding training material of the speaker improves the representation of the natural classes as well, because the distance between [l/ɹ] and the natural classes changes without the model being trained on the ambiguous sounds.

Figure 6 shows that, for the AmbL model, the distance between [l/ɹ] and the natural [l] category becomes increasingly smaller deeper into the network: The line showing hidden layer 1 (triangles) is almost always on top, and the line showing layer 5 (diamonds) is almost always at the bottom. Interestingly, there is a downward slope from the first to the last bin, indicating that with increasing numbers of [l/ɹ] training examples labeled as [l], the distance between [l/ɹ] and natural [l] continues to decrease, even though there are no natural [l] tokens in the retuning data. This continual decrease

---

[1] We repeated this experiment using a Recurrent Neural Network (RNN) model trained under the Connectionist Temporal Classification (CTC) [14] criterion. The network architecture was different from the DNN architecture used in this paper, and consisted of two convolutional layers on the raw spectrogram, followed by six layers of stacked RNN. Despite the vastly different architecture, our new model showed highly similar behavior in terms of classification rate over the time course of incremental retuning. Most interestingly, both models seemed to have forgotten what a natural [l] sounds like.

in distance between [l/ɹ] and natural [l] seems to be correlated with the continual increase in classification of the ambiguous sound as [l] for the later bins in Fig. 3, and might indicate further adaptation of the representation of the ambiguous sound towards the natural [l].

In the AmbR model (Fig. 7), the ratio of distance([l/ɹ],[l]) over distance([l/ɹ],[ɹ]) increases from layer 1 to layer 5, indicating that the neural embedding of [l/ɹ] becomes more [ɹ]-like deeper in the network. So, like the AmbL model, the AmbR model also shows lexical retuning: The speech representation of [l/ɹ] becomes increasingly closer to that of the natural [ɹ] deeper into the model. The effect of increasing amounts of adaptation material is however not as clear-cut as for the AmbL model. The distance ratio rises until bin 2 (8 [l/ɹ] training examples), then falls until bin 5, then rises until bin 7, then falls again. This inconsistency is also found in the classification scores of [l/ɹ] as [ɹ] in Fig. 4 but to a lesser extent, which suggest that the increase in the distance between the [l/ɹ] and [ɹ] categories is not large enough to substantially impact classification results.

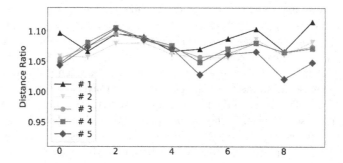

Fig. 5. Ratio of distance([l/ɹ],[l])/distance([l/ɹ],[ɹ]) for the Baseline model.

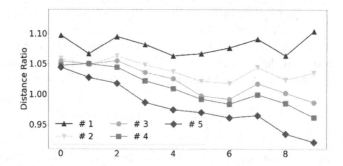

Fig. 6. Ratio of distance([l/ɹ],[l])/distance([l/ɹ],[ɹ]) for the AmbL model.

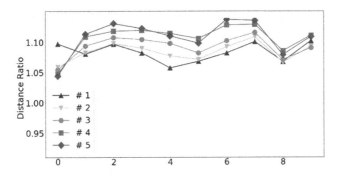

Fig. 7. Ratio of distance([l/ɹ],[l])/distance([l/ɹ],[ɹ]) for the AmbR model.

As the classification rates make a significant jump after just seeing the first bin of words for all three experimental sets, which indicates very fast adaptation, in the second experiment, we investigate how the CGN-only model adapts to a single bin of retuning data over the training course in the very first time step. Similar to the procedure above, we evaluate the classification rates by training the CGN-only model using the first training bin (training bin 0) from each experiment set (natural, AmbL, AmbR) for 30 epochs. Before the first epoch of training ($t = 0$), and after each epoch of training ($1 \leq t \leq 30$), we record the percentage of [l], [ɹ], and ambiguous [l/ɹ] sounds from the second test bin (test bin 1) that are classified as either [l] or [ɹ] (a total of 31 time points, $0 \leq t \leq 30$). Figure 8 shows the classification rates for the Baseline model: both [l] and [ɹ] sounds show immediate adaptation after the first epoch (correct response rate increases by about 20% from $t = 0$ to $t = 1$). The [ɹ] sound shows the highest accuracy over 30 epochs, but the number of [ɹ]'s correctly recognized only increases very slightly after the fifth epoch. After reaching a peak by the first epoch, the classification rate for [l] decreases until the third epoch, and then flatlines (with some small oscillations). Interestingly, while ambiguous [l/ɹ] sounds are not present in the training data, more and more [l/ɹ] get classified as [ɹ] as training progresses, meaning

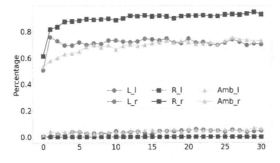

Fig. 8. Proportion of [l] and [ɹ] responses by the baseline model over 30 epochs for the first bin.

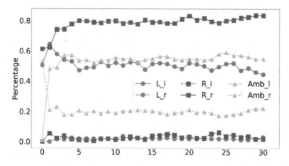

Fig. 9. Proportion of [l] and [ɹ] responses by the AmbL model over 30 epochs for the first bin.

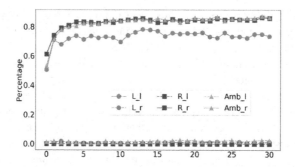

Fig. 10. Proportion of [l] and [ɹ] responses by the AmbR model over 30 epochs for the first bin

that the bias of [l/ɹ] toward [ɹ] somehow increases without the model seeing any ambiguous sounds.

Figure 9 shows the classification rates over 30 epochs for the AmbL model using stimuli from the first training bin with ambiguous sounds labeled as [l]. The classification rates at $t = 0$ are the same in Figs. 8 and 9, because they are based on the same model; it is only after the first training epoch ($t = 1$) that their rates diverge. Similar to Fig. 8, the accuracy for [ɹ] reaches 80% within 5 epochs, with a large jump at the second epoch. The accuracy for natural [l] also jumps up after the first epoch, even though there are no [l] tokens in the training data, but beginning with the second epoch, the model starts to forget how to correctly classify natural [l] tokens. The most important observation comes with the ambiguous [l/ɹ] sound. After just a single epoch on a single bin of data, the percentage of [l/ɹ] sounds classified as [l] goes from 0% to a little below 50%. However, after 5 epochs, the accuracy for [l/ɹ] as [l] flatlines around 50%, meaning that the model has reached its limit of perceptual learning by seeing only one training bin.

Figure 10 shows the classification rates over 30 epochs for the AmbR model using stimuli from the first training bin with ambiguous sounds labeled as [ɹ]. While no natural [ɹ] is present in this experiment set, the accuracy for natural [ɹ] gradually increases until the fifth epoch, meaning that perceptual learning on ambiguous sounds as [ɹ] also helps the model learn a natural [ɹ]. The ambiguous [l/ɹ] sound is classified as [ɹ] 50% of the time at $t = 0$, i.e., with no training; the $t = 0$ case is identical to those shown in Figs. 8 and 9. After just one epoch of training, using one bin of ambiguous sounds labeled as [ɹ], the model learns to perform this classification with 70% accuracy, and accuracy increases until the fifth epoch.

It is worthwhile, at this point, to remind the reader what is meant by "one epoch" in the training of a neural net. Each epoch of training consists of three stages: (1) a direction vector d is chosen; in the first epoch, this is just the negative gradient of the error; (2) a search procedure is used to choose the scale, g; (3) the neural network weights are updated as $w = w + gd$. Each epoch of training can only perform a constant shift of the previous network weights. Figures 2, 3, 4, 8, 9 and 10 show that most of the DNN adaptation occurs in the first epoch on the first bin of the adaptation material, i.e., on the first update of the direction, therefore most of the DNN adaptation can be characterized as a constant shift in the network weights. This makes sense since the model is just learning about 4 additional training tokens (one adaptation bin) — with only 4 tokens, while it is not possible to learn a very complicated modification of the boundary, learning a boundary shift is indeed possible and very likely the case here.

In a deep neural network, a constant shift of the network weights is not the same thing as a constant shift of the classification boundary, but in practice, the revision of w after the first epoch is usually not much more complicated than a shifted boundary. The finding that inter-talker adaptation can be accomplished by a constant shift in cepstral space is not new; it has previously been reported by [16]. The finding that a comparable constant shift is sufficient to learn distorted sounds, like the ambiguous [l/ɹ] sound, has never previously been reported.

4   Discussion and Concluding Remarks

Inspired by the fast adaptation of human listeners to ambiguous sounds (e.g., [1–6]), we investigated the time-course of phoneme category adaptation in a DNN, with the ultimate aim to investigate the DNN's ability to serve as a model of human perceptual learning. We based our investigation on the time-course of adaptation of the human perceptual learning experiment in [5]. In the first experiment, we provided the DNN with an increasing number of the original ambiguous acoustic stimuli from [5] as retraining tokens (in 9 bins of 4 ambiguous items), compared classification accuracy on the ambiguous items in an independent, held-out test set for the different bins, and calculated the ratio of the distance between the [l/ɹ] category and the natural [l] and [ɹ] categories, respectively, for the five hidden layers of the DNNs and for the 9 different bins. In the second experiment, the amount of training was investigated by calculating the classification rates over 30 epochs when only one bin is used for retuning.

Results (both presented here and the unpublished results with a CTC-RNN model) showed that, similar to human listeners, DNNs quickly learned to interpret the ambiguous sound as a "natural" version of the sound. After only 4 examples of the ambiguous sound, the DNN showed perceptual learning, with little further adaptation for subsequent training examples, although a slight further adaptation was observed for the model which learned to interpret the ambiguous sound as [l]. In fact, perceptual learning could already clearly be seen after only one epoch of training on those 4 examples, and showed very little improvement after the fifth epoch. This is in line with human lexically-guided perceptual learning; human listeners have been found to need 10–15 examples of the ambiguous sound to show the same type of step-like function [5, 6]. We should note, however, that it is not evident how to compare the 4 examples needed by the DNN with the 10–15 examples of the human listener. We know of no way to define the "learning rate" of a human listener other than by adjusting the parameters of a DNN until it matches the behavior of the human, which is an interesting avenue for further research into the DNN's ability to serve as a model of human perceptual learning. Nevertheless, both DNNs and human listeners need very little exposure to the ambiguous sound to learn to normalize it.

Retuning took place at all levels of the DNN. In other words, retuning is not simply a change in decision at the output layer but rather seems to be a redrawing of the phoneme category boundaries to include the ambiguous sound. This is again exactly in line with what has been found for human listeners [17].

This paper is the first to show that, similar to inter-talker adaptation, adaptation to distorted sounds can be accomplished by a constant shift in cepstral space. Moreover, our study suggests that DNNs are more like humans than previously believed: In all cases, the DNN adapted to the deviant sound very fast and after only 4 presentations, with little or no adaptation thereafter. Future research will aim to test, in perceptual experiments with human listeners, the prediction of the DNN that the speech representations of the ambiguous sound and the natural [l] and [ɹ] change very little once the category adaptation has taken place.

Acknowledgements. The authors thank Anne Merel Sternheim and Sebastian Tiesmeyer with help in earlier stages of this research, and Louis ten Bosch for providing the forced alignments of the retraining material. This work was carried out by the first author under the supervision of the second and third author.

References

1. Samuel, A.G., Kraljic, T.: Perceptual learning in speech perception. Atten. Percept. Psychophys. **71**, 1207–1218 (2009)
2. Norris, D., McQueen, J.M., Cutler, A.: Perceptual learning in speech. Cogn. Psychol. **47**, 204–238 (2003)
3. Scharenborg, O., Weber, A., Janse, E.: The role of attentional abilities in lexically-guided perceptual learning by older listeners. Atten. Percept. Psychophys. **77**(2), 493–507 (2015). https://doi.org/10.3758/s13414-014-0792-2
4. Scharenborg, O.: Janse, E: Comparing lexically-guided perceptual learning in younger and older listeners. Atten. Percept. Psychophys. **75**(3), 525–536 (2013). https://doi.org/10.3758/s13414-013-0422-4
5. Drozdova, P., van Hout, R., Scharenborg, O.: Processing and adaptation to ambiguous sounds during the course of perceptual learning. In: Interspeech 2016, San Francisco, CA, pp. 2811–2815 (2016)
6. Poellmann, K., McQueen, J.M., Mitterer, H.: The time course of perceptual learning. In: Proceedings of ICPhS (2011)

7. Gales, M.J.: Maximum likelihood linear transformations for HMM-based speech recognition. Comput. Speech Lang. **12**(2), 75–98 (1998)
8. Liao, H.: Speaker adaptation of context dependent deep neural networks. In: Proceedings of ICASSP, pp. 7947–7951 (2013)
9. Scharenborg, O., Tiesmeyer, S., Hasegawa-Johnson, M., Dehak, N.: Visualizing phoneme category adaptation in deep neural networks. In: Proceedings of Interspeech, Hyderabad, India (2018)
10. Scharenborg, O.: Modeling the use of durational information in human spoken-word recognition. J. Acoust. Soc. Am. **127**(6), 3758–3770 (2010)
11. Karaminis, T., Scharenborg, O.: The effects of background noise on native and non-native spoken-word recognition: a computational modelling approach. In: Proceedings of the Cognitive Science conference, Madison, WI, USA (2018)
12. Oostdijk, N.H.J., et al.: Experiences from the spoken Dutch Corpus project. In: Proceedingd of LREC – Third International Conference on Language Resources and Evaluation, Las Palmas de Gran Canaria, pp. 340–347 (2002)
13. Povey, D., et al.: The Kaldi speech recognition toolkit. In: IEEE Workshop on Automatic Speech Recognition and Understanding, Hawaii, US (2011)
14. Graves, A., Fernandez, S., Gomez, F., Schmidhuber, J.: Connectionist temporal classification: labelling unsegmented sequence data with recurrent neural networks. In: Proceedings of the 23rd international conference on Machine learning, Pittsburgh, Pennsylvania, USA (2006)
15. https://github.com/laurensw75/kaldi_egs_CGN
16. Pitz, M., Ney, H.: Vocal Tract normalization equals linear transformation in cepstral space. IEEE Trans. Speech Audio Process. **13**(5), 930–944 (2005)
17. Clarke-Davidson, C., Luce, P.A., Sawusch, J.R.: Does perceptual learning in speech reflect changes in phonetic category representation or decision bias? Percept. Psychophys. **70**, 604–618 (2008)

Dialogue and Spoken Language Understanding

Towards Pragmatic Understanding of Conversational Intent: A Multimodal Annotation Approach to Multiparty Informal Interaction – The EVA Corpus

Izidor Mlakar[✉], Darinka Verdonik, Simona Majhenič,
and Matej Rojc

Faculty of Electrical Engineering and Computer Science,
University of Maribor, Maribor, Slovenia
{izidor.mlakar,darinka.verdonik,simona.majhenic,
matej.rojc}@um.si

Abstract. The present paper describes a corpus for research into the pragmatic nature of how information is expressed synchronously through language, speech, and gestures. The outlined research stems from the 'growth point theory' and 'integrated systems hypothesis', which proposes that co-speech gestures (including hand gestures, facial expressions, posture, and gazing) and speech originate from the same representation, but are not necessarily based solely on the speech production process; i.e. 'speech affects what people produce in gesture and that gesture, in turn, affects what people produce in speech' ([1]: 260). However, the majority of related multimodal corpuses 'ground' non-verbal behavior in linguistic concepts such as speech acts or dialog acts. In this work, we propose an integrated annotation scheme that enables us to study linguistic and paralinguistic interaction features independently and to interlink them over a shared timeline. To analyze multimodality in interaction, a high-quality multimodal corpus based on informal discourse in a multiparty setting was built.

Keywords: Corpora and language resources · Multimodal corpus · Multimodal technologies · Natural language understanding. pragmatics · Annotation · Conversational intelligence

1 Introduction

In social and spoken interaction, language is not used in isolation and does not occur in a vacuum [2]. Embodied behavior adds more than 50 percent of non-redundant information to the common ground of the conversation [3]. The sharing of information or the exchange of information in human social interactions is far more complex than a mere exchange of words. It is multilayered and includes attitude and affect, utilizes bodily resources (embodiment) as well as a physical environment in which the discourse takes place [4].

Effective communication requires the following conditions to be fulfilled: (i) the communicator must make his or her intention to communicate recognizable and (ii) the

© Springer Nature Switzerland AG 2019
C. Martín-Vide et al. (Eds.): SLSP 2019, LNAI 11816, pp. 19–30, 2019.
https://doi.org/10.1007/978-3-030-31372-2_2

propositional content or conceptual or ideational meaning (e.g. semantic information) that they wish the recipient to receive must be represented effectively [5].

In interpersonal discourse, verbal signals carry a symbolic or semantic interpretation of information through linguistic and paralinguistic properties, while non-verbal signals (i.e. embodiments) orchestrate speech [6]:4. Non-verbal concepts, such as prosody, embodiments, emotions, or sentiment are multi-functional and operate on the psychological, sociological, and biological level and in all time frames. These signals represent the basis of cognitive capabilities and understanding [7, 8]. Embodied behavior in particular, effectively retains the semantics of the information, helps in providing suggestive influences, and gives a certain degree of cohesion and clarity to the overall discourse [9, 10]. Non-verbal behavior, although not bound by grammar, co-aligns with language structures and compensates for the less articulated verbal expression model [2, 11]. It also serves interactive purposes, such as content representation or expression of one's mental state, attitude, and social functions [12–16].

The main motivations for the work presented in this paper is driven by the goal of enabling machine 'sensing' and more natural interaction with virtual agents. Despite the considerable interest in this topic and significant progress reported, automatically understood and machine-generated information from a set of evidence is, in general, still far from perfect or natural [11, 17]. Moreover, not only speech and language affect embodiment but embodied signals also affect what people produce through language and speech [1].

This paper presents a multimodal approach to generating 'conversational' knowledge and modeling of the complex interplay among conversational signals, based on a concept of data analytics (mining) and information fusion. Our work outlines a novel analytical methodology and a model to annotate and analyze conversational signals in spoken multi-party discourse. Moreover, the results of our annotation process (i.e. the corpus) applied to a multi-party discourse setting in Slovenian are represented. In addition to capturing language-oriented signals, naïve to modern corpus linguistics, the model also provides a very detailed description of non-verbal (and paralinguistic) signals. These disparate phenomena are interconnected through the notion of co-occurrence (e.g. timeline).

2 Background

One of the main issues in sentic computing is misinterpretation of conversational signals and non-cohesive responses. As a result, 'multimodality in interaction' became one of the fundamental concepts in corpus linguistics. Especially in interactional linguistics and conversation analysis, a significant focus was shifted to embodied behavior (an overview of such research can be found in [11, 18]). The semantic domain is particularly well-suited when investigating co-verbal alignment. Research studies show how humans 'map' semantic information onto linguistic forms [10, 19, 20]. Linguistic approaches in general tend to observe embodied behavior in discourse on a linguistic basis (i.e. language and grammar). However, as argued by Birdwhistell [21], what is conveyed through the body does not meet the linguist's definition of language. Therefore, the apparent grammatical interface between language and gestures seems to be limited ([2]). In terms of creating conversational knowledge, such association operates in very narrow contexts and opens limited and highly focused opportunities to explore the interplay between verbal and non-verbal signals [8, 22].

In contrast, the researchers in [6, 14–16], among others, propose to involve additional modalities, such as sound and image, and investigate the functional nature of embodiments during discourse. The widely adopted approach to multimodality in interaction is Pierce's semiotic perspective (i.e. the 'pragmatics on the page'), which explores the meaning of images and the intrinsic visual features of written text. In [23], for instance, the authors correlated hand shapes (and their trajectories) with semiotic class based on a broader context of the observed phenomena. Although the approaches oriented towards semiotics (i.e. [24–26]) go beyond semantics and do not restrict embodiments to linguistic rules, they still restrict themselves functionally, that is to a specific phenomenon and a narrow discourse context.

In contrast to the aforementioned approaches inspired by linguistics, Feyaerts et al. [27] authors build on the cognitive-linguistic enterprise and equally incorporate all relevant dimensions of how events are utilized, including the trade-off between different semiotic channels. However, the discourse setting is limited to an artificial setting. Due to the challenging nature of informal, especially multiparty discourse, researchers tend to establish artificial settings [28]. These settings introduce laboratory conditions with targeted narration and discourse concepts between collocutors which focus on a specific task. Such data sources therefore clearly reveal the studied phenomena but hinder 'interference' of other, non-observed signals that would appear in less restricted settings. Furthermore, in most cases, a wider scope of conversational signals is intentionally left out of the conversational scenario [29, 30]. Following [27], we observe discourse as a multimodal phenomenon, in which each of the signals represents an action item, which must be observed in its own domain and under its own restrictions. We focus on corpus collection, structuring, and analysis. Instead of 'artificial' scenarios we utilize a rich data source based on an entertaining evening TV talk show in Slovene, which represents a good mixture of institutional discourse, semi-institutional discourse, and casual conversation.

3   Data Collection and Methodology: The EVA Corpus

3.1   Data Source

In this research, we used the EVA Corpus [31] which consists of 228 min in total, including 4 video and audio recordings, each 57 min long, with corresponding orthographic transcriptions. The discourse in all four recordings is a part of the entertaining evening TV talk show *A si ti tut not padu*, broadcast by the Slovene commercial TV in 2010. In total, 5 different collocutors are engaged in each episode. The conversational setting is relaxed and unrestricted. It is built around a general scenario, focused on day-to-day concepts. The discourse involves a lot of improvisation and is full of humor, sarcasm, and emotional responses. Moreover, although sequencing exists and general discourse structuring (e.g. role exchange, topic opening, grounding, etc.) applies, it is performed highly irregularly. Table 1 outlines the general characteristics of the recording used to define the proposed model of conversational expression. The utterances in the EVA Corpus are mostly single short sentences, on average consisting of 8 words. The discourse contains 1,801 discourse markers (counting only those with a minimum frequency of 10). The corpus includes a lot of non-verbal interactions: 1,727 instances in which 'movement' was attributed to convey meaning (e.g., a gesture performed with an intent) were classified.

Table 1. General characteristics of discourse in the EVA Corpus.

Utterances	
Total	1,516
AVG per speaker	303
Sentences	
Total	1,999
AVG per speaker	399.8
AVG per statement	1.32
Words	
Total	10,471
AVG per speaker	2094
AVG per sentence	7.9
Metadiscourse	
discourse markers (n > 10)	1,801
AVG per speaker	599
Non-verbal behavior	
Total number of semiotic intents	1,727

The data in Table 1 clearly outline that contributors are active and that the discourse involves short statements (i.e. under 5 s) with a significant amount of overlapping speech. Individual sentence duration ranges from 0.5 s to 5 s and 2.8 s on average. Together, all participants generate roughly 93 min of spoken content in a 57-min recording. The statements are interchanging rapidly among the collocutors and with high density.

3.2   Annotation Topology

In order to realize the aforementioned 'conversational model' and observe each conversational expression in greater detail, a multimodal annotation approach typically used in conversational analysis was adopted. For this purpose, an annotation topology with various levels, as outlined in Fig. 1, was defined. The scheme applies a two-layered analysis of the conversational episode.

In the first layer (i.e. symbolics/kinesics), signals that are primarily evident in the formulation of an idea and identify the communicative intent were observed and annotated. As outlined in Fig. 1, this layer annotates linguistic and paralinguistic signals

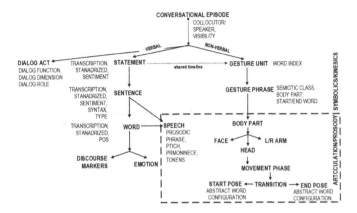

Fig. 1. The topology of annotation in the EVA Corpus: the levels of annotation describing verbal and non-verbal contexts of conversational episodes

(e.g. kinesics [21]). The second layer (i.e. articulation/prosody) is oriented towards the form and is concerned with how an abstract concept (i.e. communicative intent) is physically articulated through auditory and visual channels. It provides detailed descriptions of the structure of verbal and non-verbal components and on how the articulators (verbal and non-verbal ones) are modeled, moved, and put to use.

The material was annotated by an annotator with a background in linguistics and experience in annotation of multimodal materials. The annotations were performed in ELAN (EUDICO Linguistic Annotator) and WebAnno, converged into a single data source, specified as JSON, and visualized. The currently available annotations were performed over a nine-month period and in separate trials for each conversational concept and, in some cases, even for each signal.

3.3   Annotation Procedure and Inter-annotator Agreement

Five annotators, two with linguistic background, and three with technical background in machine interaction were involved in this phase of annotations. Annotations were performed in separate sessions, each session describing a specific signal. The annotation was performed in pairs, i.e. two or three annotators annotated the same signal. After the annotation, consensus was reached by observing and commenting on the values where the was no or little annotation agreement among multiple annotators (including those not involved in the annotation of the signal). The final corpus was generated after all disagreements were resolved. Procedures for checking inconsistencies were finally applied by an expert annotator.

Before starting with each session, the annotators were given an introductory presentation defining the nature of the signal they were observing and the exact meaning of the finite set of values they could use. An experiment measuring agreement was also performed. It included an introductory annotation session in which the preliminary inconsistencies were resolved. For the less complex signals, influenced primarily by a single modality (i.e. pitch, gesture unit, gesture phrase, body-part/modality, sentence type, etc.), the annotators' agreement measured in terms of Cohen's *kappa* [32] was high, namely between 0.8 and 0.9. As summarized by Table 2, for the more complex signals that involve all modalities for their comprehension (including speech, gestures, and text) the interpretation was less unanimous.

Table 2. Results of the preliminary inter-coder agreement experiment.

Signal	Kappa score
Sentiment	0.67
Dialog function	0.64
Dialog dimension	0.71
Intent (semiotic class)	0.48
Emotion label	0.51
Gesture unit	0.75
Movement phase	0.66

The figures indicate that agreement was 0.63 on average. Given the complexity of the task and the fact that the values in Table 2 also cover cases with possible duality of meaning, the level of agreement is acceptable and comparable to other multimodal corpus annotation tasks [25]. For *Intent* the possible duality of interpretation was surprisingly common. The annotators in general agreed on the major class and would have a difference in opinion in the minor sub-class.

3.4 Transcription and Segmentation

The audio data was transcribed in original colloquial transcriptions (verbatim), and in their standardized transcriptions (standardized Slovenian spelling). The colloquial transcriptions also include meta information transcribed in brackets '[]' (e.g., [:laugher], [gap], [incident], [:voice]). All transcriptions are segmented into statements, sentences and words while also considering the temporal domain. The boundaries for colloquial and standardized statements match completely. The conversations are split into 5 sessions, in which each session contains information and annotation levels for each individual speaker. Additionally, each word was POS tagged following the JOS specifications.

3.5 Discourse Management and Structuring

Following the ISO 24617-2 [33] guidelines, dialogue acts (DA) in the EVA Corpus were annotated as an independent concept and some adjustments to the ISO scheme were added. The definition of the ISO functional segments as the basic unit of annotation and their several layers of information (sender, addressee, dimension, and communicative function) were retained. Some non-task dimensions were merged into a single cover dimension, the social obligation dimension was generalized into social management. The results of the annotation are listed in Table 3.

Table 3. Results of DA annotation in the EVA Corpus

DA		Dialog dimensions > 200	
Total acts	3,465	Total dimensions	3,465
With 1 dimension	2,144	Task	1,960
With 2 dimensions	1,175	Communication management	608
With 3 or more dimensions	146	Feedback	445
Dialog functions			
Total functions	3,479		
Functions with frequency > 25			
inform: 982, stalling: 291, ownComprehensionFB: 272, setQuestion: 176, answer: 163, checkQuestion: 135, retraction: 112, feedbackElicitation: 108, agreement: 104, instruct: 95, confirm: 93, positive: 78, interaction Structuring: 68, negative: 65, backchannel: 64, disagreement: 48, opening: 46, argument: 43, completion: 39, request: 38, partner ComprehensionFB: 35, turnTake: 32, suggest: 31, emphasis: 28, flattery: 26			

The most common dimension was task (e.g. information providing, agreement, confirmation, instructing) which accounted for more than half of the DAs. Communication management (stalling, retraction, etc.) was the second most frequently assigned dimension. This reflects a high level of spontaneity in dialogue. The third most frequent dimension was feedback, which can be explained with a high level of interaction and informal character of the dialogue.

3.6 Discourse Markers

The present research draws on previous work on Slovene DMs [34], which includes a vast set of expressions ranging from connective devices such as *and* and *or* to the interactional *yes* and *y'know* and to production markers such as *uhm*. Altogether 121 different expressions were tagged as DMs; however, only DMs with a minimum frequency of 10 were analyzed and classified into the following groups:

DM-s (speech formation markers): *eee* 'um' (316), *eem* 'uhm' (15), *mislim* 'I mean' (24), *v bistvu* 'actually'[1] (10)

DM-d (dialogue markers):

- **DM-d(c)** (contact): *veš* 'y'know' (14), *a veš* 'y'know' (24), *glej* 'look' (23), *daj* 'come on' (17), *ne* 'right?' (183), *a ne* 'right?' (21), *ti* 'you' (10), *ej* 'hey' (14)
- **DM-d(f)** (feedback): *aja* 'I see' (18), *mhm* 'mhm' (20), *aha* 'oh' (53), *ja* 'yes' (409), *fajn* 'nice' (14)
- **DM-d(s)** (dialogue structure): *dobro* 'alright' (39), *no* 'well' (79), *ma* 'well' (10), *zdaj* 'now' (21), *čakaj* 'wait' (22)

---

[1] *It is impossible to provide exact English equivalents for the Slovenian discourse markers examined in this paper as there are no one-to-one equivalents. The translations provided here are therefore only informative, giving the general meaning of each discourse marker.*

DM-c (connectives): *in* 'and' (65), *pa* 'and' (48), *ker* 'because' (13), *ampak* 'but' (16), *tako* 'so' (20), *a* 'but' (117), *pač* 'just' (16).

Altogether 1,651 DMs were annotated which accounts for 15.8% of all spoken content (i.e. 10,471 words).

3.7   Emotion

Emotional attitude in discourse primarily pertains to the way people feel about the conversational episode, the interlocutor, or the content of the ongoing conversation. For the annotation of emotions, Plutchik's three dimensional [35] model was applied. It has the capacity to describe complex emotions and how they interact and change over time and in a broader, social context. The results are listed in Table 4.

Table 4. Cross-speaker distribution of annotated emotions in the EVA Corpus

Emotion	Instances	Emotion	Instances
Anticipation: interest	1,239	Delight	19
Trust: acceptance	671	Trust: admiration	19
Joy	349	Boredom	15
Joy: serenity	221	Sadness	15
Disapproval	137	Contempt	14
Joy: ecstasy	92	Pensiveness	12
Surprise	69	Anger: annoyance	10
Amazement	49	Pride	10
Anticipation: vigilance	43	Alarm	7
Cynicism	29	Fear: apprehension	7
Disgust	23	Optimism	7
Distraction	23	Shame	7
Curiosity	22		

In the EVA corpus, 3,312 instances of emotional attitude were identified. The 'Anticipation: interest', 'Trust: acceptance' and 'Joy' category were identified as dominant emotions.

3.8   Classification of Embodied Behavior Through Semiotic Intent

This research focuses only on 'meaningful' movement defined through an extension of semiotics as the basis for symbolic interpretation of body language in human-human interaction. We applied the classification proposed in [31], which leverages between semiotics and kinesics, and also includes functions of discourse management (i.e. [15, 16]). The following classes of semiotic intent (SI) were distinguished:

- illustrators (I), with the subclasses: outlines (I_O), ideographs (I_I), dimensional illustrators (I_D), batons (I_B);

- regulators/adapters (R), with the subclasses: self-adaptors (R_S), communication regulators (R_C), affect regulators (R_A), manipulators (R_M), social function and obligation regulators (R_O);
- deictics/pointers (D), with the subclasses: pointers (D_P), referents (D_R), and enumerators (D_E), and
- symbols/emblems (S).

Table 5. The usage of embodied behavior in the EVA Corpus

SI class	SI subclass	Frequency	Total
I	I_O	20	178
	I_I	68	
	I_D	11	
	I_B	80	
R	R_A	105	1,194
	R_C	717	
	R_M	16	
	R_O	27	
	R_S	329	
D	D_P	40	275
	D_R	219	
	D_E	16	
S	S	37	37
(undetermined)	U	43	43
Total			**1,727**

As visible in Table 5, the EVA Corpus contains 1,727 instances of SIs generated during the discourse. The distribution of SIs shows that most of the observed embodied movement correlates with regulation and adaptation of discourse (SI class R). Among regulators, communication regulators (R_C) and self-adapters (R_S) were the most utilized non-verbal mechanism. Symbols (S) and illustrators (I) exhibit the most significant linguistic link and even a direct semantic link. In most cases, they are accompanied by a speech referent, although symbols do provide a clear meaning even without a referent in speech. In the EVA Corpus, they were classified as the least frequent non-verbal mechanism, which is also in line with non-prepared discourse.

3.9 Form and Structure of Non-verbal Expressions

From the perspective of kinesics, gestures and non-verbal expressions are considered body communication generated through movement, i.e. facial expressions, head movement, or posture. The approach outlined in [36] and the definition of the annotation of form, as represented in [37], were adopted for the description of non-verbal expressions (shape and motion). The distribution of non-verbal expressions based on modality (i.e. body parts) as represented in the EVA corpus is outlined in Table 6.

Table 6. Non-verbal patterns across all speakers in the EVA Corpus.

Modality	Total	Mean per participant
FACE	53	10.6
HEAD	704	140.8
HEAD+FACE	717	143.4
LARM	34	6.8
LARM+FACE	4	0,8
LARM+HEAD	289	57.8
LARM+HEAD+FACE	230	46
LARM+RARM	74	14.8
LARM+RARM+FACE	19	3.8
LARM+RARM+HEAD	789	157.8
ALL MODALITIES	476	95.2
RARM	57	11.4
RARM+FACE	2	0.4
RARM+HEAD	428	85.6
RARM+HEAD+FACE	323	64.6

4   Conclusion

This paper presents the first Slovene multimodal corpus, the EVA Corpus. Its aim is to better understand how verbal and non-verbal signals correlate with each other in naturally occurring speech and to help improve natural language generation in embodied conversational agents. The various annotation levels incorporate and link linguistic and paralinguistic, verbal and non-verbal features of conversational expressions as they appear in multiparty informal conversations.

The concept proposed in this paper builds on the idea that a 'multichannel' representation of a conversational expression (i.e. an idea) is generated by fusing language ('what to say') and articulation ('how to say it'). On the cognitive level (i.e. the symbolic representation), an idea is first formulated through the symbolic fusion of language and social/situational context (i.e. the interplay between linguistic and paralinguistic signals interpreted as the communicative intent). On the representational level, one utilizes non-linguistic channels (i.e. gestures, facial expressions), verbal (i.e. speech) and non-verbal prosody (i.e. movement structure) to articulate the idea and present it to the target audience.

Acknowledgments. This work is partially funded by the European Regional Development Fund and the Ministry of Education, Science and Sport of the Republic of Slovenia; the project SAIAL (research core funding No. ESRR/MIZŠ-SAIAL), and partially by the Slovenian Research Agency (research core funding No. P2-0069).

References

1. Kelly, S.D., Özyürek, A., Maris, E.: Two sides of the same coin: speech and gesture mutually interact to enhance comprehension. Psychol. Sci. **21**(2), 260–267 (2010)
2. Couper-Kuhlen, E.: Finding a place for body movement in grammar. Res. Lang. Soc. Interact. **51**(1), 22–25 (2018)
3. Cassell, J.: Embodied conversational agents: representation and intelligence in user interfaces. AI Mag. **22**(4), 67 (2001)
4. Davitti, E., Pasquandrea, S.: Embodied participation: what multimodal analysis can tell us about interpreter-mediated encounters in pedagogical settings. J. Pragmat. **107**, 105–128 (2017)
5. Trujillo, J.P., Simanova, I., Bekkering, H., Özyürek, A.: Communicative intent modulates production and comprehension of actions and gestures: a Kinect study. Cognition **180**, 38–51 (2018)
6. McNeill, D.: Why We Gesture: The Surprising Role of Hand Movements in Communication. Cambridge University Press, Cambridge (2016)
7. Church, R.B., Goldin-Meadow, S.: So how does gesture function in speaking, communication, and thinking? Why Gesture?: How the hands function in speaking, thinking and communicating, vol. 7,p. 397 (2017)
8. Poria, S., Cambria, E., Bajpai, R., Hussain, A.: A review of affective computing: from unimodal analysis to multimodal fusion. Inf. Fusion **37**, 98–125 (2017)
9. Esposito, A., Vassallo, J., Esposito, A.M., Bourbakis, N.: On the amount of semantic information conveyed by gestures. In: 2015 IEEE 27th International Conference on Tools with Artificial Intelligence (ICTAI), pp. 660–667. IEEE (2015)
10. Lin, Y.L.: Co-occurrence of speech and gestures: a multimodal corpus linguistic approach to intercultural interaction. J. Pragmat. **117**, 155–167 (2017)
11. Keevallik, L.: What does embodied interaction tell us about grammar? Res. Lang. Soc. Interact. **51**(1), 1–21 (2018)
12. Vilhjálmsson, H.H.: Representing communicative function and behavior in multimodal communication. In: Esposito, A., Hussain, A., Marinaro, M., Martone, R. (eds.) Multimodal Signals: Cognitive and Algorithmic Issues. LNCS (LNAI), vol. 5398, pp. 47–59. Springer, Heidelberg (2009). https://doi.org/10.1007/978-3-642-00525-1_4
13. Arnold, L.: Dialogic embodied action: using gesture to organize sequence and participation in instructional interaction. Res. Lang. Soc. Interact. **45**(3), 269–296 (2012)
14. Kendon, A.: Semiotic diversity in utterance production and the concept of 'language'. Phil. Trans. R. Soc. B **369**, 20130293 (2014). https://doi.org/10.1098/rstb.2013.0293
15. McNeill, D.: Gesture in linguistics (2015)
16. Allwood, J.: A framework for studying human multimodal communication. In: Rojc, M., Campbell, N. (eds.) Coverbal Synchrony in Human-Machine Interaction. CRC Press, Boca Raton (2013)
17. Navarro-Cerdan, J.R., Llobet, R., Arlandis, J., Perez-Cortes, J.C.: Composition of constraint, hypothesis and error models to improve interaction in human-machine interfaces. Inf. Fusion **29**, 1–13 (2016)
18. Nevile, M.: The embodied turn in research on language and social interaction. Res. Lang. Soc. Interact. **48**(2), 121–151 (2015)
19. Hoek, J., Zufferey, S., Evers-Vermeul, J., Sanders, T.J.: Cognitive complexity and the linguistic marking of coherence relations: a parallel corpus study. J. Pragmat. **121**, 113–131 (2017)

20. Birdwhistell, R.L.: Introduction to Kinesics: An Annotation System for Analysis of Body Motion and Gesture. Department of State, Foreign Service Institute, Washington, DC (1952)
21. Adolphs, S., Carter, R.: Spoken Corpus Linguistics: From Monomodal to Multimodal, vol. 15. Routledge, London (2013)
22. Navarretta, C.: The automatic annotation of the semiotic type of hand gestures in Obama's humorous speeches. In: Proceedings of the Eleventh International Conference on Language Resources and Evaluation (LREC 2018), pp. 1067–1072 (2018)
23. Han, T., Hough, J., Schlangen, D.: Natural language informs the interpretation of iconic gestures: a computational approach. In Proceedings of the Eighth International Joint Conference on Natural Language Processing (Volume 2: Short Papers), Vol. 2, pp. 134–139 (2017)
24. Brône, G., Oben, B.: Insight interaction: a multimodal and multifocal dialogue corpus. Lang. Resour. Eval. 49(1), 195–214 (2015)
25. Paggio, P., Navarretta, C.: The Danish NOMCO corpus: multimodal interaction in first acquaintance conversations. Lang. Resour. Eval. 51(2), 463–494 (2017)
26. Lis, M., Navarretta, C.: Classifying the form of iconic hand gestures from the linguistic categorization of co-occurring verbs. In: Proceedings from the 1st European Symposium on Multimodal Communication University of Malta; Valletta, 17–18 October 2013, no. 101, pp. 41-50. Linköping University Electronic Press (2014)
27. Feyaerts, K., Brône, G., Oben, B.: Multimodality in interaction. In: Dancygier, B. (ed.) The Cambridge Handbook of Cognitive Linguistics, pp. 135–156. Cambridge University Press, Cambridge (2017)
28. Chen, L., et al.: VACE multimodal meeting corpus. In: Renals, S., Bengio, S. (eds.) MLMI 2005. LNCS, vol. 3869, pp. 40–51. Springer, Heidelberg (2006). https://doi.org/10.1007/11677482_4
29. Knight, D.: Multimodality and Active Listenership: A Corpus Approach: Corpus and Discourse. Bloomsbury, London (2011)
30. Bonsignori, V., Camiciottoli, B.C. (eds.): Multimodality Across Communicative Settings, Discourse Domains and Genres. Cambridge Scholars Publishing, Cambridge (2017)
31. Rojc, M., Mlakar, I., Kačič, Z.: The TTS-driven affective embodied conversational agent EVA, based on a novel conversational-behavior generation algorithm. Eng. Appl. Artif. Intell. 57, 80–104 (2017)
32. Cohen, J.: A coefficient of agreement for nominal scales. Educ. Psychol. Meas. 20(1), 37–46 (1960)
33. Mezza, S., Cervone, A., Tortoreto, G., Stepanov, E.A., Riccardi, G.: ISO-standard domain-independent dialogue act tagging for conversational agents (2018)
34. Verdonik, D.: Vpliv komunikacijskih žanrov na rabo diskurznih označevalcev. In: Vintar, Š. (ed.) Slovenske korpusne raziskave, (Zbirka Prevodoslovje in uporabno jezikoslovje). 1, 88–108. Znanstvena založba Filozofske fakultete, Ljubljana (2010)
35. Plutchik, R.: The nature of emotion. Am. Sci. 89, 344–350 (2001)
36. Kipp, M., Neff, M., Albrecht, I.: An annotation scheme for conversational gestures: how to economically capture timing and form. Lang. Resour. Eval. 41(3–4), 325–339 (2007)
37. Mlakar, I., Rojc, M.: Capturing form of non-verbal conversational behavior for recreation on synthetic conversational agent EVA. WSEAS Trans. Comput. 11(7), 218–226 (2012)

Lilia, A Showcase for Fast Bootstrap of Conversation-Like Dialogues Based on a Goal-Oriented System

Matthieu Riou$^{(\boxtimes)}$, Bassam Jabaian, Stéphane Huet, and Fabrice Lefèvre

LIA/CERI, Avignon University, 339 Chemin des Meinajaries,
84140 Avignon, France
{matthieu.riou,bassam.jabaian,stephane.huet,
fabrice.lefevre}@univ-avignon.fr

Abstract. Recently many works have proposed to cast human-machine interaction in a sentence generation scheme. Neural networks models can learn how to generate a probable sentence based on the user's statement along with a partial view of the dialogue history. While appealing to some extent, these approaches require huge training sets of general-purpose data and lack a principled way to intertwine language generation with information retrieval from back-end resources to fuel the dialogue with actualised and precise knowledge. As a practical alternative, in this paper, we present Lilia, a showcase for fast bootstrap of conversation-like dialogues based on a goal-oriented system. First, a comparison of goal-oriented and conversational system features is led, then a conversion process is described for the fast bootstrap of a new system, finalised with an on-line training of the system's main components. Lilia is dedicated to a chit-chat task, where speakers exchange viewpoints on a displayed image while trying collaboratively to derive its author's intention. Evaluations with user trials showed its efficiency in a realistic setup.

Keywords: Spoken dialogue systems · Chatbot ·
Goal-oriented dialogue system · On-line learning

1 Introduction

While a new avenue of research on end-to-end deep-learning-based dialogue systems has shown promising results lately [18,24,27], the need of a huge quantity of data to efficiently train these models remains a major hindrance. In the reported studies, systems are typically trained with large corpora of movie subtitles or forum data, which are suitable for modelling long, open-domain dialogues. But then, systems' developments rely on a small set of reference datasets that may be unavailable for all languages (publicly available corpora are usually in English [4,25]), or for all new domains of interest. Another difficulty is that they cannot handle entity matching between a knowledge source and utterances.

© Springer Nature Switzerland AG 2019
C. Martín-Vide et al. (Eds.): SLSP 2019, LNAI 11816, pp. 31–43, 2019.
https://doi.org/10.1007/978-3-030-31372-2_3

Despite some recent propositions to extend the range of applications of the end-to-end neural-network-based framework to task-oriented systems [10,24], the way to connect the external information to inner representation remains fundamentally unsolved [23,27].

As a consequence, classical modular architectures are still useful in many cases. They basically can be seen as a pipeline of modules processing the audio information from the user; downstream progressive treatments aim to first extract the content (speech recognition), then the meaning (semantic parsing, SP), to finally combine it with previous information (including grounding status) from the dialogue history (belief tracking). In this last module, a policy can decide from a dialogue state representation the next best action to perform according to some global criteria (generally dialogue length and success in reaching user's goal). This in-depth step of dialogue management (DM) can then supply the stream to convey the information back to the user: conversion of the dialogue manager action into utterances by the natural language generation (NLG) module followed by speech synthesis. The HIS architecture [26] offers such a setup, plunged into a global statistical framework accounting for the relationships between the data handled by the main modules of the system. Among other things it allows reinforcement learning of the DM policy. In this system some of the most sample-efficient learning algorithms had been implemented and tested [6], while on-line learning with direct interactions with the user had also been proposed [9]. Even more recently on-line learning has been generalised to the lower-level modules, SP and NLG, with protocols to control the cost of such operations during the system development (as in [8,15,16,20,27]).

HIS is meant to handle goal-oriented vocal interactions. It allows a system to exchange with users in order to address a particular need in a clearly identified field (make a hotel reservation, consult train timetables, troubleshooting, etc.). Goal-oriented dialogue systems require a database to be able to support domain specific tasks. In order to formulate system responses, entities of the database are matched with the information collected through the dialogue. The DM is responsible for making appropriate dialogue decisions according to the user goal and taking into account some uncertain information (e.g. speech recognition errors, misunderstood speech, etc.). The Partially Observable Markov Decision Process (POMDP) model [12] has been successfully employed in the Spoken Dialogue System (SDS) field [22,26] as well as in the Human Robot Interaction (HRI) context [14], due to its capacity to explicitly handle parts of the inherent uncertainty of the information which the system has to deal with (e.g. erroneous speech transcripts, falsely recognised gestures, etc.). In this setup, the agent maintains a distribution over possible dialogue states, referred to as the belief state in the literature, and interacts with its perceived environment using a reinforcement learning (RL) algorithm so as to maximise the expected cumulative discounted reward [21].

In this paper, we report on our investigations of the fast adaptation of such a system to handle conversation-like dialogues. Our underlying goal in this endeavour is to develop a system intended to be used in a neuroscience experiment.

From inside an fMRI system, users will interact with a robotic platform, vocally powered by our system, which is live-recorded and displayed inside the head-antenna. Users discuss with the system about an image and they try jointly to elaborate on the message conveyed by the image (see Sect. 3 for further details). Considering that well-performing solutions can be used directly off-the-shelf for speech recognition and synthesis, the study focuses on adapting the spoken semantic parsing and dialogue management modules only.

The remainder of this paper is organised as follows. After presenting a comparison between a goal-oriented dialogue and a conversation in Sect. 2, we present some design guidelines, forming a recipe to convert the goal-oriented dialogue system to a conversational one in Sect. 3. Section 4 provides an experimental study with human evaluations of the proposed approach and we conclude in Sect. 5.

2 Comparison of Goal-Oriented Vs Conversational Agents

On the one hand, goal-oriented dialogue agents are designed for a few particular tasks and set up to have highly-focused interactions to get information from the user to help complete the task at stake, by helping her to reach a defined goal (such as making a reservation). On the other hand, conversational systems are designed to mimic the unstructured conversational or 'chats' characteristics of human interactions [11]. The review hereafter intends to outline the most important differences between the two situations.

Of course, one must be aware that most of natural human spoken interactions are in fact a composition of goal-driven and open-minded interleaved turns. The latter in this case generally play a role of social glue between speakers, as pointed out by conversational analysis studies (e.g. in [19]). So the presentation below is somewhat artificial and solely aims at making things clearer in the purpose of the implementation of an artificial interactive system.

The most obvious difference lies in the domain covered by the interactions. In principle, goal-oriented interactions suppose a limited single-domain backdrop. Nevertheless these domains have been largely extended in the recent years, and even some possibilities exist to design multi-domain applications (see for instance [3,7]). On the contrary, conversational systems are supposed to have no limitation on the discussed topics. No such system has been built so far and this remains a research objective, mainly due to the limited understanding abilities of extant systems. It is worth mentioning here that a conversation can also happen in a restricted domain (such as banter about the weather forecast for instance). And then the distinction should be operated at other levels.

First of them, goal-oriented systems can be characterised by the existence of a back-end that the user wants to access to. It will generally be a database, but can be generalised to any knowledge source from which informational entities can be retrieved. During a conversation it is supposed that the user has no specific desire to know a particular piece of information. Even though it is not contradictory with getting to know things in a casual way, there is no incentive to do so. While conversing users are mainly interested in answers allowing them to pursue their own logic, some general knowledge is sufficient to produce responses that make sense in light of users' turns, most of the time. That is how some conversational systems could be built using a huge quantity of movie subtitles [17]. Not surprisingly, learning how to interact with a user based on movie excerpts does not end up with very coherent and purposeful reactions on behalf of the system, even when some contextual information is added [10].

Another major practical difference between goal-driven and chit-chat discussions lies in the timing. While goal-oriented systems are expected to reach the goal in the shortest possible time, it is almost the opposite for conversational ones. For these latter, the dialogue is supposed to go on as long as the speakers find some interest and motivation in the discussion (and they have available time to spend together). It arises a difficulty in using an automatic process to train such systems as one constraint guiding the learning progress (length penalty) is removed. Indeed, most of recent approaches to train DM by means of RL algorithms relied on two constraints: reach the goal (and collect a reward associated with it) and do it fast (and avoid rewards penalising each turn). Therefore, with only one remaining constraint, the kind of strategy is unclear at the end of the learning process.

Finally, a very important discrepancy between the two interaction types is the role-playing innuendo. In goal-oriented systems, the slave-to-master relationship between the user and the system is implicitly implemented, whereas when conversation is at stake, both speakers are expected to intervene at an equal level. So the conversational system becomes truly mix-initiative (in comparison to user or system-initiative systems), but this implies much more complex exchanges. More information can potentially be conveyed at each turn, with less prior expectation of what can be said. On top of all that, the conversation situation implies that the system might be able to participate in the argumentation about the discussed topics. The argumentative capacity, despite recent very promising works based on the use of game theory [2,13], remains very difficult to implement in the current architecture. This is a point that our endeavours to convert a goal-oriented system to a conversational interface have not yet been able to address appropriately, as discussed in the next section.

3   Conversion from Goal-Oriented to Conversational

In this section we present some principles behind the conversion of a goal-oriented system to its conversational version. The propositions are exemplified in the context of a new system, Lilia, intended to be used for neuroscience experiments in a near future.

The experiments presented in the paper concern a chit-chat dialogue system framed in a goal-oriented dialogue task. In this context, users discuss with the system about an image (out of a small predefined set of 6), and they tried jointly to discover the message conveyed by the image, as described in [5]. The discussion is expected to last only one minute at most. An example of a successful dialogue for the task is reported, from user trials, in Table 1, with the presented image. The dialogue is translated from French and the user inputs are given after automatic speech recognition (_asr) processing.

Table 1. Example of a successful dialogue (translated from French). Automatic transcripts of the user's inputs are given.

S Hello, my name is Lilia. Today, we've got to try to find this image's message. Maybe we could start by discussing its characteristics.
U_asr *and a head-down apple*
S And what do you feel about this fruit?
U_asr *it looks like a superhero with eyes*
S This apple looks like a hero with eyes. Concerning its look, we can say it is upside-down.
U_asr *it seems appetising and nice-looking*
S This apple with eyes really looks like a hero. Concerning its appearance, I would say nice looking.
U_asr *image want to eat more fruits and natural products superheros mean something to kids*
S I think that may be related to GMOs. In order to prove that they create strong fruits and vegetables, finally healthy.
U_asr *you're right*
S Yes, or more simply it is about recommending youth to eat more fruits in general?
U_asr *yes may be, good bye*

In order to use a goal-oriented system for such a task, the principle which has been followed is to construct, as the system's back-end, a database containing several hundreds of possible combinations of characteristics of the image. Each of them is associated with a hypothesis about the conveyed message. The overall recommended process is summarised in Algorithm 1 in its general terms, and its application to the development of Lilia is detailed hereafter.

During its interaction with the system, the user is expected to progressively provide elements about the image, which will make the system select a small subset of matching entity descriptions in the database. From this subset, it can pick other characteristics and present them as its opinion or ultimately select a pre-defined message to return as a plausible explanation of the image purpose. This would allow the user to speak rather freely about the image for several tens of seconds before arguing briefly about the message. Formally no argumentation is possible from the system's side, it can only propose canned messages. Yet by providing variants of surface forms for each of them, it is possible to simulate a progression in the system's idea of the message. For example, in the dialogue displayed in Table 1, the last two system turns are in fact issued from the same

DM dialogue act ("inform(message=GMO)") but are converted to two different utterances which give the illusion to respond to each other. Although this a very limited mechanism to mimic an argumentative capacity on behalf of the system, it appeared to work quite well during user trials, as the next section will show.

So a paramount starting point for designing the new system is to elaborate a dedicated new ontology. It should be built based not only on the expected topics but also on the targeted discussion structure. We illustrate this process for our 'image commentary' domain. The concepts chosen to describe an image have been elicited on the expectation of what a user could say about them. Here we ensure the ontology contains the elements to unroll the first part of the conversation on exchanging impressions about image characteristics. The ontology has been kept simple and generic as it is mainly based on the following concepts:

- **Is** describes physical characteristics with the following values: "nice looking", "rotten", "upside-down", etc.
- **Possesses** describes attributes of the fruit, such as: "arm", "bumps", etc.
- **Looks like** describes a resemblance of the fruit, with the following values: "human", "batman", etc.
- **Seems** describes an emotion or a feeling coming off the fruit: "sad", "tired", "appetising", etc.
- **Color** describes the main colour of the fruit.

Algorithm 1. Design guidelines for conversation-like dialogues

---

1: Enumerate possible objects of discussion → **ontology, top slot and values**
2: Elaborate a (small) set of common characteristics → **ontology, leaf slots**
3: Enumerate slot values for each object → **ontology, flat list of slot/value pairs for each object**
4: Tailor ontology to enforce dialogue structure: tag a concluding slot (possible final message of the discussion), and tag several slots as compulsory (the message can be delivered only after users have provided them) → **ontology, structure tags**
5: Generate Cartesian product of all slot/value pairs per object → produce **backend DB**
6: Use ontology to bootstrap semantic parser: keyword spotting with values (or more elaborate, as for instance using ZSSP framework [8]) → **SP**
7: Use ontology to bootstrap a set of generation templates (concatenation of single-slot templates or composition of multi-slot templates) → **NLG**
8: Multi-criteria objective function: → **reward function** for DM RL training
   - final step (e.g. informing of a particular final slot after exchanging at least several other slots, see ontology tags)
   - length penalty, to preserve global coherence
9: Train system components: → **trained SP and DM policy**
   - collect WoZ or human-human data first and batch train or
   - direct online training

For the second part of conversation, delivering a message, it has been observed two sets of images: one with damaged poor-looking fruits with human characteristics (arms, legs, eyes) and another with fruit disguised as superheros looking rather strong (as the apple in Table 1). A dedicated message has been conceived for each group: first the author's intention was to convince children that even poor-looking rotten fruits were healthy and good to eat, or fruits and vegetables in general are strong and healthy companions, as superheros are usually (for some versions of the message it has even been suggested that it could be a campaign in favour of GMO crops, see Table 1).

Those description concepts induce the system to discuss several characteristics of the image with the user, but their usage also presents some pitfalls. Firstly, when the system is discussing one concept, for example requesting about "Is", and the user answers with a characteristic of a different concept, the system may keep repeating its request while the user thinks it has answered it. Secondly, the characteristics of a given concept do not necessarily exclude each other. For example, a same fruit can have both characteristics "is=nice looking" and "is=upside-down". To implement that in the goal-oriented system, the back-end database is built as the mere Cartesian product of all the values of the ontology's slots. In the previous case this will result in two distinct DB entities for the same fruit in the database, one having the "nice looking" characteristic, and the other having the "upside-down" one.

The SP module also has to be adapted to the new task. As our goal-oriented system relies on the use of an on-line trained SP module (such as in [8]) no further manual modifications have been necessary at this step. The ontology as described above is instantiated in the model, and each concept is associated with a set of values. In Lilia, 9 concepts are considered for a total of 51 values (so 5.7 values/slot on average). Only the concept of message has been specifically addressed. As the purpose of the dialogue system is to ultimately deliver a message, the message concept can only be requested by the user. Therefore all user inputs proposing a message are labelled as a request, whatever it is said about it, to drive the system to suggest its own message in return. For all concepts the openness of the system will derive from a loose match between surface forms and concept values (the opposite of what is generally required for goal-oriented systems). SP being trained on-line, see below, it was possible to provide the trainers with instructions on how to strive to connect their vocal inputs with the ontology elements: no need to be precise as long as it allows the system to unroll its entity matching process through the turns until the final delivery of the image's message.

On the side of the DM module, the goal-oriented dialogue system was designed to receive only one dialogue act for each user input. This act could carry several concepts (for example "inform(fruit=apple,seems=strong)"), but it could not inform and request at the same time. The most essential act was extracted from the SP outputs and it was the only one to be sent to the dialogue manager. In a conversational-like dialogue, the user is very likely to produce several acts in one sentence. To handle that, all the acts are sent to the dialogue manager as if they were multiple user turns, before the system is asked to respond. As the last user input act is used by the dialogue manager as a feature to choose the next answer, the acts are reordered to have the most important at the end. Here is the complete list of acts priority, from the most important to the least. First the acts which allow the user to request something to the system and expect an immediate answer, in this order: "help", "repeat", "restart", "request", "request alternatives", "request more", "confirm". Then the acts used by the user to inform the system, on which the system would have to bounce back: "negate","deny", "inform", "affirm", "acknowledge". Finally, pragmatic acts related to the overall dialogue management: "bye", "hello", "thank you".

To allow a fast development of the system, an online RL training approach has been retained for the DM. Several instructions have been given to the expert trainers to define its reward function (how she will penalise or compliment the system for its actions, with numerical values). A conversation, by definition, is not supposed to have a precise goal. However, to be able to train the system, we made explicit the notion of success of a dialogue in this case (associated to a strong positive reward). This is a key aspect of the conversion proposed here, to be able to tag a dialogue as successful or not. So it has been proposed to consider a dialogue objectively successful when a message has been said by the system and at least two description concepts have been discussed (no matter who introduced them in the dialogue). To handle difficult examples, the users are prompted to deem a dialogue failed whenever they notice anything they consider bad (too abnormal or unnatural).

This definition of success imposes a minimal dialogue length. In order to avoid unnecessary and redundant turns, a (-1) penalty reward is given at each turn during the DM policy training. And although a conversation has no time limit, generally speaking, the assumption is made that keeping a mechanism to favour the dialogues reaching their goal swiftly is relevant.

This is coherent with a specificity of the task which is that the system does not need to learn to end the dialogue. In final experiments, the dialogue will automatically be interrupted after 1 min. In both on-line training and test phases, users were asked to end the dialogue themselves by saying bye as soon as it was successful, or when it had lasted too long already. So in a more general view this property can be preserved with an upper bound on the dialogue duration after which the system could decide to hang up.

Since the NLG module has a huge impact on user appreciation, we started with handcraft rules. Each possible dialogue act has one or a few sentence templates, for a total of roughly 80 seed templates in total. Adding different variations for a single act leads to reduce the impression of repetitions. The outputs have been specifically designed to induce the user behaviour. A small reminder of the goal is given at the start of the dialogue.

4   Evaluation

The evaluation of the converted system is presented in this section. In order to evaluate the interest of the on-line learning process, two complementary versions of the system are proposed in comparison. First, **handcraft** is a baseline version of the system without on-line learning; it uses the initial SP module (zero training) and a handcrafted (H) dialogue manager policy. Then, in order to effectively learn on-line the dialogue system, the system's developer needs to be able to both improve the SP and DM models. Therefore, an enhanced version of the system, referred to as **trained** hereafter, is obtained by replacing the initial SP module and the handcrafted dialogue manager policy by on-line learnt ones. The learning protocol proposed to achieve it, referred to as **on-line training** below, directly juxtaposes an adversarial bandit to learn the SP module and a Q-learner reinforcement learning approach to learn the dialogue manager policy following our prior work [16]. The knowledge base of the SP module as well as the DM policy are adapted after each dialogue turn.

In the experiments reported here a GUI interface has been used (a porting to the FurHat robot head platform [1] is planned for the next series in the fMRI context). The platform could rely on the I/O capacities of the Google Chrome web browser for automatic speech recognition and synthesis. Due to the cost of transcribing the user trials, no precise measure of the actual word error rate has been made; our estimation is less than 20% (with surprising variations depending on the period of the day during which the trials were carried out). The synthesis is of good quality, but cannot be used to add prosody information. So it can be perceived as a bit 'flat' every now and then, but not really disturbing, as most of the users noticed.

For **on-line training**, an expert user communicated with the system to train it. Using sample-efficient reinforcement learning algorithms allows us to converge pretty fast in terms of cumulated rewards and success rate. In our case the training session has been limited to 140 dialogues. Then a group of (mostly) naive users tested each model (48 dialogues each, so a total of 12 dialogues performed by each of our 8 users). At the end of each session, the users were asked to give a rating on a scale of 0 (worst) to 5 (best) to the understanding and generation perceived qualities of the system. The number of training dialogues, as well as the number of test sets for each configuration are recalled in Table 2, along with the results.

Table 2. Evaluation of the proposed approach with and without training

Model	Train (#dial)	Test (#dial)	Success (%)	Avg cum. Reward	Sys. Underst. Rate	Sys. Gener. Rate
Handcraft	0	94	31	−1.7	1.3	4.1
On-line training	140	96	78	9.3	2.9	4.5

The difference in performance between handcraft and on-line training models (+47% absolute in success rate) shows the impact of the SP adaptation on the overall success of the conversation, along with a better understanding (1.3 for handcraft vs. 2.9 for on-line training). The average cumulated reward rate on the test is directly correlated to the success rate and comes in confirmation of the previous observations. Also, due to a well-tuned template-based generation system, the system generation rate is high (>4) for all configurations.

From Table 3, it is possible to observe the gap in performance between the initial version of the SP module and after on-line training. For this evaluation a set of 100 utterances were randomly extracted from the user trials and their semantic annotation manually corrected. It was then possible to estimate the precision and recall of the SP outputs w.r.t. their references, and derive an overall F-measure. The measures were compared using or not the concept values in the scoring. It can be observed that after training, SP is more robust to value errors, as the gap of 5% with initial SP (65.5% vs 70.7%) is reduced to 3% (81% vs 84%). But more generally if the performance of the initial low-cost SP (65.5%) was well below standard for such system, the gap is filled after training where an 81% F-score is reached.

Table 3. Semantic Parser module evaluation: initial vs post-on-line training

Model	Complete act			Without value		
	F-Score	Precision	Recall	F-Score	Precision	Recall
Handcraft SP	65.5	60.0	72.1	70.7	65.0	77.6
Online training SP	81.0	76.3	86.5	84.0	78.9	89.8

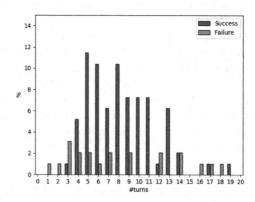

Fig. 1. Distribution of the dialogues w.r.t. the number of turns

It is worth mentioning that in complementary experiments from our prior work [16] the results obtained after **on-line training** seem to suffer of great variability, depending on the choices made by the expert training the system. The experts have a large margin of action in how they train their system: for instance, they can decide to locally reward only the correct actions (positively), or reversely, only the bad ones (negatively) or ideally, but more costly, both. Also they are free of the inputs used to train the system with: very simple to ensure a steep learning curve or more realistic to try to immediately reach the interesting regions of the DM policy state space. In any case it is worth noticing that the system performance has been shown to increase globally with time in all cases, and so a system can always be further improved to a certain level.

Some more detailed results are given in Table 4. The objective here was to determine if succeeded dialogues and failed ones have distinct features that would allow us to better handle and prevent failed cases in the future. For instance, it was hypothesised that failure could occur from more complex and long interactions from the user. But figures in Table 4 show that there is no such discrepancy between good and bad occurrences: average numbers of turns are very close (8.3 vs 8.1); the same statement applies to time durations (125 s vs 130 s), or the number of words or concepts by sentence, which are not different enough to give some clues for the reasons of failure.

Table 4. Comparison of successful and unsuccessful dialogues

Success	#dial	Avg #turns	Avg duration (seconds)	Avg #words by sentence	Avg #words by dialogue	Avg #concepts by sentence
Success	75	8.3	124.9	7.1	55.4	2.2
Failure	21	8.1	130.0	5.9	45.5	2.0
All	96	8.3	126.0	6.9	53.2	2.1

This tendency is further confirmed by looking at how succeeded and failed dialogues spread over the number of turns, as shown in the histograms of Fig. 1. The two populations, represented in two distinct series, can be compared (while we did not re-normalise the percentage at each number of turns, to make obvious the difference in population size). It can be observed that success is pretty uniformly spread in [4, 13] and failure alike, in a slightly larger interval [3, 14], with small peaks in both cases (5 for success and 3 for failure). By the way, the targeted duration of the dialogues (60 s) is on average doubled. Though departing from the instructions, it should be seen as a good point as it tends to show that users are willing to chat with the system, and are not expeditious as they could be if they had respected their guidelines giving a minute as an objective duration.

5   Conclusion

In this paper a conversion of a goal-oriented human-agent interaction system into a conversational system has been presented. After reviewing the main differences between the two types of interactions, some considerations to redesign a goal-oriented system have been proposed to handle conversation-like dialogues. This fast bootstrap of a goal-oriented system for conversation-liked dialogues is affordable in terms of development cost, and has shown an unexpected good level of performance. The user trials on a chit-chat task in French present a success rate as high as 78%, with very good perceptual ratings from the users (system's understanding and generation quality).

Acknowledgments. This workshop has been partially supported by grants ANR-16-CONV-0002 (ILCB), ANR-11-LABX-0036 (BLRI).

References

1. Al Moubayed, S., Beskow, J., Skantze, G., Granström, B.: Furhat: a back-projected human-like robot head for multiparty human-machine interaction. In: Esposito, A., Esposito, A.M., Vinciarelli, A., Hoffmann, R., Müller, V.C. (eds.) Cognitive Behavioural Systems. LNCS, vol. 7403, pp. 114–130. Springer, Heidelberg (2012). https://doi.org/10.1007/978-3-642-34584-5_9
2. Barlier, M., Perolat, J., Laroche, R., Pietquin, O.: Human-machine dialogue as a stochastic game. In: SIGDIAL, September 2015
3. Budzianowski, P., et al.: Sub-domain modelling for dialogue management with hierarchical reinforcement learning. arXiv.org (2017)
4. Casanueva, I., et al.: A benchmarking environment for reinforcement learning based task oriented dialogue management. arXiv.org, November 2017
5. Chaminade, T.: An experimental approach to study the physiology of natural social interactions. Interact. Stud. **18**(2), 254–276 (2017)
6. Daubigney, L., Geist, M., Chandramohan, S., Pietquin, O.: A comprehensive reinforcement learning framework for dialogue management optimization. Sel. Top. Sign. Proces. **6**(8), 891–902 (2012)
7. Ekeinhor-Komi, T., Bouraoui, J.L., Laroche, R., Lefèvre, F.: Towards a virtual personal assistant based on a user-defined portfolio of multi-domain vocal applications. In: 2016 IEEE SLT Workshop, pp. 106–113 (2016)
8. Ferreira, E., Jabaian, B., Lefèvre, F.: Online adaptative zero-shot learning spoken language understanding using word-embedding. In: IEEE ICASSP (2015)
9. Ferreira, E., Lefèvre, F.: Reinforcement-learning based dialogue system for human-robot interactions with socially-inspired rewards. Comput. Speech Lang. **34**(1), 256–274 (2015)
10. Ghazvininejad, M., et al.: A knowledge-grounded neural conversation model. arXiv.org (2017)
11. Jurafsky, D., Martin, J.H.: Speech and Language Processing, Chap. 29, 3rd edn., pp. 201–213. Prentice-Hall Inc., Upper Saddle River (2017). Dialogue Systems and Chatbots
12. Kaelbling, L.P., Littman, M.L., Cassandra, A.R.: Planning and acting in partially observable stochastic domains. Artif. Intell. **101**(1–2), 99–134 (1998)
13. Lewis, M., Yarats, D., Dauphin, Y.N., Parikh, D., Batra, D.: Deal or no deal? End-to-end learning for negotiation dialogues. arXiv.org (2017)
14. Lucignano, L., Cutugno, F., Rossi, S., Finzi, A.: A dialogue system for multimodal human-robot interaction. In: ICMI (2013)
15. Riou, M., Jabaian, B., Huet, S., Lefèvre, F.: Online adaptation of an attention-based neural network for natural language generation. In: INTERSPEECH (2017)

16. Riou, M., Jabaian, B., Huet, S., Lefèvre, F.: Joint on-line learning of a zero-shot spoken semantic parser and a reinforcement learning dialogue manager. In: IEEE ICASSP (2019)
17. Serban, I.V., et al.: A deep reinforcement learning Chatbot. arXiv.org (2017)
18. Serban, I.V., Sordoni, A., Bengio, Y., Courville, A., Pineau, J.: Building end-to-end dialogue systems using generative hierarchical neural network models. In: AAAI (2016)
19. Sidnell, J.: Conversation Analysis: An Introduction. Wiley-Blackwell, West Sussex (2010)
20. Su, P.H., et al.: On-line active reward learning for policy optimisation in spoken dialogue systems. In: ACL, pp. 2431–2441, Berlin, Germany, August 2016
21. Sutton, R.S., Barto, A.G.: Reinforcement learning: an introduction. IEEE Trans. Neural Netw. **9**(5), 1054 (1998)
22. Thomson, B., Young, S.: Bayesian update of dialogue state: a POMDP framework for spoken dialogue systems. Comput. Speech Lang. **24**(4), 562–588 (2010)
23. Wen, T.H., Miao, Y., Blunsom, P., Young, S.: Latent intention dialogue models. In: ICML 2017, p. 10 (2017)
24. Wen, T.H., et al.: A network-based end-to-end trainable task-oriented dialogue system. In: ACL (2017)
25. Williams, J., Raux, A., Ramachandran, D., Black, A.: The dialog state tracking challenge. In: SIGDIAL (2013)
26. Young, S., et al.: The hidden information state model: a practical framework for POMDP-based spoken dialogue management. Comput. Speech Lang. **24**(2), 150–174 (2010)
27. Zhao, T., Eskenazi, M.: Zero-shot dialog generation with cross-domain latent actions. arXiv.org, May 2018

Recent Advances in End-to-End Spoken Language Understanding

Natalia Tomashenko[1]([⊠]), Antoine Caubrière[2], Yannick Estève[1], Antoine Laurent[2], and Emmanuel Morin[3]

[1] LIA, University of Avignon, Avignon, France
{natalia.tomashenko,yannick.esteve}@univ-avignon.fr
[2] LIUM, University of Le Mans, Le Mans, France
{antoine.caubriere,antoine.laurent}@univ-lemans.fr
[3] LS2N, University of Nantes, Nantes, France

Abstract. This work investigates spoken language understanding (SLU) systems in the scenario when the semantic information is extracted directly from the speech signal by means of a single end-to-end neural network model. Two SLU tasks are considered: named entity recognition (NER) and semantic slot filling (SF). For these tasks, in order to improve the model performance, we explore various techniques including speaker adaptation, a modification of the connectionist temporal classification (CTC) training criterion, and sequential pretraining.

Keywords: Spoken language understanding (SLU) · Acoustic adaptation · End-to-end SLU · Slot filling · Named entity recognition

1 Introduction

Spoken language understanding (SLU) is a key component of conversational artificial intelligence (AI) applications. Traditional SLU systems consist of at least two parts. The first one is an automatic speech recognition (ASR) system that transcribes acoustic speech signal into word sequences. The second part is a natural language understanding (NLU) system which predicts, given the output of the ASR system, named entities, semantic or domain tags, and other language characteristics depending on the considered task. In classical approaches, these two systems are often built and optimized independently.

Recent progress in deep learning has impacted many research and industrial domains and boosted the development of conversational AI technology. Most of the state-of-the art SLU and conversational AI systems employ neural network models. Nowadays there is a high interest of the research community in end-to-end systems for various speech and language technologies. A few recent papers [5,11,16,18,21,24] present ASR-free end-to-end approaches for SLU tasks and show promising results. These methods aim to learn SLU models from acoustic signal without intermediate text representation. Paper [5] proposed an audio-to-intent architecture for semantic classification in dialog systems. An encoder-decoder framework [26] is used in [24] for

© Springer Nature Switzerland AG 2019
C. Martín-Vide et al. (Eds.): SLSP 2019, LNAI 11816, pp. 44–55, 2019.
https://doi.org/10.1007/978-3-030-31372-2_4

domain and intent classification, and in [16] for domain, intent, and argument recognition. A different approach based on the model trained with the connectionist temporal classification (CTC) criterion [13] was proposed in [11] for named entity recognition (NER) and slot filling. End-to-end methods are motivated by the following factors: possibility of better information transfer from the speech signal due to the joint optimization on the final objective criterion, and simplification of the overall system and elimination of some of its components. However, deep neural networks and especially end-to-end models often require more training data to be efficient. For SLU, this implies the demand of big semantically annotated corpora. In this work, we explore different ways to improve the performance of end-to-end SLU systems.

2 SLU Tasks

In SLU for human-machine conversational systems, an important task is to automatically extract semantic concepts or to fill in a set of *slots* in order to achieve a goal in a human-machine dialogue. In this paper, we consider two SLU tasks: named entity recognition (NER) and semantic slot filling (SF). In the NER task, the purpose is to recognize information units such as names, including person, organization and location names, dates, events and others. In the SF task, the extraction of wider semantic information is targeted. These last years, NER and SF where addressed as word labelling problems, through the use of the classical BIO *(begin/inside/outside)* notation. For instance, "*I would like to book three double rooms in Paris for tomorrow*" will be represented for the NER and SF task as the following BIO labelled sentences:

- NER: "*I::$\emptyset$ would::$\emptyset$ like::$\emptyset$ to::$\emptyset$ book::$\emptyset$ three::B-amount double::$\emptyset$ rooms::$\emptyset$ in::$\emptyset$ Paris::B-location/city for::$\emptyset$ tomorrow::B-time/date*".
- SF: "*I::B-command would::I-command like::I-command to::I-command book::I-command three::B-room/number double::B-room/type rooms::I-room/type in::$\emptyset$ Paris::B-location/city for::$\emptyset$ tomorrow::B-time/date*".

In this paper, similarly to [11], the BIO representation is abandoned in profit to a chunking approach. For instance for NER, the same sentence will be presented as:

- NER: "*I would like to book < amount three > double rooms in < location/city Paris > for < time/date tomorrow >*".

In this study, we train an end-to-end neural model to reproduce such textual representation from speech. Since our neural model emits characters, we use specific characters corresponding to each opening tag (one by named entity category or one by semantic concept), while the same symbol is used to represent the closing tag.

3 Model Training

End-to-end training of SLU models is realized through the recurrent neural network (RNN) architecture and CTC loss function [13] as shown in Fig. 1. A spectrogram of power normalized audio clips calculated on 20 ms windows is used as the input features for the system. As shown in Fig. 1, it is followed by two 2D-invariant (in the time and-frequency domain) convolutional layers, and then by five BLSTM layers with sequence-wise batch normalization. A fully connected layer is applied after BLSTM layers, and the output layer of the neural network is a softmax layer. The model is trained using the CTC loss function. The neural architecture is similar to the Deep Speech 2 [1] for ASR.

The outputs of the network depend on the task. For ASR, the outputs consist of graphemes of a corresponding language, a *space* symbol to denote word boundaries and a *blank* symbol. For NER, in addition to ASR outputs, we add outputs corresponding to named entity types and a closing symbol for named entities. In the same way, for SF, we use all ASR outputs and additional tags corresponding to semantic concepts and a closing symbol for semantic tags.

In order to improve model training, we investigate speaker adaptive training (SAT), pretraining and transfer learning approaches. First, we formalize the *-mode*, that proved its effectiveness in all our previous and current experiments.

3.1   CTC Loss Function Interpretation Related to *-mode*

The CTC loss function [13] is relevant to train models for ASR without Hidden Markov Models. The *-mode* can be seen as a minor modification of the CTC loss function.

CTC Loss Function Definition. By means of a many-to-one $\mathcal{B}$ mapping function, CTC transforms a sequence of the network outputs, emitted for each acoustic frame, to a sequence of final target labels by deleting repeated output labels and inserting a *blank* (*no label*) symbol. The CTC loss function is defined as:

$$\mathcal{L}_{CTC} = - \sum_{(\mathbf{x},l)\in Z} \ln P(l|\mathbf{x}), \qquad (1)$$

where $\mathbf{x}$ is a sequence of acoustic observations, l is the target output label sequence, and Z the training dataset. $P(l|\mathbf{x})$ is defined as:

$$P(l|\mathbf{x}) = \sum_{\pi \in \mathcal{B}^{-1}(l)} P(\pi|\mathbf{x}), \qquad (2)$$

where π is a sequence of initial output labels emitted by the model for each input frame. To compute $P(\pi|\mathbf{x})$ we use the probability of the output label π_t emitted by the neural model for frame t to build this sequence. This probability is modeled by the value $y_{\pi_t}^t$ given by the output node of the neural model related to the label π_t. $P(\pi|\mathbf{x})$ is defined as $P(\pi|\mathbf{x}) = \prod_t^T y_{\pi_t}^t$, where T denotes the number of frames.

CTC Loss Function and *-mode*. In the framework of the *-mode*, we introduce a new symbol, "$\star$", that represents the presence of a label (the opposite of the *blank* symbol) that does not need to be disambiguated. We expect to build a model that is more discriminant on the important task-specific labels. For example, for the SF SLU task important labels are the ones corresponding to semantic concept opening and closing tags, and characters involved in the word sequences that support the value of these semantic concepts (*i.e* characters occurring between an opening and a closing concept tag). In the CTC loss function framework, the *⋆-mode* consists in applying another kind of mapping function before $\mathcal{B}$. While $\mathcal{B}$ converts a sequence π of initial output labels into the final sequence l to be retrieved, we introduce the mapping function $\mathcal{S}$ that is applied to each final target output label. Let C be the set of elements l_i included in subsequences $l_a^b \subset l$ such as l_a is an opening concept tag and l_b the associated closing tag; i, a and b are indexes that handle positions in sequence l, and $a \leq i \leq b$. Let V be the vocabulary of all the symbols present in sequences l in Z, and let consider the new symbol $\star \notin V$. Let define $V^\star = V \cup \{\star\}$, and L (resp. $L^\star$) the set of all the label sequences that can be generated from V (resp. $V^\star$).

Considering n as the number of elements in l, m an integer such as $m \leq n$, we define the mapping function $\mathcal{S} : L \rightarrow L^\star, l \mapsto l'$ in two steps:

$$1.\ \forall l_j \in l \quad \begin{cases} l_j \notin C \Rightarrow l'_j = \star \\ l_j \in C \Rightarrow l'_j = l_j \end{cases}$$
$$2.\ \forall l'_j \in l' \quad l'_{j-1} = \star \Rightarrow l'_j = \emptyset \tag{3}$$

By applying $\mathcal{S}$ on the last example sentence used in Sect. 2 for NER, this sentence is transformed to:

- sentence: "*I would like to book < amount three > double rooms in < location/city Paris > for <time/date tomorrow >*".
- $\mathcal{S}$(sentence): "*⋆ < amount three > ⋆ < location/city Paris > ⋆ < time/date tomorrow >*".

To introduce *⋆-mode* in the CTC loss function definition, we modify the formulation of $P(l|\mathbf{x})$ in formula (2) by introducing the $\mathcal{S}$ mapping function applied to l:

$$P(l|\mathbf{x}) = \sum_{\pi \in \mathcal{B}^{-1} \circ \mathcal{S}(l)} P(\pi|\mathbf{x}). \tag{4}$$

3.2 Speaker Adaptive Training

Adaptation is an efficient way to reduce the mismatches between the models and the data from a particular speaker or channel. For many years, acoustic model adaptation has been a key component of any state-of-the-art ASR system. For end-to-end approaches, speaker adaptation is less studied, and most of the first end-to-end ASR systems do not use any speaker adaptation and are built on spectrograms [1] or filterbank features [2]. However, some recent works [7,28] have demonstrated the effectiveness of speaker adaptation for end-to-end models.

For SLU tasks, there is also an emerging interest in the end-to-end models which have a speech signal as input. Thus, acoustic, and particularly speaker, adaptation for such models can play an important role in improving the overall performance of these systems. However, to our knowledge, there is no research on speaker adaptation for end-to-end SLU models, and the existing works do not use any speaker adaptation.

One way to improve SLU models which we investigate in this paper is speaker adaptation. We apply i-vector based speaker adaptation [23]. The proposed way of integration of i-vectors into the end-to-end model architecture is shown in Fig. 1. Speaker i-vectors are appended to the outputs of the last (second) convolutional layer, just before the first recurrent (BLSTM) layer. In this paper, for better initialization, we first train a model with *zero pseudo i-vectors* (all values are equal to 0). Then, we use this pretrained model and fine-tune it on the same data but with the real i-vectors. This approach was inspired by [6], where an idea of using zero auxiliary features during pretraining was implemented for language models and in our preliminary experiments it demonstrated better results than direct model training with i-vectors [27].

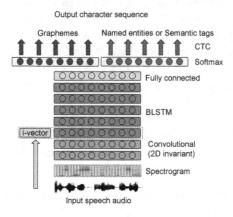

Fig. 1. Universal end-to-end deep neural network model architecture for ASR, NER and SF tasks. Depending on the task, the set of the output characters consists of: (1) ASR: graphemes for a given language; (2) NER: graphemes and named entity tags; (3) SF: graphemes and semantic SF tags.

3.3   Transfer Learning

Transfer learning is a popular and efficient method to improve the learning performance of the target predictive function using knowledge from a different source domain [19]. It allows to train a model for a given target task using available out-of-domain source data, and hence to avoid an expensive data labeling process, which is especially useful in case of low-resource scenarios.

In this paper, for SF, we investigate the effectiveness of transfer learning for various source domains and tasks: (1) ASR in the target and out-of-domain languages; (2) NER in the target language; (3) SF. For all the tasks, we used similar model architectures (Sect. 4.2 and Fig. 1). The difference is in the text data preparation and output targets. For training ASR systems, the output targets correspond to alphabetic characters and a *blank* symbol. For NER tasks, the output targets include all the ASR targets and targets corresponding to named entity tags. We have several symbols corresponding to named entities (in the text these characters are situated before the beginning of a named entity, which can be a single word or a sequence of several words) and a one tag corresponding to the end of the named entity, which is the same for all named entities. Similarly, for SF tags, we use targets corresponding to the semantic concept tags and one tag corresponding to the end of a concept. Transfer learning is realized through the chain of consequence model training on different tasks. For example, we can start from training an ASR model on audio data and corresponding text transcriptions. Then, we change the softmax layer in this model by replacing the targets with the SF targets and continue training on the corpus annotated with semantic tags. Further in the paper, we denote this type of chain as $ASR{\rightarrow}SF$. Models in this chain can be trained on different corpora, that can make this method useful in low-resource scenario when we do not have enough semantically annotated data to train an end-to-end model, but have sufficient amount of data annotated with more general concepts or only transcribed data. For NER, we also investigates the knowledge transfer from ASR.

Table 1. Corpus statistics for ASR, NER and SF tasks.

Task	Corpora	Size, h	#Speakers
ASR train	EPAC, ESTER 1,2, ETAPE, REPERE, DECODA, MEDIA, PORTMEDIA	404.6	12518
NER train	EPAC, ESTER 1,2, ETAPE, REPERE	323.8	7327
NER dev	ETAPE (dev)	6.6	152
NER test	ETAPE (test), Quaero (test)	12.3	474
SF train	1. MEDIA (train),	16.1	727
	2. PORTMEDIA (train)	7.2	257
SF dev	MEDIA (dev)	1.7	79
SF test	MEDIA (test)	4.8	208

4 Experiments

4.1 Data

Several publicly available corpora have been used for experiments (see Table 1).

ASR Data. The corpus for ASR training was composed of corpora from various evaluation campaigns in the field of automatic speech processing for French. The EPAC [9], ESTER 1,2 [10], ETAPE [14], REPERE [12] contain transcribed speech in French from TV and radio broadcasts. These data were originally in the microphone channel and for experiments in this paper were downsampled from 16 kHz to 8 kHz, since the test set for our main target task (SF) consists of telephone conversations. The DECODA [3] corpus is composed of dialogues from the call-center of the Paris transport authority. The MEDIA [4,8] and PORTMEDIA [17] are corpora of dialogues simulating a vocal tourist information server. The target language in all experiments is French. For experiments with transfer learning from ASR built in a different source language to SF in the target language, we used the TED-LIUM corpus [22]. This publicly available dataset contains 1495 TED talks in English that amount to 207 h of speech data from 1242 speakers.

NER Data. To train the NER system, we used the following corpora: EPAC, ESTER 1,2, ETAPE, and REPERE. These corpora contain speech with text transcriptions and named entity annotation. The named entity annotation is performed following the methodology of the Quaero project [15]. The taxonomy is composed of 8 main types: *person, function, organization, location, product, amount, time*, and *event*. Each named entity can be a single word or a sequence of several words. The total amount of annotated data is 112 h. Based on this data, a classical NER system was trained using *NeuroNLP2*[1] to automatically extract named entities for the rest 212 h of the training corpus. This was done in order to increase the amount of the training data for NER. Thus, the total amount of audio data to train the NER system is about 324 (112 + 212) h. The development part of the ETAPE corpus was used for development, and as a test set we used the ETAPE test and Quaero test datasets.

SF Data. The following two French corpora, dedicated to semantic extraction from speech in a context of human/machine dialogues, were used in the current experiments: MEDIA and PORTMEDIA. The corpora have manual transcription and conceptual annotation [8,29]. The MEDIA corpus is related to the hotel booking domain, and its annotation contains 76 semantic tags: *room number, hotel name, location, date*, etc. The PORTMEDIA corpus is related to the theater ticket reservation domain and its annotation contains 35 semantic tags which are very similar to the tags used in the MEDIA corpus. For joint training on these corpora, we used a combined set of 86 semantic tags.

4.2 Models

We used the *deepspeech.torch* implementation[2] for training speaker independent (SI) models, and our modification of this implementation to integrate speaker adaptation. The open-source *Kaldi* toolkit [20] was used to extract 100-dimensional speaker i-vectors. All models had similar topology (except for the number of outputs) shown in Fig. 1 for SAT models. SI models were trained in the same way, but without i-vector integration. Input features are spectrograms. They are followed by two 2D-invariant (in the time and-frequency domain) convolutional layers[3], and then by five 800-dimensional BLSTM layers with sequence-wise batch normalization. A fully connected layer is applied after BLSTM layers, and the output layer of the neural network is a softmax layer. The size of the output layer depends on the task (see Sect. 4.3). The model is trained using the CTC loss function.

4.3 Tasks

The target tasks for us are NER and SF. For each of this task, other tasks can be used for knowledge transfer. To train NER, we use ASR for transfer learning. To train SF, we use ASR on French and English, NER and another auxiliary SF task for transfer learning. Hence, we consider the following set of tasks:

---

[1] https://github.com/XuezheMax/NeuroNLP2.
[2] https://github.com/SeanNaren/deepspeech.pytorch.
[3] With parameters: kernel size = (41, 11), stride = (2, 2), padding = (20, 5).

- ASR_F – French ASR with 43 outputs {French characters, *blank* symbol}.
- ASR_E – English ASR with 28 outputs {English characters, *blank* symbol}.
- NER – French NER with 52 outputs {43 outputs from ASR_F, 8 outputs corresponding to named entity tags, 1 output corresponding to the closing tag for all named entities}.
- SF_1 – target SF task with 130 outputs {43 outputs from ASR_F, 86 outputs for semantic slot tags, 1 output for the closing tag}; trained on the training part of the MEDIA corpus.
- SF_{1+2} – auxiliary SF task; trained on the MEDIA plus PORTMEDIA training corpora.
- $NER^\star$, $SF_1^\star$ – for the target tasks NER and SF_1, we also considered $\star$-mode (Sect. 3.1).

4.4   Results for NER

Performance of NER was evaluated in terms of *precision*, *recall*, and *F-measure*. Results for different training chains for speaker-independent (SI) and speaker adaptive training models (SAT) are given in Table 2. We can see, that pretraining with ASR_F task does not lead to significant improvement in performance. When the $NER^\star$ is added to the training chain, it improves all the evaluation measures. In particular, F-measure is increased by 1.9% absolute. For each training chain, we trained a corresponding chain with speaker adaptation. Results for SAT models are given in the right part of Table 2. For all training chains, SAT models outperform SI models. The best result with SAT (F-measure 71.8%) outperforms the best SI result by 1.1% absolute.

Table 2. NER results on the test dataset in terms of Precision (P,%), Recall (R,%) and F-measure (F, %) for SI and SAT models.

Model training	SI			SAT		
	P	R	F	P	R	F
NER	78.9	60.7	68.6	80.9	60.9	69.5
$ASR_F \rightarrow NER$	80.5	60.0	68.8	80.2	61.7	69.7
$ASR_F \rightarrow NER \rightarrow NER^\star$	82.1	62.1	70.7	**83.1**	**63.2**	**71.8**

4.5   Results for SF

SF performance was evaluated in terms of *F-measure, concept error rate* (CER) and *concept value error rate* (CVER).

Results for different training chains for speaker-independent (SI) models on the test set are given in Table 3 (#1–8). The first line SF_1 shows the baseline result on the test MEDIA dataset for the SF task, when a model was trained directly on the target task using in-domain data for this task (training part of the MEDIA corpus). The second line SF_{1+2} corresponds to the case when the model was trained on the auxiliary SF task. Other lines in the table correspond to different training chains described in Sect. 3.3. In #4, we can see a chain that starts from training an ASR model for English. We can

Table 3. SF performance results on the MEDIA test dataset for end-to-end SF models trained with different transfer learning approaches. Results are given in terms of F-measure (F), CER and CVER metrics (%); SF_1 – target task; SF_{1+2} – auxiliary task; **F** and **E** refer to the languages. For the best models, the results in blue correspond to decoding using beam search with a LM.

Model training	SI				SAT			
	#	F	CER	CVER	#	F (LM)	CER (LM)	CVER (LM)
SF_1	1	72.5	39.4	52.7				
SF_{1+2}	2	73.2	39.0	50.1				
$SF_{1+2} \rightarrow SF_1$	3	77.4	33.9	44.9				
$ASR_E \rightarrow SF_{1+2} \rightarrow SF_1$	4	81.3	28.4	37.3				
$ASR_F \rightarrow SF_{1+2} \rightarrow SF_1$	5	85.9	21.7	28.4	9	87.5	19.4	25.4
$NER \rightarrow SF_{1+2} \rightarrow SF_1$	6	86.4	20.9	27.5	10	87.3	19.5	26.0
$ASR_F \rightarrow SF_{1+2} \rightarrow SF_1^\star$	7	85.9	21.2	27.9	11	**87.7 (89.2)**	18.8 (16.5)	25.5 **(20.8)**
$NER \rightarrow SF_{1+2} \rightarrow SF_1^\star$	8	87.1	19.5	27.0	12	87.6 **(89.2)**	**18.6 (16.2)**	**24.6 (20.8)**

Table 4. SF performance results on the MEDIA test dataset for different systems.

Systems in literature:	CER	Systems in this paper:	CER
Pipeline: ASR+SLU, [25]	19.9	—greedy mode	18.6
End-to-end, [11]	27.0	—beam search with LM	16.2

observe that using a pretrained ASR model from a different language can significantly (16.2% of relative CER reduction) improve the performance of the SF model (#4 vs #3). This result is noticeable since it shows that we can take benefit from linguistic resources from another language in case of lack of data for the target one. Using an ASR model trained in French (#5) provides better improvement: 36.0% of relative CER reduction (#5 vs #3). When we start the training process from a NER model (#6) we can observe slightly better results. Further, for the best two model training chains (#5 and 6) we trained corresponding models in $\star$-mode (#7 and 8). Results with speaker adaptation for four best models are shown in the right part of Table 3 (#9–12). We can see that SAT models show better results than SI ones. For CVER, we can observe a similar tendency. The results for the best models using beam search and a 4-gram LM are shown in brackets in blue. The LM was built on the texts including "$\star$". Finally, Table 4 resumes our best results (in greedy and beam search modes) and shows the comparison results on the MEDIA dataset from other works [11,25]. We can see, that the reported results significantly outperform the results reported in the literature for the current task.

Error Analysis. In the training corpus, different semantic concepts have different number of samples, that may impact the SF performance. Figure 2 demonstrates the relation between the concept error rate (CER) of a particular semantic concept and its frequency in the training corpus. Each point in Fig. 2 corresponds to a particular semantic concept. For rare tags, the distribution of errors has larger variance and means than for more frequent tags. In addition, we are interested in the distribution of different types

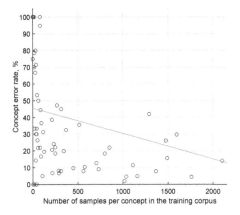

Fig. 2. Concept error rate (CER,%) results on the MEDIA test dataset for different concepts depending on the number of corresponding concepts in the training corpus. The CER results are given for the SAT model (#12), decoding with beam search and a 4-gram LM.

of SF errors (*deletions*, *insertions* and *substitutions*), which is shown in the form of a confusion matrix in Fig. 3. For better representation, we first ordered the concepts in descending order by the total number of errors. Then, we chose the first 36 concepts which have the biggest number of errors. The total amount of errors of the chosen 36 concepts corresponds to 90% of all the errors for all concepts in the test MEDIA dataset.

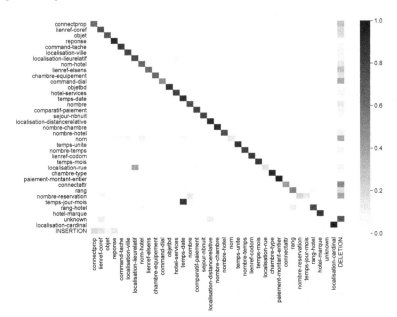

Fig. 3. Confusion matrix for concepts on the MEDIA test dataset. The last row and last column represent insertion and deletion errors correspondingly. The CER results are given for the SAT model (#12), decoding with beam search and a 4-gram LM.

The diagonal corresponds to the correctly detected concepts and other elements (except for the last row and last column) correspond to the substitution errors. The final raw represents insertion errors and the final column – deletions. Each element in the matrix shows the total number of the corresponding events (*correctly recognized concept, substitution, deletion or insertion*) normalized by the total number of such events in the row. The most frequent errors are deletions (50% of all errors), then substitutions (32.3%) and insertions (17.7%).

5 Conclusions

In this paper, we have investigated several ways to improve the performance of end-to-end SLU systems. We demonstrated the effectiveness of speaker adaptive training and various transfer learning approaches for two end-to-end SLU tasks: NER and SF. In order to improve the quality of the SF models, during the training, we proposed to use knowledge transfer from an ASR system in another language and from a NER in a target language. Experiments on the French MEDIA test corpus demonstrated that using knowledge transfer from the ASR in English improves the SF model performance by about 16% of relative CER reduction for SI models.

The improvement from the transfer learning is greater when the ASR model is trained on the target language (36% of relative CER reduction) or when the NER model in the target language is used for pretraining. Another contribution concerns SAT training for SLU models – we demonstrated that this can significantly improve the model performance for NER and SF.

Acknowledgements. This work was supported by the French ANR Agency through the ON-TRAC project, under the contract number ANR-18-CE23-0021-01, and by the RFI Atlanstic2020 RAPACE project.

References

1. Amodei, D., et al.: Deep speech 2: end-to-end speech recognition in English and Mandarin. In: International Conference on Machine Learning, pp. 173–182 (2016)
2. Bahdanau, D., Chorowski, J., Serdyuk, D., Brakel, P., Bengio, Y.: End-to-end attention-based large vocabulary speech recognition. In: ICASSP, pp. 4945–4949. IEEE (2016)
3. Bechet, F., Maza, B., Bigouroux, N., Bazillon, T., El-Beze, M., et al.: DECODA: a call-centre human-human spoken conversation corpus. In: LREC, pp. 1343–1347 (2012)
4. Bonneau-Maynard, H., Ayache, C., Bechet, F., et al.: Results of the French Evalda-Media evaluation campaign for literal understanding. In: LREC (2006)
5. Chen, Y.P., Price, R., Bangalore, S.: Spoken language understanding without speech recognition. In: ICASSP (2018)
6. Deena, S., et al.: Semi-supervised adaptation of RNNLMs by fine-tuning with domain-specific auxiliary features. In: INTERSPEECH, pp. 2715–2719. ISCA (2017)
7. Delcroix, M., Watanabe, S., Ogawa, A., Karita, S., Nakatani, T.: Auxiliary feature based adaptation of end-to-end ASR systems. In: INTERSPEECH, pp. 2444–2448 (2018)
8. Devillers, L., et al.: The French MEDIA/EVALDA project: the evaluation of the understanding capability of spoken language dialogue systems. In: LREC (2004)

9. Estève, Y., Bazillon, T., Antoine, J.Y., Béchet, F., Farinas, J.: The EPAC corpus: manual and automatic annotations of conversational speech in French broadcast news. In: LREC (2010)
10. Galliano, S., et al.: The ESTER 2 evaluation campaign for the rich transcription of French radio broadcasts. In: Interspeech (2009)
11. Ghannay, S., Caubrière, A., Estève, Y., et al.: End-to-end named entity and semantic concept extraction from speech. In: SLT, pp. 692–699 (2018)
12. Giraudel, A., Carré, M., Mapelli, V., Kahn, J., Galibert, O., Quintard, L.: The REPERE corpus: a multimodal corpus for person recognition. In: LREC, pp. 1102–1107 (2012)
13. Graves, A., et al.: Connectionist temporal classification: labelling unsegmented sequence data with recurrent neural networks. In: Proceedings of the 23rd International Conference on Machine Learning, pp. 369–376. ACM (2006)
14. Gravier, G., Adda, G., Paulson, N., et al.: The ETAPE corpus for the evaluation of speech-based TV content processing in the French language. In: LREC (2012)
15. Grouin, C., Rosset, S., Zweigenbaum, P., Fort, K., Galibert, O., Quintard, L.: Proposal for an extension of traditional named entities: from guidelines to evaluation, an overview. In: Proceedings of the 5th Linguistic Annotation Workshop, pp. 92–100 (2011)
16. Haghani, P., et al.: From audio to semantics: approaches to end-to-end spoken language understanding. arXiv preprint arXiv:1809.09190 (2018)
17. Lefèvre, F., et al.: Robustness and portability of spoken language understanding systems among languages and domains: the PortMedia project (in French), pp. 779–786 (2012)
18. Lugosch, L., Ravanelli, M., Ignoto, P., Tomar, V.S., Bengio, Y.: Speech model pre-training for end-to-end spoken language understanding. arXiv preprint arXiv:1904.03670 (2019)
19. Pan, S.J., Yang, Q.: A survey on transfer learning. IEEE Trans. Knowl. Data Eng. **22**(10), 1345–1359 (2010)
20. Povey, D., Ghoshal, A., et al.: The Kaldi speech recognition toolkit. In: ASRU (2011)
21. Qian, Y., Ubale, R., et al.: Exploring ASR-free end-to-end modeling to improve spoken language understanding in a cloud-based dialog system. In: ASRU, pp. 569–576 (2017)
22. Rousseau, A., Deléglise, P., Esteve, Y.: Enhancing the TED-LIUM corpus with selected data for language modeling and more ted talks. In: LREC, pp. 3935–3939 (2014)
23. Saon, G., Soltau, H., Nahamoo, D., Picheny, M.: Speaker adaptation of neural network acoustic models using i-vectors. In: ASRU, pp. 55–59 (2013)
24. Serdyuk, D., Wang, Y., Fuegen, C., Kumar, A., Liu, B., Bengio, Y.: Towards end-to-end spoken language understanding. arXiv preprint arXiv:1802.08395 (2018)
25. Simonnet, E., et al.: Simulating ASR errors for training SLU systems. In: LREC 2018 (2018)
26. Sutskever, I., Vinyals, O., Le, Q.V.: Sequence to sequence learning with neural networks. In: Advances in Neural Information Processing Systems, pp. 3104–3112 (2014)
27. Tomashenko, N., Caubrière, A., Estève, Y.: Investigating adaptation and transfer learning for end-to-end spoken language understanding from speech. In: Interspeech, Graz, Austria (2019)
28. Tomashenko, N., Estève, Y.: Evaluation of feature-space speaker adaptation for end-to-end acoustic models. In: LREC (2018)
29. Vukotic, V., Raymond, C., Gravier, G.: Is it time to switch to word embedding and recurrent neural networks for spoken language understanding? In: Interspeech (2015)

Language Analysis and Generation

A Study on Multilingual Transfer Learning in Neural Machine Translation: Finding the Balance Between Languages

Adrien Bardet[(✉)], Fethi Bougares[(✉)], and Loïc Barrault[(✉)]

LIUM, Le Mans University, Le Mans, France
{adrien.bardet,fethi.bougares,loic.barrault}@univ-lemans.fr

Abstract. Transfer learning is an interesting approach to tackle the low resource languages machine translation problem. Transfer learning, as a machine learning algorithm, requires to make several choices such as selecting the training data and more particularly language pairs and their available quantity and quality. Other important choices must be made during the preprocessing step, like selecting data to learn subword units, the subsequent model's vocabulary. It is still unclear how to optimize this transfer. In this paper, we analyse the impact of such early choices on the performance of the systems. We show that systems performance are depending on quantity of available data and proximity of the involved languages as well as the protocol used to determined the subword units model and consequently the vocabulary. We also propose a multilingual approach to transfer learning involving a universal encoder. This multilingual approach is comparable to a multi-source transfer learning setup where the system learns from multiple languages before the transfer. We analyse subword units distribution across different languages and show that, once again, preprocessing choices impact systems overall performance.

Keywords: Transfer learning · Machine translation · Languages proximity ·
Data quantity · Subwords distribution · Languages balance · Data balance ·
Multilingual

1 Introduction

Some major technical advances have allowed neural systems to become the most efficient approach to machine translation when a large amount of data is available [1, 19]. However, when small amounts of training data are available, neural systems struggle to obtain good performance [13]. Transfer learning consists in training a first neural machine translation (NMT) system on another language pair where a larger quantity of data is available. This already trained system is then adapted to the new data from the low resource language pair with the aim of getting better performance than a system trained only on few data.

Transfer learning in machine translation results in learning a first system called "parent" system on abundant data. Then use this learned system as a basis for the "children" systems. This basis allows the system to learn the new data. It is comparable to domain

ⓒ Springer Nature Switzerland AG 2019
C. Martín-Vide et al. (Eds.): SLSP 2019, LNAI 11816, pp. 59–70, 2019.
https://doi.org/10.1007/978-3-030-31372-2_5

adaptation where the domain is another language. The transfer generally improves the results of the child system which benefit from knowledge learned during the training of the parent system [20]. In this paper we show that the data used to train the parent system significantly impact the resulting child system. Many factors must be taken into account like data quantity, languages proximity, data preprocessing, etc. Several studies deal with quantities of data used as well as proximity of involved languages, and conclusions diverge [5, 12].

2   Related Work

Currently, neural systems require a large body of training corpus to achieve good performance, which by definition is problematic for low resource language pairs. Phrase-based statistical machine translation systems appears then as a relevant alternative [13].

Several multilingual automatic translation approaches have been developed to translate texts from low resource language pairs [7, 10, 15]. The use of universal encoders and decoders allowed [10] to design a learning system that manages several language pairs in parallel and achieves better results, especially for less-endowed languages. Specific symbols (e.g. <2s>) are used to control the output language of the universal decoder. This kind of model even makes it possible to translate pairs of languages that are not seen during training (so-called *zero-shot learning*). However, performance in such cases remains relatively low. In the same line, [17] explore different parameter sharing schemes in a multilingual model.

The choice of level of representation of words is important. [6, 16] have shown that the use of sub-lexical symbols shared between languages of the parent and child models results in an increase of transfer performance. In our work, we use this method by exploring different amounts of symbols (i.e. different vocabulary sizes).

Transfer learning tries to overcome the problem by relying on a parent system (trained on a large amount of data) that serves as the basis for learning a child system [20]. The transfer is more effective when languages are shared between the parent and the child. In this direction, [5] highlights the importance of proximity of languages in order to obtain a transfer of better quality. These observations are contradicted in [12] where better results are obtained with more distant but better endowed language pairs.

The work presented in this paper extends those of [12] and [5] on several points. Like [12], we try to evaluate the performance of the child system according to the data used in the parent system, still considering the criteria of proximity of language and quantity of data. In this study, we also consider a parent system consisting of a universal encoder (trained over several languages). We study different choices to be made for preprocessing data and the parameters of the translation model and we will try to determine the best configuration.

The objective is to better understand the correlation between the impact of the amount of data and the proximity of languages on the performance of the child system. We will see that our experiences contradict some of the conclusions of the articles mentioned above.

3 Data

In order to perform transfer learning, we need multiple language pairs. We selected the Estonian→English language pair as our low resource language pair. Our goal is to have the best possible results for this language pair, thus we need another language pair to train the parent system for the transfer.

3.1 Data Selection

We use data provided in WMT2018 machine translation evaluation campaign [2]. 2.5 million (41M words source side and 52M target side) parallel sentences are available for the Estonian→English language pair. We can not consider this as a low resource language pair, but this quantity remains low for training an NMT system that achieves good results. To assess the impact of language proximity in the parent system, we use two pairs of different languages with different proximity to Estonian. The first pair is Finnish→English. Finnish is close to Estonian since both are Finno-Ugric languages. 5 million parallel sentences are available for this language pair which represent 78M words source side and 114M target side. As distant language pair we have chosen German→English. German is a Germanic language further away from Finnish and Estonian for which 40 million parallel sentences are available corresponding to 570M words source side and 607M target side. This will allow us to evaluate the impact of the quantity of data. Both pairs have English as target so the transfer will be from close (or distant) language on source side and target language is fixed. We want to exhibit whether this significant difference in terms of quantity of data will compensate for the language distance and result in a distant (German→English) parent system from which the transfer is as effective as a close (Finnish→English) parent system.

3.2 Data Preprocessing

We use subword SPM units [14]. Systems using subword units are the current state of the art in neural machine translation. There is also a correlation in transfer quality depending on the number of subword units in common [16].

Two separate models of subword units are learned, one trained on source languages and another trained on the target language (English). Both are used in parent and child systems. The corresponding source and target vocabularies are created from tokenized data. Consequently, there is a direct correlation between data used to train the SPM models and resulting vocabularies in the NMT systems.

We take this into account by learning subword models for the source side with the data used to learn the parent **and** child systems. This subword model is then applied to all source side data, for both parent and child data. The goal is to not change the vocabulary during the parent/child transition since this would require to get representations for units that are not seen during training of the parent model.

Sentences with length less than 3 subword units and more than 100 subword units are filtered out.

Finally, subword units that occur at least 5 times in our training corpus are kept in the vocabularies while the others are matched to an unknown unit (<unk>). This process is necessary in our case since SPM can not guarantee exhaustive coverage of the corpus.

3.3 SPM Model Study

In this section we will describe 3 different SPM models that were created for the multilingual parent system approach.

We tried to combine the proximity of Finnish and Estonian and to take advantage of large amount of data from the German→English pair. For that, we built a system with universal encoder and decoder [8] with both Finnish→English and German→English corpora used as training data. One advantage of the universal approach is the capability to add one or more languages to our system without having to change the architecture. We can thus always have really comparable Estonian→English children, whereas we now have a multilingual system as a parent. [10] showed that parallel learning of multiple language pairs with a universal architecture has a positive impact on translation results. We want to verify if this is also the case for transfer learning.

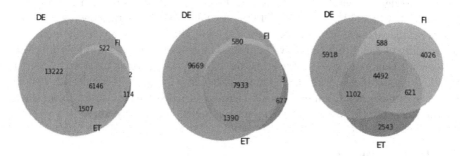

Fig. 1. Subword distribution for different SPM models built with different data distributions. Left is using 10M DE tokens only, center is using 5M DE tokens, 3M FI and 2.5M ET, and right is using 3M DE, 3M FI and 2.5M ET tokens.

We designed 3 different SPM data distributions to train separate SPM models for the source side of our universal systems.

For the first one, the model has been trained on German data only, resulting in subword units that are specific this language but are used for Finnish and Estonian. The result is that the vocabulary contains many short subword units covering the Finnish and the Estonian text. The Finnish and Estonian words end up being heavily split, which might complicate the subsequent modelling. This model is referred to as the 10-0-0 SPM. The second SPM model is using 5M German sentences, 3M in Finnish and 2.5M in Estonian. This is an intermediate model, with a more balanced data distribution across languages. It is referred to as the 5-3-2.5 SPM. The last one is made of 3M sentences in German, 3M in Finnish and 2.5M in Estonian. We force the data to be balanced for this SPM model despite the data quantity imbalance. It is referred to as the 3-3-2.5 SPM.

Figure 1 describes subword units distributions in the vocabulary of our systems obtained with the three different SPM models presented above. Figure 1 (left) shows the distribution for the 10-0-0 SPM model. This distribution is very unbalanced as expected. The Finnish specific units and Estonian specific units in this figure are unseen by the SPM model composed on German data. Figure 1 (center) shows that, even when greatly reducing the quantity of German data, the vocabulary remains mainly composed of German-specific units, however, we notice more common subword units than with the 10-0-0 SPM.

It is important to keep in mind that the distribution of subword units in the vocabulary does not reflect the actual coverage in the corpus (the number of occurrences is not taken into account).

The Fig. 1 (right) shows a balanced distribution of subword units across the 3 languages.

In this case, every language has a set of specific units that will be learned during training of the parent and/or child NMT model. We want to verify whether this will lead to a better transfer for NMT.

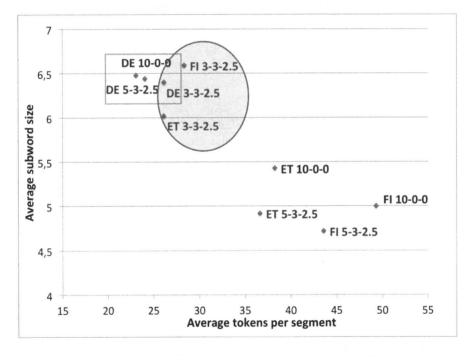

Fig. 2. Average subword units size (in characters) against average number of tokens per segment for the different SPM models. (Color figure online)

Figure 2 shows the relationship between the average subword token size (in characters) and the number of tokens per segment. This graph highlights that all the ratios for German are close to each other regardless the data distribution used to train them (they are all in the pink rectangle). It also shows that the Finnish and Estonian ratios of each SPM model are close to each other in each distribution. There is a particularity observed only with the 3-3-2.5 SPM: all 3 languages ratio are close to each other (circled in green in Fig. 2), emphasizing the balance between the 3 languages.

4 Architecture

To carry out our experiments we based our approach on the principle of trivial transfer learning of [12]. The principle is simple and consists in using an architecture that does not change between the learning of the parent system and the child system. Only training data are changed between parent and child system learning.

We use a standard end-to-end encoder/decoder architecture with an attention mechanism [1, 19]. This architecture is composed of a bi-directional encoder and a decoder based on gated recurrent units, Bi-GRU [4] of size 800. The embeddings size is set to 400. We apply a 0.3 *dropout* [18] on the embeddings, on the context before being provided to the attention mechanism and on the output before softmax. Weights are initialized according to [9] and we use Adam [11] as optimizer. The initial learning rate is set to $1.10e-4$ and size of a batch is 32. This architecture is the only configuration used for all systems presented in the next sections. They were implemented with the nmtpytorch[1] toolkit [3].

5 Experiments

All results of the systems presented here are calculated on the development dataset of the *News* translation task of WMT2018 evaluation campaign.

Table 1. Results in BLEU of the Estonian→English language pair without transfer learning with vocabularies containing only subword units coming from this language pair (source and target side separated).

# subword units	ET-EN 2.5M	ET-EN 200k
8k	14.12	10.69
16k	14.17	10.70
32k	13.60	10.10

The Estonian→English system presented in Table 1 is compared to our system using transfer learning. We can notice that the differences are small and negligible. Note that the number of source and target side subword units is the same.

We can see in Table 2, that for learning a child system Estonian→English, the models based on the subword units including German get worse results than those including Finnish. Since Finnish and Estonian are close languages, it is likely that they share more

---

[1] https://github.com/lium-lst/nmtpytorch.

Table 2. Results in BLEU of Estonian→English systems without transfer learning using subword units from the different SPM models used in transfer learning afterwards. This emphasises the impact of the subword units and vocabulary used.

# subword units	DE+ET 2.5M	DE+ET 200k	FI+ET 2.5M	FI+ET 200k
8k	10.64	-	14.47	-
16k	11.55	9.27	15.08	10.66
32k	12.52	-	13.87	-

subword units than with German, which explains the results. The hypothesis is that they coexist better in the vocabulary. This is confirmed by the results of the Estonian-German SPM model, which increases as number of subword units increases. While for the Estonian-Finnish SPM, results decrease when using 32k units compared to using 16k units. Therefore, it seems that a greater number of subword units is more favourable for German-Estonian system whereas 16k units are sufficient for the Finnish-Estonian system.

Table 3. Results in BLEU of different parent models in their respective languages pairs.

Language pair	40M	20M	10M	5M	2.5M
FI-EN	-	-	-	18.03	16.16
DE-EN	20.22	10.46	11.01	11.11	10.95

The results of the standard systems in Table 3 give us an idea of the performance obtained by parent systems in their respective language pairs. We selected 5M sentences from the German→English corpus to have a similar size to the Finnish→English corpus. We can then effectively compare the 2 different source languages of the parent systems with the same amount of data to train on. This allows us to evaluate the impact of language proximity. We observed that the performance of the German→English parent system using only 5M of randomly selected data is much lower than that of the system using all available data. One can thus expect a loss of performance when the parent system is trained with a smaller amount of data.

We had to make a compromise when defining architecture size. We want the biggest possible architecture to effectively learn the parent system, but we also want a reasonable size to avoid overfitting the child's system afterwards. For the upcoming experiments, we chose to use 16k subword units because this quantity led to the best performance for the Estonian→English system.

5.1 Results

In Table 4, we expose the results of the Estonian→English child systems that were learned from the different parent systems presented in Table 3.

Using all data for our 3 systems we get an improvement compared to our Estonian→English baseline which is 14.17 BLEU (see Table 1). Results are close but we can see that the results with the DE+ET SPM are worse, which corresponds to results in Table 2. The best result is obtained with the Finnish→English parent.

We performed several experiments with different quantities of German→English data (from 40M to 1.25M) and Finnish→English data (from 5M to 1.25M). Results show that with the same amount of data, results differ greatly among architectures. This difference is explained by the proximity of languages used to train the parent system. Finnish, which is closer to Estonian, offers a better transfer than the more distant German, confirming results in [5]. [12] shows that the quality of the parent system is important to ensure a good transfer to a child. The low performance of the German→English parent using 5M of data explains the poor results of the later learned child system.

We also tried our multilingual parent approach with universal encoder as described in Sect. 3.3.

We use a different SPM model from previous ones because this time it contains German and Finnish from the parent system, in addition to Estonian from the child system for source side.

The assumption is that by combining these two factors we should get a parent who will provide a better transfer to our child systems. The results show that this is not so obvious (see Table 4); performance is worse than German→English or Finnish→English as the only parent. One hypothesis is that the imbalance of amounts of data between the two source languages of the parent is an obstacle to learning a good quality parent.

Table 4. Results in BLEU of Estonian→English child models with different parent models used to transfer.

Parent language pair	45M (40M DE + 5M FI)	40M	20M	10M	5M	2.5M
FI-EN	-	-	-	-	**16.55**	**16.55**
DE-EN	-	**16.10**	10.46	11.28	10.92	11.18
FI+DE-EN 10-0-0 SPM	**15.71**		14.06	14.44	14.37	14.53	14.47
FI+DE-EN 5-3-2.5 SPM	13.20		13.45	13.86	14.05	14.01	13.77
FI+DE-EN 3-3-2.5 SPM	14.09		14.51	14.22	14.64	14.52	14.71

We tried different data quantity distribution between the two languages.

In the 40M column there is 35M from German→English with 5M from Finnish→English, in the 20M column it is 15M German→English with 5M Finnish→English, in the 10M column it is 5M German→English with 5M Finnish→English, in the 5M column it is 2.5M form both and in the last column it is 1.25M from both. This way we have a better vision of the impact of data quantity from the German→English language pair.

With our 3 different SPM models applied to our multilingual parent, we observe some interesting results.

The first of them is the 15.71 BLEU obtained when using the full data available on both parent pairs.

This result is surprising considering the others scores. This SPM model learned only on German→English data reveals an interesting behaviour of non DE source side data. Indeed, words are "overly" split into subword units. Surprisingly, this particularity seems to provide good results in Estonian→English. Our hypothesis is that, thanks to the quite large architecture to train the Estonian→English system, the system overfits on the small subword units. The average number of subword units per line and average subword unit size of this particular SPM model applied to our data can confirm it. Thus, for the training of this child model, the system keeps improving and stopped during the 10th epoch, while most of the other child systems presented barely passed the 5th. These observations are in line with our hypothesis of overfitting but we keep investigating about this result.

Overall the results from the 5-3-2.5 SPM are the least interesting. Results seem to increase slightly as we reduced data quantity used by a small margin. This might be related to the quantity of German→English data used for training the model.

Finally, with the more balanced SPM model (3-3-2.5), the results are quite stable with only slight changes. The results are also better at each data quantity than the 5-3-2.5 SPM.

We see an improvement thanks to the transfer for the Estonian→English child systems. However, we also want to apply this transfer in the case where few resources are available. To simulate this lack of data, we kept only 200k sentences from the Estonian→English corpus to learn new children with the same parent systems.

Table 5. Results in BLEU of the Estonian→English language pair using only 200k sentence pairs to train the child model (artificially simulated low resourced)

Parent language pair	45M (40M DE + 5M FI)	40M	20M	10M	5M	2.5M
FI-EN	-	-	-	-	**13.03**	**12.24**
DE-EN	-	**11.12**	6.87	6.99	7.10	6.96
FI+DE-EN 10-0-0 SPM	**11.05**	10.41	11.29	11.68	11.54	11.72
FI+DE-EN 5-3-2.5 SPM	10.26	9.79	10.52	11.00	10.85	10.65
FI+DE-EN 3-3-2.5 SPM	**12.19**	12.05	11.89	**12.56**	11.93	**12.45**

Results of Table 5 show us that when we have few training data for the child system, the proximity of languages is the most important feature.

Finnish→English parent system outperforms the others in this configuration.

This time our multilingual approach results are as good as with the German→English pair. Results coming from the balanced SPM outperforms it on all data quantities.

Compared to previous results with all the data on the 10-0-0 SPM system, we observe that the results are not anymore outperforming the others. The system obtains 11.05 BLEU which is not better than the previous results. This can confirm that the overfiting of the subword units works less well when less data is available like in this setup. In general the results are consistent without dependency to data quantity.

With the 5-3-2.5 SPM model, the results are consistent as before: they are still worse than the two others SPM models.

Our 3-3-2.5 SPM models now outperforms the others as well as the German→English parent systems. Their results even get close to the Finnish→English parent. We believe that the balance of the subword units across the 3 source languages involved is particularly effective in this case where few data are available for the child system.

6   Conclusion

In this paper, we showed that transfer learning for NMT depends on the quantity of available data and the proximity of involved languages. Also, carefully training the subword models can lead to a better language equilibrium in the vocabulary leading to better translation results.

These parameters are therefore to be taken into account for the choice of parent systems.

Our results are in line with those obtained by [20] and [5]; proximity of languages used for transfer learning is more important than data quantities. With equivalent amounts of data, parent systems using pairs of closer languages perform better, but the quality of the parent systems in question should not be neglected and should be taken into account in the results of the child systems. The token distribution in the vocabulary is also of greater importance and have an impact on system performance.

Our universal multilingual approach end up showing some interesting results, especially in low resource context. We presented an analysis of the subword units distribution and the importance of the preprocessing steps ahead of the training process. We showed that the balance between the different languages involved in the system is extremely relevant for the performance of the child systems afterwards.

In the future we want to keep investigating subword units distribution with different examples to better explain the relation between those factors and the systems performance results.

Acknowledgments. This work was supported by the French National Research Agency (ANR) through the CHIST-ERA M2CR project, under the contract number ANR-15-CHR2-0006-017.

References

1. Bahdanau, D., Cho, K., Bengio, Y.: Neural machine translation by jointly learning to align and translate. CoRR abs/1409.0473 (2014). http://arxiv.org/abs/1409.0473
2. Bojar, O., et al.: Findings of the 2018 conference on machine translation (WMT18). In: Proceedings of the Third Conference on Machine Translation, Volume 2: Shared Task Papers, pp. 272–307. Association for Computational Linguistics, Belgium, Brussels, October 2018. http://www.aclweb.org/anthology/W18-6401

3. Caglayan, O., García-Martínez, M., Bardet, A., Aransa, W., Bougares,F., Barrault, L.: NMTPY: a flexible toolkit for advanced neural machine translation systems. Prague Bull. Math. Linguist. **109**, 15–28 (2017). https://doi.org/10.1515/pralin-2017-0035, https://ufal. mff.cuni.cz/pbml/109/art-caglayan-et-al.pdf
4. Cho, K., van Merrienboer, B., Gülçehre, Ç., Bougares, F., Schwenk, H., Bengio, Y.: Learning phrase representations using RNN encoder-decoder for statistical machine translation. CoRR abs/1406.1078 (2014). http://arxiv.org/abs/1406.1078
5. Dabre, R., Nakagawa, T., Kazawa, H.: An empirical study of language relatedness for transfer learning in neural machine translation. In: Proceedings of the 31st Pacific Asia Conference on Language, Information and Computation, pp. 282–286. The National University (Philippines) (2017). http://aclweb.org/anthology/Y17-1038
6. Durrani, N., Dalvi, F., Sajjad, H., Belinkov, Y., Nakov, P.: One size does not fit all: comparing NMT representations of different granularities. In: Proceedings of the 2019 Conference of the North American Chapter of the Association for Computational Linguistics: Human Language Technologies, Volume 1 (Long and Short Papers), pp. 1504–1516. Association for Computational Linguistics, Minneapolis, Minnesota, June 2019. https://www.aclweb.org/ anthology/N19-1154
7. Gu, J., Wang, Y., Chen, Y., Li, V.O.K., Cho, K.: Meta-learning for low-resource neural machine translation. In: Proceedings of the 2018 Conference on Empirical Methods in Natural Language Processing, pp. 3622–3631. Association for Computational Linguistics, Brussels, Belgium, October–November 2018. https://www.aclweb.org/anthology/D18-1398
8. Ha, T., Niehues, J., Waibel, A.H.: Toward multilingual neural machine translation with universal encoder and decoder. CoRR abs/1611.04798 (2016). http://arxiv.org/abs/1611.04798
9. He, K., Zhang, X., Ren, S., Sun, J.: Delving deep into rectifiers: surpassing human-level performance on ImageNet classification. CoRR abs/1502.01852 (2015). http://arxiv.org/abs/ 1502.01852
10. Johnson, M., et al.: Google's multilingual neural machine translation system: enabling zeroshot translation. CoRR abs/1611.04558 (2016). http://arxiv.org/abs/1611.04558
11. Kingma, D.P., Ba, J.: Adam: a method for stochastic optimization. CoRR abs/1412.6980 (2014). http://arxiv.org/abs/1412.6980
12. Kocmi, T., Bojar, O.: Trivial transfer learning for low-resource neural machine translation. In: Proceedings of the Third Conference on Machine Translation: Research Papers, WMT 2018, Belgium, Brussels, 31 October–1 November 2018, pp. 244–252 (2018). https://aclanthology. info/papers/W18-6325/w18-6325
13. Koehn, P., Knowles, R.: Six challenges for neural machine translation. In: Proceedings of the First Workshop on Neural Machine Translation, pp. 28–39. Association for Computational Linguistics, Vancouver, August 2017. https://doi.org/10.18653/v1/W17-3204, https://www. aclweb.org/anthology/W17-3204
14. Kudo, T., Richardson, J.: Sentencepiece: a simple and language independent subword tokenizer and detokenizer for neural text processing. CoRR abs/1808.06226 (2018). http://arxiv. org/abs/1808.06226
15. Lakew, S.M., Erofeeva, A., Negri, M., Federico, M., Turchi, M.: Transfer learning in multilingual neural machine translation with dynamic vocabulary. In: IWSLT 2018, October 2018
16. Nguyen, T.Q., Chiang, D.: Transfer learning across low-resource, related languages for neural machine translation. CoRR abs/1708.09803 (2017). http://arxiv.org/abs/1708.09803
17. Sachan, D., Neubig, G.: Parameter sharing methods for multilingual self-attentional translation models. In: 3rd Conference on Machine Translation (WMT), Brussels, Belgium, October 2018. https://arxiv.org/abs/1809.00252
18. Srivastava, N., Hinton, G., Krizhevsky, A., Sutskever, I., Salakhutdinov, R.: Dropout: a simple way to prevent neural networks from overfitting. J. Mach. Learn. Res. **15**, 1929–1958 (2014). http://jmlr.org/papers/v15/srivastava14a.html

19. Sutskever, I., Vinyals, O., Le, Q.V.: Sequence to sequence learning with neural networks. CoRR abs/1409.3215 (2014). http://arxiv.org/abs/1409.3215
20. Zoph, B., Yuret, D., May, J., Knight, K.: Transfer learning for low-resource neural machine translation. CoRR abs/1604.02201 (2016). http://arxiv.org/abs/1604.02201

A Deep Learning Approach to Self-expansion of Abbreviations Based on Morphology and Context Distance

Daphné Chopard$^{(\boxtimes)}$ (ID) and Irena Spasić (ID)

School of Computer Science and Informatics, Cardiff University,
5 The Parade, Cardiff CF24 3AA, UK
{ChopardDA,SpasicI}@cardiff.ac.uk

Abstract. Abbreviations and acronyms are shortened forms of words or phrases that are commonly used in technical writing. In this study we focus specifically on abbreviations and introduce a corpus-based method for their expansion. The method divides the processing into three key stages: abbreviation identification, full form candidate extraction, and abbreviation disambiguation. First, potential abbreviations are identified by combining pattern matching and named entity recognition. Both acronyms and abbreviations exhibit similar orthographic properties, thus additional processing is required to distinguish between them. To this end, we implement a character-based recurrent neural network (RNN) that analyses the morphology of a given token in order to classify it as an acronym or an abbreviation. A siamese RNN that learns the morphological process of word abbreviation is then used to select a set of full form candidates. Having considerably constrained the search space, we take advantage of the Word Mover's Distance (WMD) to assess semantic compatibility between an abbreviation and each full form candidate based on their contextual similarity. This step does not require any corpus-based training, thus making the approach highly adaptable to different domains. Unlike the vast majority of existing approaches, our method does not rely on external lexical resources for disambiguation, but with a macro F-measure of 96.27% is comparable to the state-of-the art.

Keywords: Natural language processing · Text normalisation · Abbreviation disambiguation · Neural networks · Corpus-based methods

1 Introduction

In recent years, text data has become ubiquitous in many critical fields. For example, it is nowadays standard practice for medical practitioners to write and rely on electronic reports when taking care of patients. As narratives are an important source of information, this growth has been accompanied by a surge in *Natural Language Processing* (NLP) applications, such as information retrieval

© Springer Nature Switzerland AG 2019
C. Martín-Vide et al. (Eds.): SLSP 2019, LNAI 11816, pp. 71–82, 2019.
https://doi.org/10.1007/978-3-030-31372-2_6

and topic modelling. While NLP systems have displayed stellar performance on numerous tasks, they rely most of the time on clean and normalised data due to the tasks' complexity.

However, as actual text data can rarely be found in canonical form, transforming text data into a unique standard representation — also called text normalization — is a key aspect of NLP pipelines. Some documents might for instance contain uncommon tokens that cannot be directly recognised by a standard NLP system and must first be resolved. The normalisation of short forms (e.g., contractions) is particularly critical in any field that involves the regular and rapid writing of reports, such as aviation or healthcare, where such forms are frequently used to speed up the writing or to ease a repetitive task. The word underlying a short form is hidden and therefore inaccessible to NLP applications, thus skewing their performance [22]. This ambiguity inevitably leads to a loss of information which weakens the system's understanding of language.

Liu et al. [13] revealed in a study conducted in 2001 that among the 163,666 short forms they retrieved from the *Unified Medical Language System* (UMLS), 33.1% of them referred to multiple full forms. Similarly, Li et al. [11] reported that the 379,918 short forms which could be found on the website AcronymFinder.com had in average 12.5 corresponding full forms. Furthermore, they noted that 37 new short forms were added daily to the website. These observations further highlight the need for an automatic method for short form expansion that is highly adaptable, preferably unsupervised and domain-independent.

Short forms can be divided into two categories: those that refer to a single word (e.g., *PT* for *patient*) and those that refer to multiple words (e.g., *DOB* for *date of birth*). (Note that, although not completely accurate from a linguistic point of view, in the remainder of this work we will refer to the former as *abbreviations* and to the latter as *acronyms* for the sake of simplicity.) The way abbreviations and acronyms relate to full forms is intrinsically different, due to their distinctive nature. Furthermore, while acronyms tend to follow pre-defined rules, new abbreviations are often created spontaneously. Consequently, we believe that these two types of short forms should be considered independently.

2   Related Works

NLP applications often use external lexical resources to expand the short forms prior to text analysis. However, short forms often correspond to multiple long forms [11], which implies that *word sense disambiguation* (WSD) is a required as part of pre-processing. Unlike acronyms, which are often standardised within a domain, authors often create ad hoc abbreviations, which may not be encoded in existing lexicons.

When they are included in specialised biomedical terminologies, it is has been shown that simple techniques, such as bag-of-words, combined with majority sense prevalence were effective in practice despite an expectation that sophisticated techniques based on biomedical terminologies, semantic types, part-of-speech and language modelling and machine learning approaches would be necessary [16]. The ShARe/CLEF eHealth 2013 challenge [17] created a reference standard of clinical short forms normalized to the *Unified Medical Language System* (UMLS) [2]. The challenge evaluated the accuracy of normalizing short forms compared to a majority sense baseline approach, which ranged from 43% to 72%. In line with findings suggested in [16], a majority sense baseline approach achieved the second-best performance. Nonetheless, machine learning approaches to clinical abbreviation recognition and disambiguation were found to be as effective with F-score over 75% [27]. However, this study focused on 1,000 most frequent abbreviations in a corpus used for evaluation, which makes it possible to successfully translate their distribution into a classification model. This also means that the given approach may not necessarily work with ad hoc abbreviations. Another problem with using supervised machine learning methods for abbreviation disambiguation in clinical texts is associated with the acquisition of training data. Manually annotating abbreviations and their senses in a large corpus is time-consuming, labour-intensive and error-prone. In addition, the learnt model may not be transferrable across domains, which makes supervised learning impractical for this particular text mining problem. With accuracy up to 90%, semi-supervised classification algorithms proved to be a viable alternative for abbreviation disambiguation [6]. Moreover, an F-score of 95% could be reached by using an unsupervised approach [9], which avoids the need to retrain a classification models or use bespoke feature engineering, which makes the approach domain independent. It also proved to be robust with respect to ad hoc abbreviations. Word embeddings provide an alternative way to represent the meaning of clinical abbreviations. Three different methods for deriving word embeddings from a large unlabelled clinical corpus have been evaluated [29]. Adding word embeddings as additional features to be used by supervised learning methods improved their performance the on clinical abbreviation disambiguation task.

All of the above mentioned systems, post-process clinical notes long after clinicians originally created them. The results show that post-processing clinical abbreviation cannot yet guarantee 100% accuracy in their identification and disambiguation. With this problem in mind, a system for real-time clinical abbreviation recognition and disambiguation has been implemented. The system interacts with an author during note generation asking them to verify correct abbreviation senses suggested automatically by the system [26]. With the accuracy of 89% and the processing time ranging from 0.630 to 1.649 milliseconds, the system incurred around 5% of total document entry time, which demonstrated the feasibility of integrating a real-time abbreviation recognition and disambiguation module with clinical documentation systems.

3   Our Method

The first step of our pipeline identifies short forms contained in a document and determines the ones referring to single words (abbreviations) rather than multiple words (acronyms). Then, for each abbreviation, a set of full form candidates is extracted. Finally, the *Word Mover's Distance* (WMD) which leverages the power of word embeddings is used to disambiguate each abbreviation.

3.1   Step 1: Abbreviation Identification

Let us assume we have a document D for which we would like to automatically resolve all abbreviations. As mentioned above, we must initially identify all abbreviations in the document: all short forms contained in the text are first extracted, then we discard all acronyms.

To extract all short forms the text is first tokenized, and we gradually discard tokens that cannot be short forms. To begin with, all tokens that are recognised as English words are discarded. For the sake of this work, abbreviations that are identical to existing English words—such as the short form *tab* for the word *tablet*—are assumed to always appear immediately followed by a period so that the two together can be identified as a single token. Indeed, in such case, the use of a full point is standard practice to mark abbreviations to avoid any confusion.

Secondly, all tokens containing less than 2 or more than 6 characters are rejected. An abbreviation should consist of at least two characters as a single character is extremely ambiguous. Moreover, since abbreviations are short per definition, we set a strict upper limit of 6 characters. This threshold is identical to the one used in previous works on abbreviations [13, 28]. This helps the system discard unknown words that are not abbreviations (e.g., misspelled words).

Although they must be discarded eventually, locations and names are rarely part of an English dictionary and are therefore still retained by the system at this point. To deal with this, a *named entity recognition* (NER) system is applied to the original document to classify the remaining tokens. Those that are labelled as *PERSON* or *LOCATION* are not retained any further.

After this simple processing, we obtain a list of short forms that includes both abbreviations and acronyms. In order to model morphological differences between the two, we develop a deep neural network that learns to distinguish between abbreviations and acronyms. Using a deep architecture rather than other rule-based or machine learning methods has the advantage of obviating the need for any manual features, giving more flexibility to the model to accommodate any type of short forms.

More specifically, the deep neural network takes as input a sequence of characters which is first processed by fully-connected layers for representation learning. Then, a recurrent neural network (RNN) sequentially reads the improved representation for structure learning. Finally, a soft-max layer predicts whether the sequence is an abbreviation or an acronym (i.e., refers to a single or multiple words).

At the end of this first step, the system yields a set A of abbreviations, namely $A = \{w \in D | w \text{ is an abbreviation}\}$.

3.2   Step 2: Full Form Candidates Identification

In the second step, a set of full form candidates Φ_α must be identified for each abbreviation $\alpha \in A$. Based on the assumption that characters appear in the exact same order in both forms, a simple rule-based solution could consist in searching for all words within the document that have as a subset all the characters of contained in the abbreviation. However, many short forms such as *xmas* and *x-mas* for the word *Christmas* contain characters that are not present in their full form. Terada *et al.* chose to address this manually by not considering specific characters such as X [23]. Unfortunately, this limits the system to follow manually engineered rules. Therefore, for maximal flexibility, we instead take advantage of a deep learning architecture to select full form candidates.

We design a siamese RNN [18] to learn how full forms relate to short forms. An illustration of the architecture is depicted in Fig. 1. Every abbreviation $\alpha \in A$ and every word w in the document D that is recognised as an English word are first encoded as a sequence of characters. Then, one by one, each abbreviation is fed along with one word to the network. The two are first processed by multiple fully-connected layers for representation learning. Each improved representation is then fed to one of two independent RNN: one that processes the sequence corresponding to the short form and one that processes the full form. Finally, the output of the two RNNs are compared by the network which must decide whether this word could potentially be referred to by the short form or not. If so, the word is added to the list of full-form candidates Φ for that abbreviation. Instead of feeding only corpus word, we could feed every English word contained in the corpus along with each abbreviation to get a set of full form candidates that is more comprehensive.

Fig. 1. Siamese RNN to select a set of full form candidates

3.3 Step 3: Abbreviation Disambiguation

In order to determine the right full form for each abbreviation $\alpha \in A$ the system must select the best of all full form candidates Φ_α. We rely on the assumption that short forms and their corresponding true full forms share a similar context in order to disambiguate each abbreviation and find the most appropriate of all full form candidates.

To compare the context of an abbreviation and its full form candidates, we propose to use the *Word Mover's Distance* (WMD), a measure that was developed by Kusner *et al.* [10] to assess similarity between two documents. It takes advantage of the semantic properties inherent to word embeddings to match documents that have a similar meaning, although they consist of very different words. First, each word i is represented by a d-dimensional word embeddings $\mathbf{x}_i$. The use of pretrained word vectors allows the model to take advantage of the linear properties of continuous space word representations [14] without the need to train it on the chosen corpus. For each document D, the n-dimensional normalized bag-of-words (BOW) vector is denoted as $\mathbf{f}^D$ with entries $f_i^D = c_i / \sum_{j=1}^n c_j$, where c_i is the number of occurrences of word i in document D, and n the vocabulary size. Finally, let $\mathbf{T} \in \mathbb{R}^{n \times n}$ be the transport matrix whose entries $\mathbf{T}_{ij}$ denote how much of f_i^D should travel to $f_j^{D'}$, where D' is a another document. Then the WMD minimizes the following linear optimization problem:

$$\min_{\mathbf{T} \geq 0} \sum_{i,j=1}^{n} \mathbf{T}_{ij} \|\mathbf{x}_i - \mathbf{x}_j\|_2, \text{ subject to } \sum_{i=1}^{n} \mathbf{T}_{ij} = f_j^{D'}, \sum_{j=1}^{n} \mathbf{T}_{ij} = f_i^D \; \forall i,j \quad (1)$$

which is an instance of the well-studied earth mover's distance problem for which many efficient solutions already exist [12, 15, 19, 21, 25].

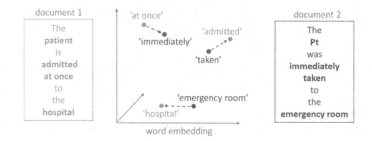

Fig. 2. WMD for abbreviation disambiguation: the minimum cumulative distance between non-stop words in the context of target abbreviation *Pt* and of full form candidate *patient* is computed

Figure 2 illustrates how this measure is used to determine the full forms that are semantically close to target short form. In this example, the abbreviation *Pt* used in the sentence *"The Pt was immediately taken to the emergency room"* refers to the word *patient*, which can be found in the sentence *"The patient is admitted at once to the hospital."*. Although the two sentences do not share a single representative word, they are semantically very close. To minimize the cumulative distance between the two sets of context words $c_1 = \{\text{admitted, at once, hospital}\}$ and $c_2 = \{\text{immediately, taken, emergency room}\}$, the words *admitted* and *taken* are matched together, and so are the words *at once* and *immediately*, and the words *hospital* and *emergency room*, because their respective word embeddings lie close together in the word space. Since the context words for the short form *Pt* have a similar meaning as the context words for the candidate *patient*, the WMD between the two sentences will be small.

For each of the abbreviations α identified in the first step, we have filtered—in the second step—a set Φ_α of potential full form candidates. Let S(w) be the set of all sentences in document D containing word w, i.e., $\text{S(w)} = \{s \in D | w \in s\}$. We must determine the best full form candidate ϕ_α^* that abbreviation α refers to in sentence s_α. We compute for each candidate $\phi \in \Phi_\alpha$ the WMD between the sentence the abbreviation appears in (i.e., s_α) and each sentence containing this candidate (i.e., $S(\phi)$). The individual WMD are then summed up and averaged to yield a disambiguation score $\sigma(\phi)$:

$$\sigma(\phi) = \sum_{\tilde{s} \in S(\phi)} \frac{\mathrm{WMD}(s_\alpha, \tilde{s})}{|S(\phi)|}. \tag{2}$$

Eventually, we consider the full form with smallest WMD as being the best one, namely $\phi_\alpha^* = \min_{\phi \in \Phi_\alpha}(\sigma(\phi))$.

4 Evaluation

4.1 Abbreviation Identification (Step 1)

Our pipeline separates the abbreviation identification step into two subsequent parts: short forms identification and abbreviation-acronym differentiation.

We first evaluate the former on discharge summaries from the 2009 i2b2 medication challenge [24] that we have manually annotated to this end. We use the python library *NLTK* [1] both for word tokenization and to find non-standard English words. More specifically, we build an English dictionary that combines all words found in the Brown, Reuters, and Words corpora. To label entities, we use the Stanford NER [5]. We succesfully identify short forms with an average F1-score of 62.20% with high precision (92.18%) but low recall (52.90%). The low recall can be explained by the presence of numerous short forms that are common enough to be part of an English vocabulary (e.g., *MD, Dr., mg, cm, tel, ID*) and, as a result, are discarded by our method. Such short forms would likely be discarded if using a different dictionary, or could easily be resolved using a standard abbreviation dictionary. When ignoring them, recall improves drastically and our approach achieves an average F1-score of **91.04%** (Precision: 95.20%, Recall: 87.38%).

Second, the network that we developed for abbreviation-acronym differentiation is trained and evaluated using distinct samples from the 32,048 unique short forms of the CARD framework [27]. The network takes as input a short form x_i and outputs a probability score $y_i \in [0, 1]$, where 0 denotes that the short form refers to multiple words (acronyms) and 1 denotes that the short form refers to a single word (abbreviation). Each short form is represented as a 6×8 matrix where each row corresponds to the binary representation of one character based on its Unicode code point—6 being the highest number of characters in a sequence. Standard characters have code point value at most 256, which allows a compact representation in only 8 dimensions. As a comparison, a one-hot encoding would require around 100 dimensions, depending on the actual number of characters in the corpus. Since short forms can sometimes refer to either an abbreviation or an acronym depending on the context, we assign each of them a label between 0 and 1 which accounts for this versatility. More precisely, the label is computed as a weighted sum of the nature of all possible expansions:

$$y_i = \frac{\sum Abb(x_i)}{\sum Abb(x_i) + \sum Acr(x_i)}, \tag{3}$$

where $Abb(x_i)$ and $Acr(x_i)$ are the sets of full forms of x_i that are single word and multiple words respectively. The deep architecture consists of 4 fully connected layers (with 16, 32, 16 and 8 nodes respectively) for representation learning followed by an RNN consisting of an LSTM cell with 48 units for morphology learning. Finally, a softmax layer outputs the probability score. The network is optimised using the Adam optimiser [8] to minimise the cross-entropy loss. The network is trained on 80% of the data, while the test set, which is used to determine the final out-of-sample performance, is composed of 20% of the short forms. Due to the high imbalance of the data set (only 19.15% of them are abbreviations), we oversample samples from this smaller class so that there are roughly the same number of examples in both classes. Since the goal is to retrieve all abbreviations, recall is the most important measure for this task. On test set we achieve a recall of **72%** (F1-score of 59%, Precision 49%). These scores could be improved by training the network on a larger data set. However, we can easily set a minimum threshold for disambiguation, which will eventually reject short forms that do not have any full form with context close enough therefore discarding any remaining acronyms.

4.2   Full Form Candidates Identification (Step 2)

For this second step, we use the 31,922 abbreviations from the CARD data set. We assign the label $y_i = 1$ to each pair $x_i = (shortform, fullform)$ contained in the data set. Negative examples are created by randomly selecting a short form from the data set and pairing it with another randomly selected full form. If the pair is not already contained in the dataset, we assign it the label $y_i = 0$ and add it to the set of samples. We repeat this process until we have as many negative as positive samples. We train the siamese RNN on 70% of the data whereas 30% are left out for testing. The deep architecture consists of 6 fully-connected layers (with 16, 32, 64, 32, 16, and 8 nodes respectively) for representation learning followed by two independent RNN – one for the abbreviation, one for the candidate. Each RNN consists of an LSTM cell with 64 units. The output of each network are then compared by stacking them and feeding them to 3 successive fully-connected layers consisting of 36, 18, and 2 nodes respectively. Prediction is achieved through a final softmax layer. We use dropout to prevent overfitting.

For comparison we implement the simple rule-based baseline suggested by Terada *et al.* [23]. It is based on the assumptions that the short form always contain less characters than the full form and the characters contained in the short form is a subset in the same order of the characters contained in a potential full form candidate (except for "X", "–", "/"). The result of the comparison is displayed in Table 1. Our approach achieves a slightly higher F1-score than the baseline, but with much higher recall. Once again, recall is the most important measure as it is crucial to select the true full form as part of the candidates, whereas false positive will be naturally dealt with in the final step.

The network handles around 5,000 samples per second for inference, which means that it needs 8 seconds for a corpus of 40,000 words and less than 40s for the entire Oxford English dictionary (i.e. 171,476 words).

Table 1. Comparison between our deep-learning based full form selection approach against a rule-based baseline [%]

Method	Prec	Rec	F1
Baseline [23]	**90.12**	64.53	75.21
Ours (siamese RNN)	75.21	**84.04**	**78.57**

4.3   Abbreviation Disambiguation (Step 3)

Due to the lack of well-established benchmarks for abbreviation disambiguation, we evaluate our approach on a subset of the MSH WSD data set, which mostly contains acronyms. We believe that although acronyms and abbreviations have different morpholgical properties, disambiguation is similar. Hence we here reproduce an experiment first conducted by Prokofyev *et al.* [20] and then later replicated by Li *et al.* [11] and Ciosici *et al.* [4] respectively. The subset of MSH WSD—a data set of abstracts from the biomedical domain created by Jimeno-Yepes *et al.* [7]—selected by Li *et al.* consists of 11,272 abstracts which contain a total of 69 ambiguous short forms each having in average 2.1 full form candidates.

To assess the performance of our approach, we compare it with the same methods as Ciosici *et al.* [4]. First, a simple baseline called *FREQUENCY* which, as its name implies, simply selects the most frequent full form candidate. Clearly, such approach completely disregards any context information and relies purely on the corpus statistics for determining the best candidate. Second, the *Surrounding Based Embedding* (SBE) model, a word embedding-based model developed by Li *et al.* [11] which first computes word embeddings of abbreviations by summing the word embeddings of the words in a window around the abbreviation before disambiguating between full form candidates using the cosine similarity. Similarly, Distr. Sim. is an approach introduced by Charbonnier and Wartena [3] which relies on word embeddings to build weighted average vectors of the context which are compared using the cosine similarity. Finally, the last benchmarks we compare with is the *Unsupervised Abbreviation Disambiguation* (UAD) method developed by Coisici *et al.* [4], which deal with disambiguation as a word prediction task. For more details on implementation of all benchmarks, please refer to the work of Coisici *et al.* [4].

We use 300 dimensional pretrained words embeddings based on part of the Google News corpus which contains about 100 billion words. Our scores are computed as average of a 3-fold cross-validation, similarly to the implementation of the benchmarks. Table 2 illustrates the performance of WMD compared to other benchmarks for short form disambiguation.

Table 2. Comparison of our disambiguation approach against different benchmarks on a subset of the MSH WSD dataset [%]

Method	Acc	Weighted			Macro		
		Prec	Rec	F1	Prec	Rec	F1
FREQUENCY	54.14	30.04	54.14	38.46	25.55	46.34	32.79
SBE [11]	82.48	83.07	82.48	82.53	82.18	82.16	81.87
Distr. Sim. [3]	80.19	80.87	80.19	80.25	79.90	80.12	79.71
UAD [4]	90.62	92.28	90.62	90.66	91.35	91.36	90.59
ours (WMD)	**96.36**	**96.38**	**96.36**	**96.36**	**95.65**	**96.97**	**96.27**

5   Conclusion

In this study, we introduced a domain-independent approach to matching abbreviations to their full form in corpus as part of text normalization. Unlike the vast majority of existing approaches to abbreviation expansion, which use an external lexicon in order to interpret an abbreviation and match it to its full form, we extract full forms from the corpus itself. This approach is based on an assumption that a full form is actually used elsewhere in the corpus. The likelihood of such an event increases with the size of a corpus, which makes the approach suitable for large-scale text mining applications. However, since our siamese RNN is able to select full form candidates from the whole English dictionary in a short amount of time, one could use publicly available resources (such as Wikipedia) to find context for the candidates that are not part of the corpus.

An advantage of using a corpus instead of a lexicon, which is typically domain-specific, makes our approach domain independent. It also avoids the need for maintaining an external lexicon up to date, while making our approach robust with respect to ad hoc abbreviations. However, one may argue that our approach still makes use of an external lexical resource. Indeed, we do use an external resource to train an RNN to model the morphological differences between acronyms and abbreviations. Assuming that these morphological properties are universal across the language rather than specific to a domain, then once trained on any representative lexicon, the model itself is readily reusable across domains and does not require to be re-trained. The same can be said about the second RNN, which models the morphological principles of word abbreviation. Once trained on any abbreviation lexicon, the model can be used to expand abbreviations that were not present in the training data (i.e. the lexicon). Lastly, the deep learning approach taken avoids the need for manual feature engineering, while the novel use of existing lexicons avoids the need for manual annotation of training data.

Finally, the use of deep learning to constrain the search space of possible matches based on the morphological structure of both abbreviations and full forms paves the way for more sophisticated approaches that can be utilised to analyse their contexts. We used Word Mover's Distance to leverage pre-trained word embeddings to measure semantic compatibility between abbreviations and full forms based on an assumption that both are used in similar contexts.

We have evaluated the different steps of our approach and achieved F1-scores of 59%, 78.57% and 96.36% respectively. These results are in line with those reported by other state-of-the art methods.

References

1. Bird, S., Klein, E., Loper, E.: Natural Language Processing with Python: Analyzing Text with the Natural Language Toolkit. O'Reilly Media, Sebastopol (2009)
2. Bodenreider, O.: The unified medical language system (UMLS): integrating biomedical terminology. Nucleic Acids Res. **32**, D267–D270 (2004)
3. Charbonnier, J., Wartena, C.: Using word embeddings for unsupervised acronym disambiguation. In: Proceedings of the 27th International Conference on Computational Linguistics, pp. 2610–2619. Association for Computational Linguistics (2018)
4. Ciosici, M.R., Sommer, T., Assent, I.: Unsupervised abbreviation disambiguation. arXiv preprint arXiv:1904.00929 (2019)
5. Finkel, J.R., Grenager, T., Manning, C.: Incorporating non-local information into information extraction systems by Gibbs sampling. In: Proceedings of the 43rd Annual Meeting on Association for Computational Linguistics, pp. 363–370. Association for Computational Linguistics (2005)
6. Finley, G.P., Pakhomov, S.V., McEwan, R., Melton, G.B.: Towards comprehensive clinical abbreviation disambiguation using machine-labeled training data. In: AMIA Annual Symposium Proceedings, vol. 2016, p. 560. American Medical Informatics Association (2016)
7. Jimeno-Yepes, A.J., McInnes, B.T., Aronson, A.R.: Exploiting MeSH indexing in MEDLINE to generate a data set for word sense disambiguation. BMC Bioinform. **12**(1), 223 (2011)
8. Kingma, D.P., Ba, J.: Adam: a method for stochastic optimization. In: International Conference on Learning Representations (2015)
9. Kreuzthaler, M., Oleynik, M., Avian, A., Schulz, S.: Unsupervised abbreviation detection in clinical narratives. In: Proceedings of the Clinical Natural Language Processing Workshop (ClinicalNLP), pp. 91–98 (2016)
10. Kusner, M., Sun, Y., Kolkin, N., Weinberger, K.: From word embeddings to document distances. In: International Conference on Machine Learning, pp. 957–966 (2015)
11. Li, C., Ji, L., Yan, J.: Acronym disambiguation using word embedding. In: Twenty-Ninth AAAI Conference on Artificial Intelligence (2015)
12. Ling, H., Okada, K.: An efficient earth mover's distance algorithm for robust histogram comparison. IEEE Trans. Pattern Anal. Mach. Intell. **29**(5), 840–853 (2007)
13. Liu, H., Lussier, Y.A., Friedman, C.: A study of abbreviations in the UMLS. In: Proceedings of the AMIA Symposium, p. 393. American Medical Informatics Association (2001)
14. Mikolov, T., Yih, W.t., Zweig, G.: Linguistic regularities in continuous space word representations. In: Proceedings of the 2013 Conference of the North American Chapter of the Association for Computational Linguistics: Human Language Technologies, pp. 746–751 (2013)

15. Monge, G.: Mémoire sur la théorie des déblais et des remblais. Histoire de l'Académie royale des sciences de Paris (1781)
16. Moon, S., McInnes, B., Melton, G.B.: Challenges and practical approaches with word sense disambiguation of acronyms and abbreviations in the clinical domain. Healthcare Inform. Res. **21**(1), 35–42 (2015)
17. Mowery, D.L., et al.: Normalizing acronyms and abbreviations to aid patient understanding of clinical texts: ShARe/CLEF eHealth Challenge 2013, task 2. J. Biomed. Semant. **7**(1), 43 (2016)
18. Mueller, J., Thyagarajan, A.: Siamese recurrent architectures for learning sentence similarity. In: Thirtieth AAAI Conference on Artificial Intelligence (2016)
19. Pele, O., Werman, M.: Fast and robust earth mover's distances. In: 12th International Conference on Computer Vision, pp. 460–467. IEEE (2009)
20. Prokofyev, R., Demartini, G., Boyarsky, A., Ruchayskiy, O., Cudré-Mauroux, P.: Ontology-based word sense disambiguation for scientific literature. In: Serdyukov, P., et al. (eds.) ECIR 2013. LNCS, vol. 7814, pp. 594–605. Springer, Heidelberg (2013). https://doi.org/10.1007/978-3-642-36973-5_50
21. Rubner, Y., Tomasi, C., Guibas, L.J.: A metric for distributions with applications to image databases. In: Sixth International Conference on Computer Vision, pp. 59–66. IEEE (1998)
22. Spasić, I.: Acronyms as an integral part of multi-word term recognition-a token of appreciation. IEEE Access **6**, 8351–8363 (2018)
23. Terada, A., Tokunaga, T., Tanaka, H.: Automatic expansion of abbreviations by using context and character information. Inf. Process. Manage. **40**(1), 31–45 (2004)
24. Uzuner, Ö., Solti, I., Cadag, E.: Extracting medication information from clinical text. J. Am. Med. Inform. Assoc. **17**(5), 514–518 (2010)
25. Wolsey, L.A., Nemhauser, G.L.: Integer and Combinatorial Optimization. Wiley, New York (2014)
26. Wu, Y., Denny, J., Rosenbloom, S., Miller, R., Giuse, D., Song, M., Xu, H.: A preliminary study of clinical abbreviation disambiguation in real time. Appl. Clin. Inform. **6**(02), 364–374 (2015)
27. Wu, Y., Denny, J.C., Trent Rosenbloom, S., Miller, R.A., Giuse, D.A., Wang, L., Blanquicett, C., Soysal, E., Xu, J., Xu, H.: A long journey to short abbreviations: developing an open-source framework for clinical abbreviation recognition and disambiguation (CARD). J. Am. Med. Inform. Assoc. **24**(e1), e79–e86 (2016)
28. Xu, H., Stetson, P.D., Friedman, C.: A study of abbreviations in clinical notes. In: AMIA Annual Symposium Proceedings. vol. 2007, p. 821. American Medical Informatics Association (2007)
29. Xu, J., Zhang, Y., Xu, H., et al.: Clinical abbreviation disambiguation using neural word embeddings. Proc. BioNLP **15**, 171–176 (2015)

Word Sense Induction Using Word Sketches

Ondřej Herman[(✉)], Miloš Jakubíček, Pavel Rychlý, and Vojtěch Kovář

Faculty of Informatics, Masaryk University,
Botanická 68a, 602 00 Brno, Czech Republic
{xherman1,jak,pary,xkovar3}@fi.muni.cz

Abstract. We present three methods for word sense induction based on Word Sketches. The methods are being developed a part of an semiautomatic dictionary creation system, providing annotators with the summarized semantic behavior of a word. Two of the methods are based on the assumption of a word having a single sense per collocation. We cluster the Word Sketch based collocations by their co-occurrence behavior in the first method. The second method clusters the collocations using word embedding model. The last method is based on clustering of Word Sketch thesauri. We evaluate the methods and demonstrate their behavior on representative words.

Keywords: Word sense induction · Word sketch · Collocations · Word embeddings

1 Word Sense Induction

The task of word sense induction (WSI) aims to identify the different senses of polysemous words from bulk text in an unsupervised setting. The problem has a long history, but none of the current solutions yield satisfactory results.

The closely related task of word sense discrimination assigns a specific occurrence of a word within its context to a predefined sense inventory.

Based on the Harris' distributional hypothesis [5], words with similar meanings appear in similar contexts, and therefore different meanings of the same word tend to be present in differing contexts. Insight into the senses of the word can be gained by investigating the contexts the word appears in.

The methods we are looking for are to be used to assist an annotator to properly describe the different senses a word can take on, therefore we would like the method to be transparent and give understandable sense clusters. For this application it is not an issue if more than one cluster consist of the same word sense, as long as a single cluster does not contain a mixture of different senses. For a speaker of the language, joining clusters is easy, while separating them is laborious. Therefore, in the following exposition we specify a higher number of word sense clusters than we expect to occur in the examples, even though the actual amount of investigated senses is likely smaller.

© Springer Nature Switzerland AG 2019
C. Martín-Vide et al. (Eds.): SLSP 2019, LNAI 11816, pp. 83–91, 2019.
https://doi.org/10.1007/978-3-030-31372-2_7

2   Spectral Clustering

The methods described below employ spectral clustering as an important building block. Spectral clustering [9] is a family of techniques which operate on
the pairwise similarities between the clustered objects, that is, on the similarity graph. Spectral clustering is based on the eigendecomposition of the graph
Laplacian L:

$$L = D - A \tag{1}$$

where D is a diagonal matrix where $d_{ii} \in D$ is the sum of weights of edges coin-
cident with the i-th vertex and A is a non-negative symmetric matrix, where
$a_{ij} = a_{ji} \in A$ is the similarity between the i-th and j-th vertex. The number
of clusters n is then chosen and the first n smallest eigenvalues and their corresponding eigenvectors are used to project L into a well-behaved space with
reduced dimension, in which clusters are easier to find. The usual choice for
clustering the reduced space is k-means.

The technique does not depend on clusters of specific shape and does not
require the similarity function to satisfy the properties of a metric. The technique
is based on standard linear algebra methods, which have been studied deeply and
can be implemented in an efficient way.

The usual formulation of spectral clustering requires the number of clusters to
be specified beforehand. One commonly used heuristic is based on the eigengap
heuristic [9], which selects the position of the first large difference between the
eigenvalues of the Laplacian ordered by magnitude as the number of clusters.
This method, while simple to implement, has no theoretical basis [9,10]. A more
robust (and complex) heuristic is described in [10].

To calculate the clustering, we use the implementation provided by the ven-
erable scikit-learn [8] library, with modifications which allow us to examine the
eigenvalues used during the computation.

3   Clustering of Word Sketch Co-occurrences

Word Sketch, as implemented in the Sketch Engine [6] is a summary of the contexts a specific word appears in, collated by different grammatical relations. The
Word Sketches are extracted from text employing a collection of regular rules,
each of which describe a collocation and the grammatical relation the collocation
appears in. The result is a collection of triples of the form (*headword, relation,
collocate*). The rules aim to trade precision for recall, so that the resulting relations of a word describe the contexts in which the word appears as completely
as possible. Triples which do not satisfy a criterion specified by a co-occurrence
metric are discarded, so that only salient triples remain. The Fig. 1 shows the
word sketch for the word *palm* calculated from the BNC corpus [3] (Fig. 2).

palm
(noun) Alternative PoS: <u>verb</u> (freq: 93)
British **National Corpus (BNC) v2.2 freq = <u>1,759</u>** (15.66 per million)

modifiers of "palm" 19.27	nouns and verbs modified by "palm" 30.13	verbs with "palm" as object 17.96	verbs with "palm" as subject 12.05	prepositional phrases
coconut 27 10.75	springs 24 10.18	grease 8 9.31	sweat 8 9.68	"palm" of ... 237 13.47
coconut palms	in palm springs	slap 12 9.04	damp 4 9	... in "palm" 110 6.25
potted 13 9.70	tree + 173 9.72	sway 8 8.91	fringe 4 8.93	... of "palm" 92 5.23
potted palms	palm trees	outstretch 7 8.64	cup 3 8.46	... into "palm" 83 4.72
sweaty 7 8.96	frond 13 9.44	sweat 4 8.09	face 40 7.97	... with "palm" 71 4.04
sago 3 8.13	palm fronds	rub 10 7.88	with palms facing upwards	... on "palm" 31 1.76
cupped 3 8	grove 20 9.16	kiss 8 7.65	leave 5 3.83	against "palm" 23 1.31
cabbage 3 7.74	palm groves	upturn 3 7.65	lie 3 3.58	"palm" in ... 22 1.25
banana 3 7.31	sunday 27 9.02	press 17 7.16	be 48 0.19	"palm" against ... 17 0.97
hand 13 6.75	on palm sunday	flatten 3 7.16	palms were	
date 12 6.37	beach 40 9.01	rest 4 6.88		
date palms	palm beach	wipe 5 6.83	**"palm" and/or ...** 13.08	
open 21 6.19	sander 9 8.91	wave 5 6.73	banana 4 8.11	
open palm	thatch 4 7.78	place 11 5.37	finger 10 8.08	
pink 4 6.19	leaf 6 7.50	lift 5 5.27	thigh 3 7.45	
warm 8 6.07	oil 22 7.46	spread 3 5.18		
left 5 5.75		push 5 5.15		

Fig. 1. Word sketch for the noun *palm*

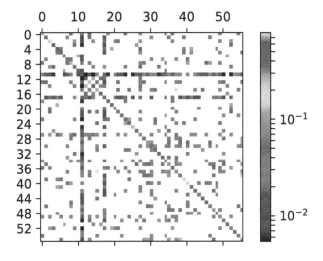

Fig. 2. Document co-occurrences of the Word Sketch collocations of the noun *palm*

The WSI method described in this section is based on the following assumptions. For a specific word,

1. each (*relation, collocate*) pair has a single sense
2. two (*relation, collocate*) pairs co-occurring in a document belong to the same sense

To identify the word senses, a $n \times n$ matrix C is constructed, where $n = |(relation, collocate)|$ and c_{ij} is the number of documents in which the i-th and

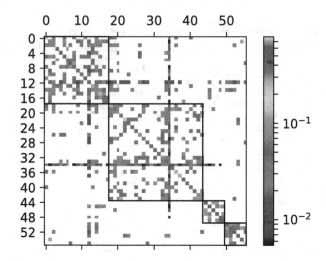

Fig. 3. Clustered document co-occurrences of the Word Sketch collocations of the noun *palm*

j-th (*relation, collocate*) pair appeared together. Only pairs with positive rank are kept and the pairs which do not occur in at least four different documents are discarded.

While C can be clustered directly, better result can be obtained by normalizing it. Using raw counts has tendency to create singleton clusters for collocates which appear in many different contexts, such as prepositions. To reduce the influence, we calculate the Dice-normalized C' as

$$C' = 2C \oslash (C + C^T) \tag{2}$$

where $\oslash$ represents element-wise matrix division. The result obtained by clustering C' using the spectral clustering algorithm with the desired number of clusters set to 4, contain the following (*relation, collocate*) pairs:

Cluster 1 (18 pairs)	Cluster 2 (26 pairs)
nouns and verbs modified by "palm-n" leaf-n	verbs with "palm-n" as object outstretch-v
verbs with "palm-n" as object sway-v	verbs with "palm-n" as object raise-v
"palm-n" and/or ... flower-n	verbs with "palm-n" as object open-v
nouns and verbs modified by "palm-n" grove-n	"palm-n" of ... hand-n
Cluster 3 (6 pairs)	**Cluster 4** (6 pairs)
modifiers of "palm-n" sweaty-j	nouns and verbs modified by "palm-n" oil-n
verbs with "alm-n" as object wipe-v	nouns and verbs modified by "palm-n" court-n
nouns and verbs modified by "palm-n" springs-n	modifiers of "palm-n" oil-n
nouns and verbs modified by "palm-n" beach-n	nouns and verbs modified by "palm-n" sunday-n

The pairs are shown in the order they appear within the clusters in the Fig. 3. The clusters obtained are mostly pure, with some exceptions, such as *palm-n springs-n* and *palm-n beach-n* appearing in the third cluster. This can be likely alleviated by modifying the construction of the similarity matrix.

The method yields reasonable results for many words and has the ability to provide extract the specific occurrences of the induced senses.

3.1   Clustering of Word Sketch Thesaurus

Thesaurus is a list of words similar a given word. As similar words appear in similar context, their word sketches will be similar, so the similarity of two words can be obtained by calculating the intersection of the word sketches of the two words.

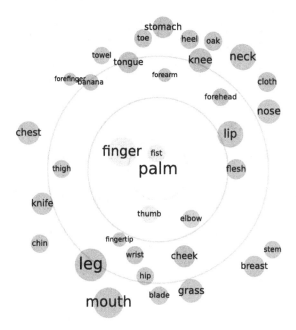

Fig. 4. Thesaurus for the noun *palm*

In Sketch Engine, the thesaurus is calculated by comparing all pairs of word sketches. For illustration, the thesaurus of the noun *palm* is visualized in the Fig. 4.

Based on the assumption that the thesaurus for a polysemous word will contain words similar to the different senses, clustering the words contained in the thesaurus based on their pairwise similarities can give insight into the senses the word can take on.

In this method, the matrix T to be clustered consists of the elements t_{ij}, which give the similarity of the i-th and j-th entries in the thesaurus of the word we are identifying the senses for, in the respective thesauri corresponding to the i-th and j-th entries. For example, the Fig. 5 shows the pairwise similarities of the elements of the thesaurus for the noun *palm*. Extracting 4 clusters yields the result shown in the Fig. 6. The most similar words grouped by cluster are:

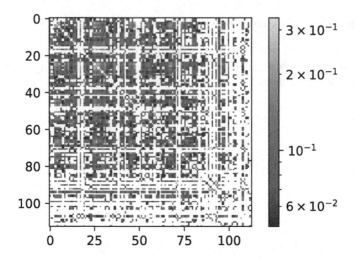

Fig. 5. Pairwise similarities of thesaurus entries for the noun *palm*

max cluster similarity	cluster representatives
.081	banana-n oak-n meadow-n fern-n vine-n
.127	fist-n finger-n thumb-n elbow-n flesh-n
.096	fingertip-n forefinger-n outward-n knuckle-n
.063	pine-n willow-n bamboo-n fig-n

This method gives easily interpretable results and seems to be more stable than clustering the word sketch co-occurrences directly and fewer occurrences of the investigated word are necessary to provide a satisfactory result. The drawback is that the information about the senses of specific occurrences in the corpus is not retained. Another issue is that the minimal similarity, for which the word sketch thesaurus items are indexed is .05, which is close to the maximum similarity in some of the obtained clusters, so the more distant or less frequent senses might be hidden below this threshold.

3.2 Clustering Context Word Embeddings

Another approach we investigated is based on clustering the contexts according to the embedding vectors. We calculated skip-gram embeddings of dimension 100 using the fastText package [4] and for every occurrence of the examined word, we calculated the average of the left and right collocate embeddings and used the HDBSCAN [7] algorithm to cluster these vectors. The length of the context to be examined turns out to be a crucial parameter. When the context is too narrow, the clusters are strongly influenced by noise. On the other hand, context which is too wide will not contain enough discriminating information for the clustering algorithm to exploit. For this experiment, we use 10 tokens to the left and 10 tokens to the right of the word.

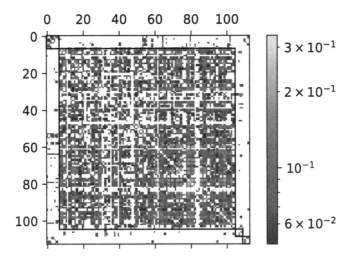

Fig. 6. Clustered pairwise similarities of thesaurus entries for the noun *palm*

HDBSCAN can determine the number of clusters automatically. For the word *palm*, 6 clusters have been found and 93 % of the total vectors had been left unclassified. Three of the clusters are similar, containing mostly repeated text:

```
BUST UPLIFT 6. With hands clasped behind, and palms facing inwards, raise the arms 30 times
BUST UPLIFT 9. With hands clasped behind and palms facing inwards, raise the arms 30 times
BUST UPLIFT 9. With hands clasped behind and palms facing inwards, raise the arms 35 times
```

The remaining three clusters consist of three salient senses: palm oil, palm trees, and palm as a body part. While the actual obtained senses are well separated, the result is very sparse and the model hard to inspect, modify and understand. Ensuring that this method works reliably on arbitrary words seems to be difficult.

3.3   Clustering Word Sketches by Word Embeddings

The previous method can be enriched using the information contained within word sketches. Instead of considering each context of the word as a candidate entering the clustering algorithm, we create a single vector for each (*relation, collocate*) pair by averaging the context vectors obtained from the obtained in the same way as in the previous method and then cluster these vectors using HDBSCAN. For the noun *palm*, when using the default configuration of HDB-SCAN, all word sketch pairs are discarded by the algorithm. Changing the *alpha* (robust single linkage distance scaling) parameter is reduced to 0.5, discards 55 % of the pairs and the rest is split into two senses, of which the most salient ones according to word sketch rank are:

Cluster 1 (77 pairs)	Cluster 2 (16 pairs)
"palm-n" of ... hand -n	modifiers of "palm-n" coconut-j
... into "palm-n" nail-n	nouns and verbs modified by "palm-n" tree-n
... into "palm-n" fist-n	modifiers of "palm-n" potted-j
... with "palm-n" forehead-n	"palm-n" from ... stone-n
nouns and verbs modified by "palm-n" springs-n	nouns and verbs modified by "palm-n" frond-n
"palm-n" on ... apron-n	nouns and verbs modified by "palm-n" grove-n
... into "palm-n" bit-n	verbs with "palm-n" as subject fringe-v
verbs with "palm-n" as subject sweat-v	verbs with "palm-n" as object sway-v
... into "palm-n" kiss-n	nouns and verbs modified by "palm-n" thatch-n
... into "palm-n" dig-v	modifiers of "palm-n" cabbage-n

While the senses are well separated, the sparsity could be an issue, as an important sense may have been missed. Additionally, this method employs not only one, but two black boxes: word embeddings and HDBSCAN.

4   Conclusion

While all of the methods give interesting results, we have not evaluated them thoroughly and have not explored the parameter space well at this time.

In addition to the described methods, we implemented the very elegant method based on sparse dictionary learning as described in [1], which aims to decompose the word vector into a weighted sum of *discourse atoms*, but on our data, it failed to yield any interesting result.

We also investigated the Adaptive Skip-gram [2], which trains a word embedding model for multiple senses for each word in a single step. The senses we obtained are reasonable, but we found the model to be too opaque and the information about the sense at a specific corpus position can be reconstructed only inexactly. We devised a method based on clustering word sketch co-occurrences, which is efficient and provides reasonable senses along with the specific corpus positions the senses for a given word appear at.

A different method based on word sketch thesaurus provides better word sense clustering, but has the drawback of not providing the specific positions of the senses.

The two word embedding based methods have shown the ability to produce great results, however they are finicky and difficult to tune and interpret.

Acknowledgments. This work has been partly supported by the Grant Agency of CR within the project 18-23891S and the Ministry of Education of CR within the OP VVV project CZ.02.1.01/0.0/0.0/16_013/0001781 and LINDAT-Clarin infrastructure LM2015071.

References

1. Arora, S., Li, Y., Liang, Y., Ma, T., Risteski, A.: Linear algebraic structure of word senses, with applications to polysemy. Trans. Assoc. Comput. Linguist. **6**, 483–495 (2018)
2. Bartunov, S., Kondrashkin, D., Osokin, A., Vetrov, D.: Breaking sticks and ambiguities with adaptive skip-gram. arXiv preprint arXiv:1502.07257 (2015)

3. Consortium, B., et al.: The British national corpus, version 2 (BNC world). Distributed by Oxford University Computing Services (2001)
4. Grave, E., Mikolov, T., Joulin, A., Bojanowski, P.: Bag of tricks for efficient text classification. In: Proceedings of the 15th Conference of the European Chapter of the Association for Computational Linguistics, EACL, pp. 3–7 (2017)
5. Harris, Z.S.: Distributional structure. Word **10**(2–3), 146–162 (1954)
6. Kilgarriff, A., Rychlý, P., Smrž, P., Tugwell, D.: Itri-04-08 the sketch engine. Inf. Technol. **105**, 113 (2004)
7. McInnes, L., Healy, J., Astels, S.: hdbscan: hierarchical density based clustering. J. Open Source Softw. **2**(11), 205 (2017)
8. Pedregosa, F., et al.: Scikit-learn: machine learning in Python. J. Mach. Learn. Res. **12**, 2825–2830 (2011)
9. Von Luxburg, U.: A tutorial on spectral clustering. Stat. Comput. **17**(4), 395–416 (2007)
10. Zelnik-Manor, L., Perona, P.: Self-tuning spectral clustering. In: Advances in Neural Information Processing Systems, pp. 1601–1608 (2005)

Temporal "Declination" in Different Types of IPs in Russian: Preliminary Results

Tatiana Kachkovskaia[(⊠)] [iD]

Saint Petersburg State University, Saint Petersburg, Russia
kachkovskaia@phonetics.pu.ru

Abstract. The paper explores temporal changes within an intonational phrase in Russian. The main question we aim to answer is whether we can speak about temporal "declination" in a similar way we speak about melodic declination. In order to answer this question, we analysed stressed vowel duration in intonational phrases (IPs) of different types using a speech corpus. We have found that (1) most intonational phrases in Russian do not have temporal "declination" or "inclination" in the pre-nuclear part: the tempo is relatively stable until the nucleus, where a noticeable lengthening is observed; (2) the rarely occurring temporal "declination" or "inclination" in certain types of IPs can be considered a specific speaker's trait; (3) the amount of lengthening on the last stressed vowel within the IP may play a role in distinguishing final and non-final IPs, rising vs. falling nuclei, but this is also speaker-specific.

Keywords: Prosody · Speech tempo · Segmental duration · Intonational phrase · Russian phonetics

1 Introduction

One of the most well-known prosodic universals is melodic declination, which is defined by Vaissiere in the following way: "In relatively long stretches of continuous speech, there is a global tendency for the F0 curve[1] to decline with time, despite successive local rises and falls" [7][p. 75]. One of the consequences of this is the reset of the baseline, which occurs at the boundary between one stretch of speech and the following one. As a result, resetting the baseline functions as a boundary marker, and declination itself—as an organizing trend which joins a number or words into a longer speech unit.

In a similar way we could attempt to speak about "temporal" declination. What is known so far is that the last word within a large prosodic unit is longer due to the phenomenon called pre-boundary (final) lengthening, which is a universal itself. Less is known about the other words within the prosodic unit:

Supported by the Government of Russia (President's grant MK-406.2019.6).

[1] fundamental frequency, or melodic, curve.

C. Martín-Vide et al. (Eds.): SLSP 2019, LNAI 11816, pp. 92–99, 2019.
https://doi.org/10.1007/978-3-030-31372-2_8

is duration evenly distributed among the other words, or we can speak about some "declination" (or "inclination") of tempo as well? The prosodic features—melody, duration, intensity—often interact with each other, either working in parallel (melody + intensity) or compensating each other (e.g., in whispered speech, where no clear melody is observed, especially for tone languages). Thus, given melodic declination, we might expect some accompanying temporal declination or inclination as well.

For Russian the melodic declination was described in detail in [3]. One of the key results of the paper is that melodic declination differs across types of utterance—almost zero in general questions (level tone), and quite significant in declaratives. Similar results were described for Dutch by van Heuven and Haan [1][p. 125]. In particular, this means that declination helps distinguish some types of utterances even before the utterance is finished—and perceptually this was proved for Russian in [6][p. 116–120], where speakers could successfully disambiguate general questions from non-final parts of declaratives before they heard the last word of the prosodic unit, probably relying on the melodic declination pattern. For speech tempo no research of this kind has been done yet.

In this paper we compared 4 types of intonational phrases (IPs) differing in the melodic movement within the nucleus and finality of the IP within the utterance: (1) low-falling, utterance-final (declaratives); (2) rising-falling[2], utterance final (general questions); (3) falling, utterance-medial (as, e.g., in cases of punctuation marks such as colon or semicolon); (4) rising-falling, utterance-medial (occurring usually with a comma or dash, or no punctuation mark at all to divide long stretches of speech). In the future more types of IPs may be analysed, but so far we have taken the ones that occur more frequently in Russian speech.

2   Materials and Methods

Temporal "declination", as well as melodic, can be calculated in several ways. When choosing a method for this paper, we aimed at obtaining results comparable with those found earlier for melodic declination in Russian as presented in [3]. In that study melodic declination was calculates by peak values for stressed syllables (upper declination). In a similar way, here we calculated temporal declination based on stressed syllables.

In Russian stressed syllables are longer than the unstressed. Within the stressed syllable, vowel duration is a more reliable measurement as consonants in the stressed syllables are not always longer than in the unstressed (see [2]). This is why we calculated temporal declination based on stressed vowel duration.

---

[2] Rise-fall—one of the most frequent types of nuclei in Russian speech—is used in general questions and non-final IPs; the rise is realized on the stressed syllable, and the melodic peak is close to its right boundary (or sometimes even later).

As vowel phonemes differ in their inherent duration, we calculated duration in z-scores using the formula suggested in [9]:

$$\tilde{d}(i) = \frac{d(i) - \mu_p}{\sigma_p}$$

where $\tilde{d}(i)$ is the normalized duration of segment i, $d(i)$ is its absolute duration, and μ_p and σ_p are the mean and standard deviation of the duration of the corresponding phone p. The mean and standard deviation values were calculated over the whole corpus for each speaker separately.

We also limited our data by only those vowels that occurred in CV syllables, in order to eliminate influence from syllable length.

The material is a subcorpus of CORPRES [5]: 20 h of segmented speech containing fictional texts recorded from 4 speakers, all native Russians with standard pronunciation. The recordings come with manual segmentation into sounds and prosodic annotation in terms of [8]. The basic large prosodic unit in the annotation is the IP defined as (1) the largest phonological chunk into which utterances are divided, (2) containing a single most prominent word (nucleus), (3) serving to join the words tightly connected with each other in terms of semantic or syntactic stricture (the definition consistent with the one given in [4][p. 311]).

Using a Python script, we retrieved all the IPs with a given type of nucleus and a given length in clitic groups (CGs). In this paper we only observe IPs with final position of the nucleus (which is not always the case in Russian, but still the vast majority). A clitic group is defined as one stressed lexical word plus (possibly) one or more adjacent unstressed clitics. Thus, the number of clitic groups within an IP equals the number of stressed syllables.

The data were analysed separately for each speaker in case individual strategies are found. Then, for each type of nucleus and each IP length average values were obtained for CG 1, CG 2, CG 3 etc. This was summarized in graphs and tables. Then each pair of adjacent words were compared using the two-tailed Student's t-test for independent datasets with unequal variances (e.g., to find out whether CG 1 is longer than CG 2 in 4-word declaratives, we compared the means for the respective stressed vowels' normalized durations using a t-test).

3   Results and Discussion

Table 1 summarizes the results of the analysis for four speakers. The table contains data for the four types of IPs, and for each type—for IPs made up of 3, 4 and 5 clitic groups (CGs). The first clitic group obtains the number 1. Thus, e.g., in an average 3-word IP of type 1 (utterance-final IP with a falling nucleus) for speaker C the first stressed vowel has the normalized duration of -0.33, the second -0.37, and the third 0.79.

The value of 1 corresponds to one standard deviation for vowel duration for the particular vowel phoneme and for the particular speaker. Standard deviation values in our data fall within the range of 23 to 35 ms. Thus, in our example the

Table 1. Mean normalized vowel duration in stressed CV syllables in IPs of different length (3 to 5 clitic groups) for each clitic group (CG) within the phrase. IP type 1 stands for utterance-final IPs with low-falling nucleus; type 2—utterance final, rising-falling nucleus; type 3—utterance-medial, falling nucleus; type 4—utterance-medial, rising-falling nucleus. Speakers C and K are females, speakers A and M are males. Asterisk (*) marks the value which is significantly different from the left neighbour; brackets around the asterisk mean that the p-value is above 0.05, but very close to this value (up to 0.065). Bold font marks the IP types where the value for the 1-st word differs significantly from the value for the penultimate word. Brackets mark a case of a small dataset. Empty cells mean a lack of a reliable dataset.

CG number		1	2	3	4	5	1	2	3	4	5
IP type	IP len										
		Speaker C					Speaker K				
1	3	−0.33	−0.37	0.79*			−0.42	−0.51	0.80*		
	4	−0.47	−0.30	−0.35	0.92*		−0.59	−0.68	−0.55	0.73*	
	5	−0.46	−0.40	−0.41	−0.46	0.86*	−0.68	−0.48	−0.59	−0.46	0.73*
2	3	−0.51	−0.64	0.43*			−0.69	−0.63	0.29*		
	4	**−0.25**	**−0.63(*)**	**−0.80**	**0.10***		−0.53	−0.71	−0.58	0.39*	
	5	(−0.18)	(−0.39)	(−0.59)	(−0.79)	(0.42)*					
3	3	−0.46	−0.49	0.82*			−0.61	−0.57	0.96*		
	4	−0.54	−0.50	−0.42	0.92*		−0.69	−0.59	−0.65	0.91*	
	5	−0.48	−0.55	−0.36	−0.34	0.99*	−0.69	−0.52	−0.43	−0.35	1.18*
4	3	−0.53	−0.53	0.12*			−0.61	−0.50*	0.36*		
	4	−0.47	−0.44	−0.49	0.14*		**−0.72**	**−0.60(*)**	**−0.54**	**0.46***	
	5	−0.60	−0.53	−0.38	−0.51	0.24*	−0.73	−0.69	−0.55	−0.51	0.34*
		Speaker A					Speaker M				
1	3	−0.26	−0.33	0.30*			−0.45	−0.39	0.18*		
	4	**−0.46**	**−0.29***	**−0.23**	**0.37***		−0.31	−0.37	−0.33	0.38*	
	5	−0.67	−0.28*	−0.40	−0.56	0.24*	**−0.71**	**−0.49**	**−0.19(*)**	**−0.40**	**0.20***
2	3	−0.50	−0.54	−0.01*			−0.36	−0.60*	1.04*		
	4	−0.47	−0.53	−0.42	−0.06*		−0.44	−0.49	−0.43	0.68*	
	5										
3	3	−0.42	−0.47	0.79*			−0.30	−0.34	0.62*		
	4	−0.57	−0.47	−0.44	0.49*		−0.42	−0.28	−0.30	0.67*	
	5	−0.62	−0.69	−0.45	−0.49	0.95*	**−0.47**	**−0.41**	**−0.31**	**−0.24**	**0.52***
4	3	−0.52	−0.50	0.30*			−0.49	−0.42	0.60*		
	4	−0.54	−0.44	−0.45	0.23*		**−0.53**	**−0.38***	**−0.34**	**0.66***	
	5	−0.61	−0.54	−0.37	−0.46	0.35*	−0.48	−0.49	−0.34	−0.36	0.54*

first and the second words have almost equal stressed vowel duration, while on the last word (which is the nucleus) we observe a noticeable change in duration— 1.16 standard deviations, i.e. more than 26 ms.

The results of statistical analysis are also shown in Table 1. An asterisk (*) marks the values that differ significantly from the left neighbour (p-value < 0.05). Those cases where the p-value was very close to 0.05 (up to 0.065) are marked by an asterisk in brackets.

In our data every last clitic group within the IP has significantly greater value; the p-values as for these cases were all below 0.002. This proves that the nucleus in phrase-final words is much longer than other stressed syllables, but this result is not new, at least for Russian (e.g., see [2]).

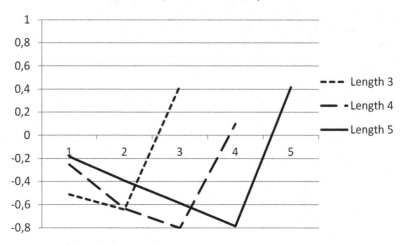

Fig. 1. Temporal pattern of a general question for speaker C showing temporal "declination". The graph presents data for IPs of different length (3–5 clitic groups); the x axis shows the index number of the clitic group within the IP.

In some cases IP-*medial* clitic groups also reveal a significant change in duration (see, e.g., 4-word IPs of type 1 for speaker A); the p-values are usually higher than for IP-final CGs (0.01 to 0.05). In most cases it is a decrease or an increase in duration on the second clitic group. In terms of IP type and length this is unsystematic.

We also analysed statistically the difference between the first CG and the penultimate CG. If the difference is significant, we may conclude that we observe temporal declination or inclination (not level tempo) in the pre-nuclear part of the IP. In Table 1 such cases are marked by bold font. We can see that there are only a few clear cases of declination, and they are very speaker-specific. Speaker C marks general questions (IP type 2) with temporal declination (see Fig. 1), while the other 3 speakers do not. Speakers K and M tend to mark non-final IPs with rising-falling nucleus with temporal inclination (see Fig. 2); speaker M also shows some temporal inclination in IPs with a falling nucleus; speaker A has a slight temporal declination in declaratives.

In the large majority of cases *no* temporal declination is observed (see, e.g., Fig. 3). This means that the temporal pattern of a typical Russian IP can be described in the following way: relatively stable tempo from the first CG to the penultimate CG, and a noticeable lengthening on the last CG.

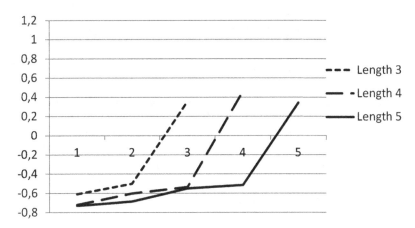

Fig. 2. Temporal pattern of a non-utterance-final IP with a rising-falling nucleus for speaker K showing a slight temporal "inclination". The graph presents data for IPs of different length (3–5 clitic groups); the x axis shows the index number of the clitic group within the IP.

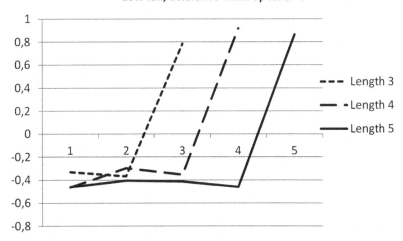

Fig. 3. Temporal pattern of a declarative for speaker C showing no temporal "declination" or "inclination". The graph presents data for IPs of different length (3–5 clitic groups); the x axis shows the index number of the clitic group within the IP.

However, the amount of lengthening on the last CG differs between IP types and between speakers. A number of individual strategies could be formulated.

1. Many speakers distinguish between IPs with falling and with rising-falling nuclei (speakers A, C, K)—falling nuclei have longer vowels.
2. Some speakers use duration to signal non-finality (speaker A)—vowels in nuclei of non-final IPs (types 3 and 4) are longer.
3. Some speakers tend to lengthen the nuclear vowel in general questions (speaker M).

The speaker-specific nature of the temporal pattern of the IP is in accordance with the findings presented in [3] for melodic declination. Still, the main result or this preliminary study is that temporal "declination" is absent in most types of IPs for most speakers. It might suggest that speech tempo is something that needs to be kept stable while other parameters are changing. But this required further proof. The next step of this ongoing research might be an analysis of those IPs which *do* have temporal "delination" or "inclination" using a series of perception experiments.

4   Conclusions

An analysis of 20 h of speech recorded from 4 speakers of Standard Russian enabled to obtain the following preliminary results.

1. Most intonational phrases in Russian do not have temporal "declination" or "inclination" in the pre-nuclear part: the tempo is relatively stable until the nucleus, where a noticeable lengthening is observed.
2. Temporal "declination" or "inclination" in certain types of IPs can be considered a specific speaker's trait.
3. The amount of lengthening on the last stressed vowel within the IP may play a role in distinguishing final and non-final IPs, rising vs. falling nuclei; this is also speaker-specific.

References

1. van Heuven, V., Haan, J.: Phonetic correlates of statement versus question intonation in dutch. In: Botinis, A. (ed.) Intonation. Text, Speech and Language Technology, vol. 15, pp. 119–144. Springer, Dordrecht (2000). https://doi.org/10.1007/978-94-011-4317-2_6
2. Kachkovskaia, T., Nurislamova, M.: Word-initial consonant lengthening in stressed and unstressed syllables in Russian. In: Karpov, A., Jokisch, O., Potapova, R. (eds.) SPECOM 2018. LNCS (LNAI), vol. 11096, pp. 264–273. Springer, Cham (2018). https://doi.org/10.1007/978-3-319-99579-3_28
3. Kocharov, D., Volskaya, N., Skrelin, P.: F0 declination in Russian revisited. In: Proceedings of 18th International Congress of Phonetic Sciences (2015)
4. Ladd, D.R.: Intonational phrasing: the case for recursive prosodic structure. Phonol. Yearb. **3**, 311–340 (1986)

5. Skrelin, P., Volskaya, N., Kocharov, D., Evgrafova, K., Glotova, O., Evdokimova, V.: CORPRES. In: Sojka, P., Horák, A., Kopeček, I., Pala, K. (eds.) TSD 2010. LNCS (LNAI), vol. 6231, pp. 392–399. Springer, Heidelberg (2010). https://doi.org/10.1007/978-3-642-15760-8_50
6. Svetozarova, N.: The Intonation System of Russian [Intonacionnaya Sistema Russkogo Yazyka]. Leningrad University (1982)
7. Vaissiere, J.: Language-independent prosodic features. In: Cutler, A., Ladd, D.R. (eds.) Prosody: Models and Measurements, Springer Series in Language and Communication, vol. 14, pp. 53–66. Springer, Heidelberg (1983)
8. Volskaya, N., Skrelin, P.: Prosodic model for Russian. In: Proceedings of Nordic Prosody X, pp. 249–260. Frankfurt am Main: Peter Lang (2009)
9. Wightman, C., Shattuck-Hufnagel, S., Ostendorf, M., Price, P.: Segmental durations in the vicinity of prosodic phrase boundaries. J. Acoust. Soc. Am. **91**, 1707–1717 (1992)

Geometry and Analogies: A Study and Propagation Method for Word Representations

Sammy Khalife[(⊠)], Leo Liberti[(⊠)], and Michalis Vazirgiannis

LIX, CNRS, Ecole Polytechnique, Institut Polytechnique de Paris,
91128 Palaiseau, France
{khalife,liberti,mvazirg}@lix.polytechnique.fr

Abstract. In this paper we discuss the well-known claim that language analogies yield almost parallel vector differences in word embeddings. On the one hand, we show that this property, while it does hold for a handful of cases, fails to hold in general especially in high dimension, using the best known publicly available word embeddings. On the other hand, we show that this property is not crucial for basic natural language processing tasks such as text classification. We achieve this by a simple algorithm which yields updated word embeddings where this property holds: we show that in these word representations, text classification tasks have about the same performance.

1 Introduction

1.1 Context and Motivations

The motivation to build word representations as vectors in a Euclidean space is twofold. First, geometrical representations can possibly enhance our understanding of a language. Second, these representations can be useful for information retrieval on large datasets, for which semantic operations become algebraic operations. First attempts to model natural language using simple vector space models go back to the 1970s, namely Index terms [22], term frequency inverse document frequency (TF-IDF) [20], and corresponding software solutions SMART [21], Lucene [10]. In recent work about word representations, it has been emphasized that many analogies such as *king* is to *man* what *queen* is to *woman*, yielded almost parallel difference vectors in the space of the two most significant coordinates [15,18], that is to say (if $d = 2$):

$$(u_i \mid 1 \leq i \leq n) \in \mathbb{R}^d \text{ being the word representations}$$

$$\text{(3,4) is an analogy of (1,2)} \Leftrightarrow \exists \epsilon \in \mathbb{R}^d \text{ s.t } u_2 - u_1 = u_4 - u_3 + \epsilon \qquad (1)$$

$$\text{where} \qquad ||\epsilon|| \ll \min(||u_2 - u_1||, ||u_4 - u_3||)$$

In Eq. (1) $||x|| \ll ||y||$ means in practice that $||x||$ is much smaller than $||y||$. Equation (1) is stricter than just parallelism, but we adopt this version because

© Springer Nature Switzerland AG 2019
C. Martín-Vide et al. (Eds.): SLSP 2019, LNAI 11816, pp. 100–111, 2019.
https://doi.org/10.1007/978-3-030-31372-2_9

it corresponds to the version the scientific press has amplified in such a way that now it appears to be part of layman knowledge about word representations [5,14,23]. We hope that our paper will help clear a misinterpretation.

Recent work leads us to cast word representations into two families: *static representations*, where each word of the language is associated to a unique element (scope of this paper), and *dynamic representations*, where the entity representating each word may change on the context (we do not consider this case in this paper).

1.2   Contributions

The attention devoted in the literature and the press to Eq. (1) might have been excessive, based on the following criteria:

o The proportion of analogies leading to the geometric Eq. (1) is small.
o The classification of analogies based on Eq. (1) or parallelism does not appear as an easy task.

Second, we present a very simple propagation method in the graph of analogies, enabling our notion of parallelism in Eq. (1). Our code is available online.[1]

2   Related Work

2.1   Word Embeddings

In the *static representations* family, after the first vector space models (Index terms, TF-IDF, see SMART [21], Lucene [10]), Skip-gram and statistical log-bilinear regression models became very popular. The most famous are Glove [18], Word2vec [15], and fastText [4]. Since word embeddings are computed once and for all for a given string, this causes polysemy for fixed embeddings. To overcome this issue, the family of *dynamic representations* have gained in attention very recently due to the increase of deep learning methods. ELmo [19], and Bert [9] representations take in account context, letters, and n-grams of each word. We do not address comparison with these methods in this paper because of the lack of analysis of their geometric properties.

There have been attempts to evaluate the semantic quality of word embeddings [11], namely:

o Semantic similarity (Calculate Spearman correlation between cosine similarity of the model and human rated similarity of word pairs)
o Semantic analogy (Analogy prediction accuracy)
o Text categorisation (Purity measure).

---

[1] Link to repository https://github.com/Khalife/Geometry-analogies.git.

However, in practice, these semantic quality measures are not preferred for applications: the quality of word embeddings is evaluated on very specific tasks, such as text classification or named entity recognition. In addition, recent work [17] has shown that the use of analogies to uncover human biases should be carried out very carefully, in a fair and transparent way. For example [7] analyzed gender bias from language corpora, but balanced their results by checking against the actual distribution of jobs between genders.

2.2   Relation Embeddings for Named Entities

An entity is a real-world object and denoted with a proper name. In the expression "Named Entity", the word "Named" aims at restricting the possible set of entities to only those for which one or many rigid designators stands for the referent. Named entities have an important role in text information retrieval [16].

For the sake of completeness, we report work on the representation of relations between entities. Indeed, an entity relation can be seen as an example of relation we consider for analogies (example: Paris is the capital of France, such as Madrid to Spain). There exist several attempts to model these relations, for example as translations [6,24], or as hyperplanes [12].

2.3   Word Embeddings, Linear Structures and Pointwise Mutual Information

In this subsection, we will focus on a recent analysis of pointwise mutual information, which aims at providing a piece of explanation of the linear structure for analogies [1,2]. This work provides a generative model with priors to compute closed form expressions for word statistics. In the following, $f = O(g)$ (resp. $f = \tilde{O}(g)$) means that f is bounded by g (resp. bounded ignoring logarithmic factors) in the neighborhood considered. The generation of sentences in a given text corpus is made under the following generative assumptions:

○ *Assumption 1*: The ensemble of word vectors consists of i.i.d samples generated by $v = s\,\hat{v}$, where $\hat{v}$ is drawn from the spherical Gaussian distribution in $\mathbb{R}^d$ and s is an integrable random scalar, always upper bounded by a constant $\kappa \in \mathbb{R}^+$.

○ *Assumption 2*: The text generation process is driven by a random walk of a vector, i.e if w_t is the word at step t, there exists a discourse vector c_t such that $\mathsf{P}(w_t = w|c_t) \propto \exp(\langle c_t, v_w \rangle)$. Moreover, $\exists \kappa \geq 0$ and $\epsilon_1 \geq 0$ such that $\forall t \geq 0$:

$$|s| \leq \kappa$$
$$\mathbb{E}_{c_{t+1}}(e^{\kappa \sqrt{d}||c_{t+1} - c_t||_2}) \leq 1 + \epsilon_1 \tag{2}$$

In the following, $\mathsf{P}(w, w')$ is the probability that two words w and w' occur in a window of size 2 (the result can be generalized to any window size), $\mathsf{P}(w)$ is the marginal probability of w. $\mathsf{PMI}(w, w')$ is the pointwise mutual information between two words w and w' [8]. Under these conditions, we have the following result [1]:

Theorem 1. *Let n denote the number of words and d denote the dimension of the representations. If Assumptions 1 and 2 are verified, then using the same notations, the following holds for any words w and w′ :*

$$\mathsf{PMI}(w, w') \triangleq \log \frac{\mathsf{P}(w, w')}{\mathsf{P}(w)\mathsf{P}(w')} = \frac{\langle v_w, v_{w'} \rangle}{d} \pm O(\epsilon)$$

$$\text{with} \quad \epsilon = \tilde{O}(\frac{1}{\sqrt{n}}) + \tilde{O}(\frac{1}{d}) + O(\epsilon_1) \tag{3}$$

Equation (3) shows that we could expect high cosine similarity for pointwise close terms (if ϵ is negligible).

The main aspect we are interested in is the relationship between linear structures and analogies. In [1], the subject is treated with an assumption following [18], stated in Eq. (4). Let χ be any set of words, and a and b words are involved in a semantic relation $\mathcal{R}$. Then there exist two scalars $v_{\mathcal{R}}(\chi)$ and $\xi_{ab\mathcal{R}}(\chi)$ such that:

$$\frac{\mathsf{P}(\chi|a)}{\mathsf{P}(\chi|b)} = v_{\mathcal{R}}(\chi) \, \xi_{ab\mathcal{R}}(\chi) \tag{4}$$

We failed to fully understand the argument made in [1,18] linking word vectors to differences thereof. However, if we assume Eq. (4), by Eq. (3) we obtain the following.

Corollary 2. *Let V be the n × d matrix whose rows are the vectors of words in dimension d. Let v_a and v_b be vectors corresponding respectively to words a and b. Assume a and b are involved in a relation $\mathcal{R}$. Let $\log(v_{\mathcal{R}})$ the element-wise log of vector $v_{\mathcal{R}}$. Then there exists a vector $\xi'_{ab\mathcal{R}} \in \mathbb{R}^n$ such that:*

$$V(v_a - v_b) = d \log(v_{\mathcal{R}}) + \xi'_{ab\mathcal{R}} \tag{5}$$

Proof. Let x a word, and a, b two words sharing a relation $\mathcal{R}$. On the one hand, taking the log of Eq. (4):

$$\log(\frac{\mathsf{P}(x|a)}{\mathsf{P}(x|b)}) = \log(v_{\mathcal{R}}(x)) + \log(\xi_{ab\mathcal{R}}(x)) \tag{6}$$

On the other hand, using Eq. (3), $\exists \epsilon_{abx} \in \mathbb{R}$ such that:

$$\log(\frac{\mathsf{P}(x|a)}{\mathsf{P}(x|b)}) = \log(\frac{\mathsf{P}(x, a)\mathsf{P}(b)}{\mathsf{P}(x, b)\mathsf{P}(a)})$$

$$= \log(\frac{\mathsf{P}(x, a)\mathsf{P}(b)\mathsf{P}(x)}{\mathsf{P}(x, b)\mathsf{P}(a)\mathsf{P}(x)})$$

$$= \mathsf{PMI}(x, a) - \mathsf{PMI}(x, b)$$

$$\log(\frac{\mathsf{P}(x|a)}{\mathsf{P}(x|b)}) = \frac{\langle v_x, v_a - v_b \rangle}{d} + \epsilon_{abx} \tag{7}$$

Combining Eqs. (6) and (7), for any x:

$$\langle v_x, v_a - v_b \rangle = d \log(v_{\mathcal{R}}(x)) + d(\log(\xi_{ab\mathcal{R}}(x)) - \epsilon_{abx}) \tag{8}$$

Let V the matrix whose rows are the word vectors. $V(v_a - v_b)$ is a vector of $\mathbb{R}^n$ whose component associated with word x is exactly $\langle v_x, v_a - v_b \rangle$. Then, let $v_{\mathcal{R}}$ be the element-wise log of vector $v_{\mathcal{R}}$, and $\xi'_{ab\mathcal{R}}$ the vectors of components $d(\log \xi_{ab\mathcal{R}}(x) - \epsilon_{abx})$. Then, Eq. (8) is exactly Eq. (5).   □

It is shown in [1] that $||V^+ \xi'_{ab\mathcal{R}}|| \leq ||\xi'_{ab\mathcal{R}}||$, where V^+ is the pseudo-inverse of V. In other words, the "noise" factor ξ' can be reduced. This reduction may not be sufficient if $\xi_{ab\mathcal{R}}$ is too large to start with. In the next section we shall propose an empirical analysis of existing embeddings with regard to analogies and parallelism of vector differences.

3   Experiments with Existing Representations

In this section, we present a list of experiments we ran on the most famous word representations.

3.1   Sanity Check

The exact meaning of the statement that analogies are geometrically character-ized in word vectors is as follows [14,18]. For each quadruplet of words involved in an analogy (a, b, c, d), consider the word vector triplet (v_a, v_b, v_c), and the difference vector $x_{ab} = v_b - v_a$. Then we run PCA on the set of word vectors to get representations in $\mathbb{R}^2$. Find the k nearest neighbours of $v_c + x_{ab}$ in the word embedding set (with k small). Finally, examine the k words and choose the most appropriate word d for the analogy $a : b = c : d$. We ran this protocol in many dimension with a corpus of analogies obtained from [13]. We display the results obtained in Fig. 1.

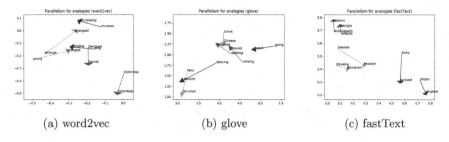

(a) word2vec	(b) glove	(c) fastText

Fig. 1. Sanity check

3.2   Analogies Protocol

In this subsection we show that the protocol we described in Sect. 3.1 for finding analogies does not really work in general. We ran it on 50 word triplets (a, b, c) as input, with $k = 10$ in the k-NN stage, but only obtained 35 correct valid analogies, namely those in Fig. 2.

'Athens:Greece=Baghdad:Iraq', 'Ottawa:Canada=Islamabad:Pakistan'
'Ashgabat:Turkmenistan=Athens:Greece', 'Beirut:Lebanon=Bern:Switzerland',
'Bujumbura:Burundi=Conakry:Guinea', 'Doha:Qatar=Hanoi:Vietnam',
'his:her=brothers:sisters', 'easy:easier=simple:simpler',
'low:lower=tight:tighter', 'strong:stronger=bad:worse',
'cold:coldest=low:lowest', 'discover:discovering=enhance:enhancing',
'play:playing=sing:singing', 'think:thinking=implement:implementing',
'Cambodia:Cambodian=Croatia:Croatian', 'Greece:Greek=Italy:Italian',
'flying:flew=jumping:jumped', 'looking:looked=screaming:screamed',
'selling:sold=taking:took', 'thinking:thought=flying:flew',
'child:children=snake:snakes', 'mouse:mice=computer:computers',

Fig. 2. Some valid analogies following Protocol 3.2

3.3   Turning the Protocol into an Algorithm

The protocol described in Sect. 3.2 is termed "protocol" rather than "algorithm" because it involves a human interaction when choosing the appropriate word out of the set of $k = 5$ nearest neighbours to $v_c + (v_b - v_a)$. Since natural language processing tasks usually concern sets of words of higher cardinalities than humans can handle, we are interested in an algorithm for finding analogies rather than a protocol. In this section we present an algorithm which takes the human decision out of the protocol sketched above. Then we show that this algorithm has the same shortcomings as the protocol, as shown in Sect. 3.2.

We first remark that the obvious way to turn the protocol of Sect. 3.2 into an algorithm is to set $k = 1$ in the k-NN stage, which obviously removes the need for a human choice. If we do this, however, we cannot even complete the famous "king:man = queen:woman" analogy: instead of "woman", we actually get "king" using glove embeddings.

Following our first definition in Eq. (1), we instead propose the notion of strong parallelism in Eq. (9):

$$||v_d - v_c - (v_b - v_a)|| \leq \tau \min(||v_b - v_a||, ||v_d - v_c||) \tag{9}$$

where τ is a small scalar. Equation (9) is a sufficient condition for quasi-parallelism between $v_d - v_c$ and $v_b - v_a$. The algorithm is very simple: given quadruplets (a, b, c, d) of words, and tag the quadruplet as a valid analogy if Eq. (9) is satisfied. We also generalize the PCA dimensional reduction from 2D to more dimensionalities. We ran this algorithm on a database of quadruplets corresponding to valid analogies, and obtained the results in Table 1. The fact that the results are surprisingly low was one of our initial motivations for this work. The failure of this algorithm indicates that the geometric relation Eq. (1) for analogies may be more incidental than systematic.

Table 1. Analogies from Eq. (9), F1-score

Dimension	word2vec		glove		fastText	
	$\tau = 0.1$	$\tau = 0.2$	$\tau = 0.1$	$\tau = 0.2$	$\tau = 0.1$	$\tau = 0.2$
2	1.08%	5.17%	3.34%	12.93%	0.97%	4.92%
10	0.00%	0.00%	0.00%	0.09%	0.00%	0.00%
20	0.00%	0.00%	0.00%	0.00%	0.00%	0.00%
50	0.00%	0.00%	0.00%	0.00%	0.00%	0.00%
100	0.00%	0.00%	0.00%	0.00%	0.00%	0.00%
300	0.00%	0.00%	0.00%	0.00%	0.00%	0.00%

3.4 Supervised Classification

The failure of an algorithm for correctly labelling analogies based on Eq. (9) (see Sect. 3.3) does not necessarily imply that analogies are not correctly labeled (at least approximately) using other means. In this section we propose a very common supervised learning approach (a simple $k-$NN).

More precisely, we trained a $5-$NN to predict analogies using vector differences, following Eq. (1). If (a, b, c, d) is an analogy quadruplet, we use the representation:

$$x_{abcd} = (v_b - v_a, v_d - v_c) \tag{10}$$

to predict the class of the quadruplet (a, b, c, d) (either no relation or being the capital of, plural, etc). If the angles between the vectors $v_b - v_a$ and $v_d - v_c$ (hint of parallelism) contain important information with respect to analogies, this representation should yield a good classification score. The dataset used is composed of 13 types of analogies, with thousand of examples in total (see Footnote 1). We considered 1000 pairs of words sharing a relation, with 13 labels (1 to 13, respectively: capital-common-countries and capital-world (merged), currency, city-in-state, family, adjective-to-adverb, opposite, comparative, superlative, present-participle, nationality-adjective, past-tense, plural, plural-verbs), and 1000 pairs of words sharing no relation (label 0). In order to generate different random quadruplets, we ran 500 simulations. Average results are in Table 2.

The results in Table 2 suggest that the representations obtained from Eq. (10) allow a good classification of analogies in dimension 10 when Euclidean geometry is used with a $5-$NN. However, in the remaining dimensions, vector differences do not encode enough information with regards to analogies.

Table 2. Multi-class F1 score classification of analogies based on representation 10 (5-nearest neighbors)

Dimension	word2vec	glove	fastText
2	62.47%	69.30%	68.74%
10	86.44%	85.62%	90.40%
20	74.74%	77.45%	80.57%
50	55.11%	61.24%	55.30%
100	50.57%	51.26%	50.56%
300	51.12%	51.72%	49.98%

4   Parallelism for Analogies with Graph Propagation

In this section we present an algorithm which takes an existing word embedding as input, and outputs a modified word embedding for which analogies correspond to a notion of parallelism in vector differences. These new word embeddings will be later used (see Sect. 5) to confirm the hypothesis that analogies corresponding to parallel vector differences does not make the word embedding better for common classification tasks.

Let us consider a family of semantic relations $(\mathcal{R}_k | 1 \leq k \leq r)$. For instance, this family can contain the plural or superlative relation. One of the relations $\mathcal{R}_k$ creates the analogy $a : b = c : d$, if and only if: $a\mathcal{R}_k b$ and $c\mathcal{R}_k d$, i.e semantic relations create quadruplets of analogies in the following sense:

$$(a, b, c, d) \text{ is an analogy quadruplet} \iff \exists k, \; a\mathcal{R}_k b \text{ and } d\mathcal{R}_k c \qquad (11)$$

A sufficient condition for relation (1) to hold for a quadruplet is for each pair a, b in the relation $\mathcal{R}_k$:

$$\exists \mu_k \in \mathbb{R}^d, \; a\mathcal{R}_k b \iff v_b = v_a + \mu_k \qquad (12)$$

Equation (12) can be generalized to other functions than summing a constant vector, namely it suffices that

$$\exists f_k : \mathbb{R}^d \longrightarrow \mathbb{R}^d, \; v_a \mathcal{R}_k v_b \iff v_b = f_k(v_a) \qquad (13)$$

Other choices of f_k might be interesting, but are not considered in this work.

In order to generate word vectors satisfying Eq. (12), we propose a routine using propagation on graphs. The first step consists in building a directed graph of words (V, E) encoding analogies:

$$(i, j) \in E \Leftrightarrow \exists k \; (i\mathcal{R}_k j) \qquad (14)$$

Therefore, we can label each edge with the type k of analogy involved (namely being the capital of, plural, etc, ...). Then, we use a graph propagation algorithm (Algorithm 1) involving Eq. (12) relation. We remark that propagation requires initial node representations.

Algorithm 1. Graph propagation for analogies

    Data: List of relations, vectors $\mu_1, ..., \mu_K \in \mathbb{R}^d$
    Result: New representations
1 Build graph G of analogies (Eq. (14));
2 Extract connected components $C_1, ..., C_c$ from G;
3 **for** $j = 1 \rightarrow c$ **do**
4   Select source node $s_1 \in C_j$;
5      $v_{s_1} \leftarrow$ Generate initial representation of s_1;
6      $s_2, ..., s_{|C_j|} \leftarrow$ Breadth first search from s_1;
7   **for** $r = 2 \rightarrow |C_j|$ **do**
8          $k \leftarrow$ index of relation between s_r and s_{r+1};
9          $v_{s_{r+1}} = v_{s_r} + \mu_k$;
10   **end**
11 **end**
12 Return $(v_i \mid 1 \leq i \leq |G|)$

Proposition 1. *Let G the graph of analogies. If G is a forest, then the representations obtained with Algorithm 1 verify Eq. (12).*

Proof. A forest structure implies the existence of a source node s for each component in G. For each component, every visited node with breadth-first search starting from s has only one parent, so the update defined Line 9 in Algorithm 1 defines a representation that verify Eq. (12) for the current node and its parent. □

However, if G is not a forest, words can have several parents. In this case, if $(parent_1, child)$ is visited before $(parent_2, child)$, our graph propagation method will not respect Eq. (12) for $(parent_1, child)$. This is the case with homonyms. For example, Peso is the currency for Argentina, but the currency for Mexico too. In practice, we selected $\mu_1, \ldots, \mu_K$ as a family of independent vectors in $\mathbb{R}^d$. We found better results in our experiments with $\forall i, \|\mu_i\| \geq d$. This can be explained by the fact that the vectors of relations needs to be non negligible when compared to difference of the words vectors.

5   Experiments with New Embeddings

In this section we present results of the experiments described in Sec. 3 with the updated embeddings obtained with the propagation Algorithm 1. We call $X++$ the new word embeddings obtained with the propagation algorithm from the word embeddings X.

5.1   Classification of Analogies

Analogies from "Parallelism": As in Sec. 3.3 using Eq. (9). Results are in Table 3. F1-scores are almost perfect (by design) in all dimensions.

Table 3. Analogies from Eq. (9) with updated embeddings, F1-score

| Dimension | word2vec++ | | glove++ | | fastText++ | |
	$\tau = 0.1$	$\tau = 0.2$	$\tau = 0.1$	$\tau = 0.2$	$\tau = 0.1$	$\tau = 0.2$
2	96.80%	96.50%	97.92%	97.15%	98.15%	97.61%
10	98.48%	98.54%	97.88%	97.88%	98.25%	98.31%
20	98.12%	98.18%	98.14%	98.43%	96.56%	96.56%
50	96.80%	96.80%	98.28%	98.36%	98.17%	98.17%
100	98.08%	98.08%	98.19%	98.19%	98.06%	98.06%
300	98.41%	98.41%	98.40%	98.40%	98.30%	98.30%

With Supervised Learning: Same experiments as in Sec. 3.4: 1000 pairs of words sharing a relation with 13 labels (1 to 13), and 1000 pairs of words sharing no relation (label 0). Results are in Table 4.

Table 4. Multi-class F1 score on classification of analogies based on relation 10 with updated embeddings (5-nearest neighbors)

Dimension	word2vec++	glove++	fastText++
2	99.73%	99.44%	99.31%
10	99.75%	99.36%	99.64%
20	99.80%	99.52%	99.94%
50	99.56%	99.63%	99.49%
100	99.89%	99.54%	99.42%
300	99.40%	99.86%	99.45%

5.2   Text Classification: Comparison Using KNN

We used three datasets: one for binary classification (Subjectivity) and two for multi-class classification (WebKB and Amazon). For reasons of time computation we used a subset of WebKB and Amazon datasets (500 samples). The implementation and datasets are available online (see Footnote 1). Results are in Table 5.

Table 5. Text classification ($d = 20$), F1-score

	word2vec	glove	fastText	word2vec++	glove++	fastText++
Subjectivity	81.69%	81.02%	82.14%	81.69%	80.38%	81.57
WebKB	71.50%	71.00%	70.50%	71.50%	72.00%	72.00
Amazon	65.20%	63.60%	60%	65.20%	61.00%	56.40

6   Conclusion

In this paper we discussed the well-advertised "geometrical property" of word embeddings w.r.t. analogies. By using a corpus of analogies, we showed that this property does not hold in general, in two or more dimensions. We conclude that the appearance of this geometrical property might be incidental rather than systematic or even likely.

This is somewhat in contrast to the theoretical findings of [1]. One possible way to reconcile these two views is that the concentration of measure argument in [1, Lemma 2.1] might yield high errors in vectors spaces having dimension as low as $\mathbb{R}^{300}$. Using very high-dimensional vector spaces might conceivably increase the occurrence of almost parallel differences for analogies. By the phenomenon of *distance instability* [3], however, algorithms based on finding closest vectors in high dimensions require computations with ever higher precision when the vectors are generated randomly. Moreover, the model of [1] only warrants approximate parallelism. So, even if high dimensional word vectors pairs were almost parallel with high probability, verifying this property might require considerable computational work related to floating point precision.

By creating word embeddings on which the geometrical property is enforced by design, we also showed empirically that the property appears to be irrelevant w.r.t. the performance of a common information retrieval algorithm (k-NN). So, whether it holds or not, unless one is trying to find analogies by using the property, is probably a moot point. We are obviously grateful to this property for the (considerable, but unscientific) benefit of having attracted some attention of the general public to an important aspect of computational linguistics.

References

1. Arora, S., Li, Y., Liang, Y., Ma, T., Risteski, A.: A latent variable model approach to PMI-based word embeddings. Trans. Assoc. Comput. Lingui. **4**, 385–399 (2016)
2. Arora, S., Li, Y., Liang, Y., Ma, T., Risteski, A.: Linear algebraic structure of word senses, with applications to polysemy. Trans. Assoc. Comput. Lingui. **6**, 483–495 (2018)
3. Beyer, K., Goldstein, J., Ramakrishnan, R., Shaft, U.: When is "Nearest Neighbor" meaningful? In: Beeri, C., Buneman, P. (eds.) ICDT 1999. LNCS, vol. 1540, pp. 217–235. Springer, Heidelberg (1999). https://doi.org/10.1007/3-540-49257-7_15
4. Bojanowski, P., Grave, E., Joulin, A., Mikolov, T.: Enriching word vectors with subword information. arXiv preprint arXiv:1607.04606 (2016)
5. Bolukbasi, T., Chang, K.W., Zou, J.Y., Saligrama, V., Kalai, A.T.: Man is to computer programmer as woman is to homemaker? debiasing word embeddings. In: Advances in Neural Information Processing Systems, pp. 4349–4357 (2016)
6. Bordes, A., Usunier, N., Garcia-Duran, A., Weston, J., Yakhnenko, O.: Translating embeddings for modeling multi-relational data. In: Advances in Neural Information Processing Systems, pp. 2787–2795 (2013)
7. Caliskan, A., Bryson, J.J., Narayanan, A.: Semantics derived automatically from language corpora contain human-like biases. Science **356**(6334), 183–186 (2017)

8. Church, K.W., Hanks, P.: Word association norms, mutual information, and lexicography. Comput. Linguist. **16**(1), 22–29 (1990)

9. Devlin, J., Chang, M.W., Lee, K., Toutanova, K.: BERT: pre-training of deep bidirectional transformers for language understanding. arXiv preprint arXiv:1810.04805 (2018)

10. Hatcher, E., Gospodnetic, O.: Lucene in Action. Manning Publications, Shelter Island (2004)

11. Jastrzebski, S., Leśniak, D., Czarnecki, W.M.: How to evaluate word embeddings? on importance of data efficiency and simple supervised tasks. arXiv preprint arXiv:1702.02170 (2017)

12. Lin, Y., Liu, Z., Sun, M., Liu, Y., Zhu, X.: Learning entity and relation embeddings for knowledge graph completion. AAAI **15**, 2181–2187 (2015)

13. Mikolov, T., Chen, K., Corrado, G., Dean, J.: Efficient estimation of word representations in vector space. arXiv preprint arXiv:1301.3781 (2013)

14. Mikolov, T., Sutskever, I., Chen, K., Corrado, G.S., Dean, J.: Distributed representations of words and phrases and their compositionality. In: Advances in Neural Information Processing Systems, pp. 3111–3119 (2013)

15. Mikolov, T., Yih, W.T., Zweig, G.: Linguistic regularities in continuous space word representations. In: Proceedings of the 2013 Conference of the North American Chapter of the Association for Computational Linguistics: Human Language Technologies, pp. 746–751 (2013)

16. Nadeau, D., Sekine, S.: A survey of named entity recognition and classification. Lingvisticae Investigationes **30**(1), 3–26 (2007)

17. Nissim, M., van Noord, R., van der Goot, R.: Fair is better than sensational: man is to doctor as woman is to doctor. arXiv preprint arXiv:1905.09866 (2019)

18. Pennington, J., Socher, R., Manning, C.: Glove: Global vectors for word representation. In: Proceedings of the 2014 Conference on Empirical Methods in Natural Language Processing (EMNLP), pp. 1532–1543 (2014)

19. Peters, M.E., Neumann, M., Iyyer, M., Gardner, M., Clark, C., Lee, K., Zettlemoyer, L.: Deep contextualized word representations. arXiv preprint arXiv:1802.05365 (2018)

20. Ramos, J., et al.: Using TF-IDF to determine word relevance in document queries. In: Proceedings of the First Instructional Conference on Machine Learning, Piscataway, NJ, vol. 242, pp. 133–142 (2003)

21. Salton, G.: The SMART retrieval system—Experiments in automatic document processing. Prentice-Hall, Inc., Upper Saddle River, NJ, USA (1971). https://dl.acm.org/citation.cfm?id=1102022

22. Salton, G., Wong, A., Yang, C.S.: A vector space model for automatic indexing. Commun. ACM **18**(11), 613–620 (1975)

23. Vylomova, E., Rimell, L., Cohn, T., Baldwin, T.: Take and took, gaggle and goose, book and read: evaluating the utility of vector differences for lexical relation learning. arXiv preprint arXiv:1509.01692 (2015)

24. Wang, Z., Zhang, J., Feng, J., Chen, Z.: Knowledge graph embedding by translating on hyperplanes. AAAI **14**, 1112–1119 (2014)

Language Comparison via Network Topology

Blaž Škrlj[1,2(✉)] and Senja Pollak[2,3]

[1] Jožef Stefan International Postgraduate School, Ljubljana, Slovenia
[2] Jožef Stefan Institute, Ljubljana, Slovenia
[3] Usher Institute, Medical School, University of Edinburgh, Edinburgh, UK
{blaz.skrlj,senja.pollak}@ijs.si

Abstract. Modeling relations between languages can offer understanding of language characteristics and uncover similarities and differences between languages. Automated methods applied to large textual corpora can be seen as opportunities for novel statistical studies of language development over time, as well as for improving cross-lingual natural language processing techniques. In this work, we first propose how to represent textual data as a directed, weighted network by the text2net algorithm. We next explore how various fast, network-topological metrics, such as network community structure, can be used for cross-lingual comparisons. In our experiments, we employ eight different network topology metrics, and empirically showcase on a parallel corpus, how the methods can be used for modeling the relations between nine selected languages. We demonstrate that the proposed method scales to large corpora consisting of hundreds of thousands of aligned sentences on an of-the-shelf laptop. We observe that on the one hand properties such as communities, capture some of the known differences between the languages, while others can be seen as novel opportunities for linguistic studies.

Keywords: Computational typology · Cross-linguistic variation · Network theory · Language modeling · Comparative linguistics · Graphs · Language representation

1 Introduction and Related Work

Understanding cross-linguistic variation has for long been one of the foci of linguistics, addressed by researchers in comparative linguistics, linguistic typology and others, who are motivated by comparison of languages for genetic or typological classification, as well as many other theoretical or applied tasks. Comparative linguistics seeks to identify and elucidate genetic relationships between languages and hence to identify language families [26]. From a different angle, linguistic typology compares languages to learn how different languages are, to see how far these differences may go, and to find out what generalizations can be made regarding cross-linguistic variation on different levels of language

© Springer Nature Switzerland AG 2019
C. Martín-Vide et al. (Eds.): SLSP 2019, LNAI 11816, pp. 112–123, 2019.
https://doi.org/10.1007/978-3-030-31372-2_10

structure and aims at mapping the languages into types [6]. The availability of large electronic text collections, and especially large parallel corpora, have offered new possibilities for computational methodologies that are developed to capture cross-linguistic variation. This work falls under computational typology [1,13], an emerging field with the goal of understanding of the differences between languages via computational (quantitative) measures. Recent studies already offer novel insights into the inner structure of languages with respect to various sequence fingerprint comparison metrics, such as for example the Jaccard measure, the intra edit distance and many other boolean distances [21]. Such comparisons represent e.g., sentences as vectors, and evaluate their similarity using plethora of possible metrics. Albeit useful, vector-based representation of words, sentences or broader context does not necessarily capture the context relevant to the task at hand and the overall structure of a text collection. Word or sentence embeddings, which recently serve as the language representation workhorse, are not trivial to compare across languages, and can be expensive to train for new languages and language pairs (e.g., BERT [8]). Further, such embeddings can be very general, possibly problematic for use on smaller data sets and are dependent on input sequence length.

In recent years, several novel approaches to computational typography have been applied. For example, Bjerva et al. [2] compared different languages based on distance metrics computed on universal dependency trees [19]. They discuss whether such language representations can model geographical, structural or family distances between languages. Their work shows how a two layer LSTM neural network [12] represents the language in a structural manner, as the embeddings mostly correlate with structural properties of a language. Their main focus is thus on explaining the structural properties of neural network word embeddings. Algebraic topology was also successfully used to study syntax properties by Port et al. [20]. Similar efforts of statistical modelling of language distances were previously presented in e.g., [14] who used Kolmogorov complexity metrics.

In contrast, we propose a different approach to modeling language data. The work is inspired by ideas of node representation as seen in contemporary geometric and manifold learning [10] and the premises of computational network theory, which studies the properties of interconnected systems, found within virtually every field of science [27]. Various granularities of a given network can be explored using approaches for community detection, node ranking, anomaly identification and similar [5,9,15]. We demonstrate that especially information flow-based community detection [7] offers interesting results, as it directly simulates information transfer across a given corpus. In the proposed approach, we thus model a corpus (language) as a single network, exposing the obtained representation to powerful network-based approaches, which can be used for language comparison (as demonstrated in this work), but also for e.g., keyword extraction (cf. [4] who used TopicRank) and potentially also for representation learning and end-to-end classification tasks.

The purpose of this work is twofold. First, we explore how a text can be transformed into a network with minimal loss of information. We believe that this powerful and computationally efficient text representation that we name *text2net*, standing for text-to-network transformation, can be used for many new tasks. Next, we show how the obtained networks can be used for cross-lingual analysis across nine languages (36 language pairs).

This work is structured as follows. In Sect. 2 we introduce the networks and the proposed text2net algorithm. Next, we discuss network-topological metrics (Sect. 3) that we use for the language comparison experiment in Sect. 4. The results are presented in Sect. 5, followed by discussion and conclusions in Sect. 6.

2   Network-Based Text Representation

First, we discuss the notion of networks, and next present our text2net approach.

2.1   Networks

We first formally define the type of networks considered in this work.

Definition 1 (Network). *A network is an object consisting of nodes, connected by arcs (directed) and/or edges (undirected). In this work we focus on directed networks, where we denote with $G = (N, A)$ a network G, consisting of a set of nodes N and a set of arcs $A \subseteq N \times N$ (ordered pairs).*

Such simple networks are not necessarily informative enough for complex, real world data. Hence, we exploit the notion of weighted directed networks.

Definition 2 (Directed weighted network). *A directed weighted network is defined as a directed network with additional, real-valued weights assigned to arcs.*

Note that assigning weights to arcs has two immediate consequences: arcs can easily be pruned (using a threshold), and further, algorithms, which exploit arc weights can be used. We continue to discuss how a given text is first transformed into a directed weighted network G.

2.2   text2net Algorithm

Given a corpus T, we discuss the mapping text2net : $T \rightarrow G$. As text is sequential, the approach captures global word neighborhood, proceeding as follows:

1. Text is first tokenized and optionally stemming, lemmatization and other preprocessing techniques are applied to reduce the space of words.
2. text2net traverses each input sequence of tokens (e.g., words, or lemmas or stems depending on Step 1), and for each token (node) stores its successor as a new node connected with the outbound arc. This step can be understood as breaking the text into triplets, where two consecutive words are connected via a directed arc (therefore preserving the sequential information).

3. During construction of such triplets, arcs commonly repeat, as words often appear in same order. Such repetitions are represented as arc weights. Weight assignment can depend on the arc type. For this purpose, we introduce a mapping $\rho(a) \to \mathbb{R}; a \in A$ (A is the set of arcs), a mapping which assigns a real value to a given arc with respect to that arc's properties.
4. Result is a weighted, directed network representing weighted token co-occurrence.

The algorithm can thus formally be stated as given in Algorithm 1. The key idea is to incrementally construct a network based on text, while traversing the corpus *only once* (after potential selected preprocessing steps).

We next discuss the text2net's computational complexity. To analyze it, we assume the following: the text corpus T is comprised of s sentences. In terms of space, the complexity can be divided into two main parts. First, the memory needed to store the sentence being currently processed and the memory for storing the network. As the sentences can be processed in small batches, we focus on the spatial complexity of the token network. Let the corpus consist of t tokens. In the worst case, all tokens are interconnected and the spatial complexity is quadratic $\mathcal{O}(t^2)$. Due to Zipf's law networks are notably smaller as each word is (mostly) connected only with a small subset of the whole vocabulary (heavy tailed node degree distribution). The approach is thus both spatially, as well as computationally efficient, and can easily scale to corpora comprised of hundreds of thousands of sentences.

In terms of hyperparameters, the following options are available (offering enough flexibility to model different aspects of a language, rendering text2net suitable as the initial step of multiple down-stream learning tasks):

Algorithm 1. text2net algorithm.

Data: Text corpus T (of documents $d_1 \ldots d_n$), empty weighted network G
Parameters : Minimum number of tokens per sentence t_s, Minimum token length t_l, word transformation function f, stopwords σ, weight prunning threshold θ, frequency weight function ρ
Result: A weighted network G

```
 1  for d ∈ T do
 2      orderedTokens := getTokens(d, tₗ,tₛ,f,σ);          ▷ Get token sequence.
 3      for qᵢ ∈ orderedTokens do
 4          arc := (qᵢ,qᵢ₊₁);                              ▷ Construct an arc.
 5          addToNetwork(G, arc);                          ▷ Construct the network.
 6          if arc ∈ current set of arcs of G then
 7              | update arc's weight via ρ;               ▷ Update weights.
 8          end
 9      end
10  end
11  G := prunenetwork(G,θ);                                ▷ Prune the network.
12  return G
```

- minimum sentence length considered for network construction (t_s),
- minimum token length (t_l),
- optional word transformation (e.g., lemmatisation) (f),
- optional stopwords or punctuation to be removed (σ),
- arc weight assignment function (ρ) (e.g., co-occurrence frequency),
- a threshold for arc prunning based on weights (θ).

3   Considered Network Topology Metrics

In this section we discuss the selected metrics that we applied to directed weighted networks. The metrics vary in their degree of computational complexity.

Number of nodes. The number of nodes present in a given network.

Number of edges. The number of edges in a given network.

InfoMap communities. The InfoMap algorithm [22] is based on the idea of minimal description length of the walks performed by a random walker traversing the network. It obtains a network partition by minimizing the description lengths of random walks, thus uncovering dense regions of a network, which represent communities. Once converged, InfoMap yields the set of a given network's nodes N partitioned into a set of partitions which potentially represent functional modules of a given network.

Average node degree. How many in- and out connections a node has on average. For this metric, networks were considered as undirected. See below:

$$\text{AvgDeg} = \frac{1}{|N|} \sum_{n \in N} \deg_{in}(n) + \deg_{out}(n).$$

Network density. The network density represents the percentage of theoretically possible edges. This metric is defined as:

$$\text{Density} = \frac{|A|}{|N|(|N| - 1)};$$

where $|A|$ is the number of arcs and $|N|$ is the number of nodes. This measure represents more coarse-grained clustering of a network.

Clustering coefficient. This coefficient is defined as the geometric average of the subnetwork edge weights:

$$\text{ClusCoef} = \frac{1}{|N|} \sum_{u \in N} \left(\frac{1}{deg(u)(deg(u) - 1)} \right) \sum_{vw} \sqrt[3]{(\hat{w}_{uv}\hat{w}_{uw}\hat{w}_{vw})};$$

here, $\hat{w}_{vw}$ for example represents the weight of the arc between nodes v and w. The $deg(u)$ corresponds to the u-th node's degree. Intuitively, this coefficient represents the number of closed node triplets w.r.t. number of all possible triplets. The higher the number, the more densely connected (clustered) the network. See [3] for detailed description of the metrics above.

4   Language Comparison Experiments

In this section we discuss the empirical evaluation setting, where we investigated how the proposed network-based text representation and network-topology metrics can be used for the task of language comparison. We use the parallel corpus (i.e., corpus of aligned sentences across different languages) from the DGT corpus, i.e. Directorate-General for Translation translation memory, provided by Joint Research Centre and available in OPUS [25]. We selected nine different languages: EN – English, ES – Spanish, ET – Estonian, FI – Finish, LV – Latvian, NL – Dutch, PR – Portugese, SI – Slovene, SK – Slovak, covering languages from different historical origins and language families: Romance languages (PT, ES), Balto-Slavic languages including Slavic (SI, SK) and Baltic (LV) language examples, Germanic langauges (EN, NL), as well as Finnic languages from Uralic family (FI, ET). The selected languages have also different typological characteristics. For example in terms of morphological typology, EN can be considered as mostly analytic, while majority of others are synthetic languages, where for example FI is considered as agglutinative, while Slavic languages are fusional as they are highly inflected.

The goal of the paper was to use the network topology metrics for langauge comparison. We considered all the pairs between the selected languages, resulting in 36 comparisons for each network-based metric. From the parallel corpus we sampled 100,000 sentences for each language, resulting in 900,000 sentences, which match across languages.

From each language, we constructed a network using text2net with following parameters: the minimum number of tokens per sentence (t_s) was set to 3, the minimum length of a given token (t_l) to 1, the word transformation function transformed words to lower-case, no lemmatisation was used, and punctuation was removed. We defined $\rho(\text{arc}) = 1$.

We compared the pairs of languages as follows. For each of the two languages, we transformed the text into a network. The discussed network topology metrics were computed for each of the two networks. Differences between the metrics' values are reported in tabular form (Table 1), as well as visualized as heatmaps (Fig. 1). In the latter, the cells are colored according to the *absolute* difference in a given metric for readability purposes. Thus, the final result of the considered analysis are differences in a selected network topology metric. The selected results were further visualized in Fig. 2.

We used NLTK [16] for preprocessing, Py3plex [23], NetworkX [11], Cytoscape [24] for network analysis and visualization and Pandas for numeric comparisions [17]. Full code is available at: https://github.com/SkBlaz/language-comparisons.

While we do not have full linguistic hypotheses about the expected mapping of the linguistic characteristics and the topological metrics, we believe that the network-based comparisons should show differences between the languages. For example, the number of nodes might capture linguistic properties, such as inflectional morphology, where we could expect that morphologically rich languages would have more nodes. Number of edges might capture linguistic properties,

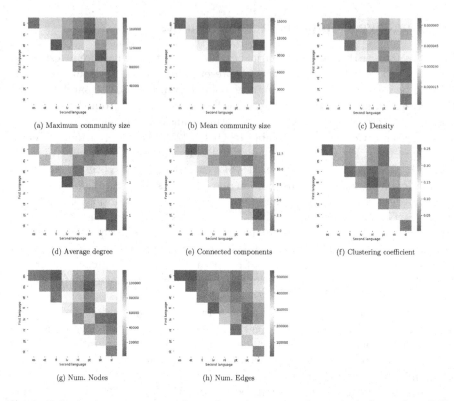

Fig. 1. Pairwise language comparison via various network-topological metrics. Cells represent the absolute differences between metrics of individual text-derived networks. Red regions represent very different networks, and blue very similar ones. (Color figure online)

such as the flexibility of the word order. The other measures are less intuitive and will be further investigated in future work. However, we believe that more complex the language (including aspects of morphology richness and word order flexibility), the richer the corresponding network's structure, while the number of connected components might offer insights into general dispersity of a given language, and could pinpoint grammatical differences if studied in more detail. Also clustering coefficient might be dependent on how fixed is the word order of a given language. None of the above has been systematically investigated, and the hypothesis is, that differences between languages will have high variability and show already known, as well as novel groupings of the languages.

5   Results

In this section we present the results of cross-lingual comparison. The inter-language differences in tabular format are given in Table 1. The measures given in the table are the differences in: #Nodes—the number of nodes, #Edges—the

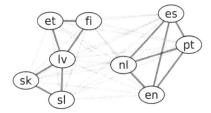

Fig. 2. Language network based on the Clustering coeff. The red links are present after the threshold of 10^{-3} was applied. Gray links represent connections that are not present given the applied threshold. We can see two groups, one formed by Balto-Slavic and Finnic languages, the other by Germanic and Romanic. (Color figure online)

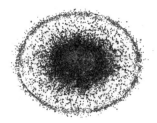

Fig. 3. Visualization of the English DGT subcorpus. This network was constructed using the proposed text2net algorithm, where each link corresponds to the *followed by* relation between a given pair of word tokens. Clustering emerges, indicating the presence of meso-scale topological structures in such networks. Different colors correspond to different communities detected using InfoMap. (Color figure online)

number of edges, Mean degree—mean node degree, Density—network density as defined in Sect. 3, MaxCom—maximum community size, MeanCom—mean community size, both computed using InfoMap communities, Clustering—clustering coefficient and CC—the number of connected components. The differences in the table are presented in L2-L1 absolute differences, while for nodes and edges we also present the differences as relative percentages of the e.g., number of nodes of the second language w.r.t the number of nodes of the first language[1]. It can be observed that some language pairs differ substantially even if only node counts are considered, where EN-FI is the pair with the largest difference, which is not surprising. English is for example an analytical language, while Finnish agglutinative with very rich morphology. Further, some of the metrics indicate groupings, which can be further investigated using heatmaps and direct visualization of language-language links.

From heatmaps shown in Fig. 1, where colors of individual cells represent differences between a given metric's values across languages, we can make several interesting observations. Based on Num. of nodes, FI and ET are very similar,

---

[1] For nodes $N_{\text{diff}} = \frac{100 \cdot |N_2|}{|N_1|}$, and for edges $E_{\text{diff}} = \frac{100 \cdot |E_2|}{|E_1|}$; the first language's values are compared against the second language's values.

Table 1. Differences between selected network-topology metrics across languages. The values are computed as L2-L1, or reported as L2 relative to L1.

Language pair	#Nodes	#Edges	Mean degree	Density ($\cdot 10^{-4}$)	MaxCom	MeanCom	Clustering	CC	N_{diff} (%)	E_{diff} (%)
en-es	10232	15251	−1.34	−0.18	11100	538.53	−0.00	5	110.90	101.71
en-et	108986	539449	−2.85	−0.62	−72382	−6578.80	−0.19	14	233.53	187.45
en-fi	117623	474376	−3.96	−0.65	114803	649.71	−0.20	12	249.01	179.60
en-lv	53162	464366	1.02	−0.34	48411	−1208.95	−0.14	8	164.35	177.99
en-nl	30786	99189	−2.30	−0.36	30839	836.35	0.06	2	138.10	115.73
en-pt	10778	56039	−0.51	−0.14	10249	−366.93	0.02	4	113.07	107.48
en-sk	59715	425657	−0.22	−0.41	60709	2757.51	−0.12	6	174.20	172.24
en-sl	46764	337421	−0.11	−0.34	−68833	−5693.41	−0.09	4	156.30	154.82
es-et	97822	518349	−1.62	−0.35	−80506	−5877.34	−0.19	7	210.57	184.29
es-fi	110493	479001	−2.70	−0.49	108672	1004.91	−0.20	6	224.53	176.58
es-lv	42062	442253	2.34	−0.16	42066	2382.83	−0.14	3	148.20	174.99
es-nl	21501	88846	−1.04	−0.20	21235	1433.40	0.06	−2	124.52	113.78
es-pt	971	43922	0.84	0.04	1232	1922.25	0.02	−2	101.96	105.67
es-sk	49382	406578	0.99	−0.20	49740	4703.95	−0.12	1	157.08	169.34
es-sl	36317	321960	1.17	−0.18	36362	6935.34	−0.10	−3	140.94	152.21
et-fi	10262	−68268	−1.32	−0.05	183810	7318.28	−0.01	5	106.63	95.82
et-lv	−57119	−80883	4.01	0.29	−51457	1237.71	0.05	−5	70.38	94.95
et-nl	−75247	−424500	0.50	0.24	−69698	−1081.51	0.25	−7	59.14	61.74
et-pt	−96441	−471464	2.45	0.49	81260	8871.81	0.21	−9	48.42	57.34
et-sk	−47901	−109523	2.56	0.20	−40340	5107.84	0.07	−7	74.60	91.89
et-sl	−61594	−194218	2.80	0.27	117767	15563.93	0.08	−11	66.93	82.60
fi-lv	−66730	−11261	5.33	0.34	−72108	−2285.02	0.06	−8	66.00	99.10
fi-nl	−89718	−393774	1.71	0.30	−89284	−3638.18	0.26	−5	55.46	64.44
fi-pt	−110479	−439797	3.60	0.54	−111720	−6939.93	0.22	−7	45.41	59.84
fi-sk	−59295	−46799	3.96	0.26	−182908	−6349.16	0.08	−10	69.96	95.90
fi-sl	−72939	−134593	4.11	0.32	−71835	8022.41	0.11	−13	62.77	86.20
lv-nl	−19634	−354516	−3.52	−0.05	−18654	−318.15	0.20	−2	84.02	65.02
lv-pt	−41716	−402441	−1.46	0.21	−36193	4468.15	0.16	−6	68.80	60.39
lv-sk	7581	−34478	−1.38	−0.08	−123658	−7706.36	0.02	−5	105.99	96.77
lv-sl	−5810	−122602	−1.21	−0.03	1014	7032.05	0.05	−6	95.10	86.99
nl-pt	−20143	−43781	1.86	0.24	−19930	−314.52	−0.04	−1	81.88	92.87
nl-sk	27385	314730	2.09	−0.04	27329	1161.44	−0.19	3	126.14	148.83
nl-sl	13810	230590	2.32	0.03	7637	−2267.56	−0.16	−3	113.18	133.78
pt-sk	48780	361817	0.12	−0.29	47881	1201.90	−0.15	5	154.06	160.25
pt-sl	35260	275981	0.32	−0.22	35831	6622.65	−0.12	0	138.23	144.05
sk-sl	−13637	−85421	0.23	0.07	−130809	−9182.12	0.03	−3	89.73	89.89

and the most different to other languages. Both are agglutinative languages and part of the Uralic language family. In terms of Num. of edges, the largest differences are between ET and EN, while the most similar are LV and FI; in pairwise comparison with EN, we can see that PT, ES and NL have similar statistics, which are all languages from Germanic (NL) or Romanic family. We believe that some measures could also indicate groupings based on morphological or other typological properties beyond the currently known ones. For example, Max. community size on one hand points FI and ET as very different, as well as SI and SK (where in both pairs the two languages are belonging to the same language family), but on the other hand PT and ES are very similar. Further, Clustering coefficient yields insights into context structure and similar properties of groupings of basic semantic units, such as words, where high similarity between ES and PT, as well as SI and SK can be observed. Finally, the number of connected components offers insights into general dispersity of a given language, and could pinpoint grammatical differences if studied in more detail. Again, we see the most remarkable differences between EN and FI and ET, but also FI and SI, while Romanic and Germanic languages are more similar. There are many open questions. E.g., which linguistic phenomena make EN-FI being quite different in Average degree, while FI-NL are relatively similar (despite EN and NL being in the same language group)?

Clustering coefficient is also shown in an alternative visualisation, i.e. in a colored network in Fig. 2. Here, we consider Clustering coefficient metric, where we adjust the color so that it represents only very similar languages (low absolute difference in the selected metric). We selected this metric, as the heatmap yielded the most block-alike structure, indicating strong connections between subsets of languages. We can see that Balto-Slavic and Finnic languages group together, while Germanic and Romanic form another group. Finally, we visualized the English corpus network in Fig. 3. Colored parts of the network correspond to individual communities. It can be observed that especially the central part of the network contains some well defined structures (blue and red). The figure also demonstrates, why various network-topological metrics were considered, as from the structure alone, no clear insights can be obtained at such scale.

6   Discussion and Conclusions

In this work, our aim was to provide one of the first large-scale comparisons of languages based on corpus-derived networks. To the best of our knowledge, the use of network topologies on sequence-based token networks are novel and it is not yet known to what characteristics the network topologies correspond. Second, we investigated whether the difference in some metrics correspond known relationships between languages, or represent novel language groupings.

We have shown that the proposed network-based text representation offers a pallete of novel opportunities for language comparison. Commonly, methods operate on sequence level, and are as such limited to one dimensional interactions with respect to a given token. In this work we attempted to lift this constraint by introducing richer, global word neighborhood. We were able to cast the language comparison problem to comparing network topology metrics, for which we show can be informative for genetic and typographic comparisons. For example, the Slovene and Slovak languages appear to have very similar global network structure, indicating comparison using communities picks up some form of evolutionary language distance. In this work we explored only very simple language networks by performing virtually no preprocessing. We believe a similar idea could be used to form networks from lemmatized text or even Universal Dependency Tags, potentially opening another dimension.

Overall, we identified the clustering coefficient as the metric, which, when further inspected, yielded some of the well known language-language relationships, such as for example high similarity between Spanish and Portugese, as well as Slovenian and Slovak languages. Similar observation was made when community structure was compared. We believe such results demonstrate network-based language comparison represents a promising venue for scalable and more informative studies of how languages, and text in general, relate to each other.

In future, we will closer connect the interpretation of network topological features with linguistic properties, also by single language metrics. Also, we believe that document-level classification tasks can benefit from exploiting the inner document structure (e.g., the Graph Aggregator framework could be leveraged instead of/in addition to conventional RNN-based approaches). The added value of graph-based similarity for classification was demonstrated e.g., in [18] for psychosis classification from speech graphs. We also believe that our cross-language analysis, could be indicative for the expected quality of cross-lingual representations. Last but not least, we plan to perform additional experiments to see if the results are stable, leading to similar findings of other corpora genres and corpora of other sizes, and also using comparable not only parallel data.

Acknowledgements. The work was supported by the Slovenian Research Agency through a young researcher grant [BŠ], core research programme (P2-0103), and project Terminology and knowledge frames across languages (J6-9372). This work was supported also by the EU Horizon 2020 research and innovation programme, Grant No. 825153, EMBEDDIA (Cross-Lingual Embeddings for Less-Represented Languages in European News Media). The results of this publication reflect only the authors' views and the EC is not responsible for any use of the information it contains.

References

1. Asgari, E., Schütze, H.: Past, present, future: a computational investigation of the typology of tense in 1000 languages. arXiv:1704.08914 (2017)
2. Bjerva, J., Östling, R., Han Veiga, M., Tiedemann, J., Augenstein, I.: What do language representations really represent? Comput. Linguist. **4**(2), 381–389 (2019)
3. Bollobás, B.: Modern Graph Theory, vol. 184. Springer, Heidelberg (2013)
4. Boudin, F.: PKE: an open source python-based keyphrase extraction toolkit. In: Proceedings of COLING 2016, the 26th International Conference on Computational Linguistics: System Demonstrations, Osaka, Japan, pp. 69–73, December 2016
5. Brandao, M.A., Moro, M.M.: Social professional networks: a survey and taxonomy. Comput. Commun. **100**, 20–31 (2017)
6. Daniel, M.: Linguistic Typology and the Study of Language, pp. 43–68. Oxford University Press, Oxford (2010)
7. De Domenico, M., Lancichinetti, A., Arenas, A., Rosvall, M.: Identifying modular flows on multilayer networks reveals highly overlapping organization in interconnected systems. Phys. Rev. X **5**(1), 011027 (2015)
8. Devlin, J., Chang, M.W., Lee, K., Toutanova, K.: Bert: pre-training of deep bidirectional transformers for language understanding. arXiv:1810.04805 (2018)
9. Fortunato, S.: Community detection in graphs. Phys. Rep. **486**(3–5), 75–174 (2010)
10. Goyal, P., Ferrara, E.: Graph embedding techniques, applications, and performance: a survey. Knowl. Based Syst. **151**, 78–94 (2018)
11. Hagberg, A., Swart, P., S Chult, D.: Exploring network structure, dynamics, and function using networkx. Technical report, LANL, Los Alamos, NM, USA (2008)

12. Hochreiter, S., Schmidhuber, J.: Long short-term memory. Neural Comput. **9**(8), 1735–1780 (1997)
13. Kepser, S., Reis, M.: Linguistic evidence: empirical, theoretical and computational perspectives, vol. 85. Walter de Gruyter (2008)
14. Kettunen, K., Sadeniemi, M., Lindh-Knuutila, T., Honkela, T.: Analysis of EU languages through text compression. In: Salakoski, T., Ginter, F., Pyysalo, S., Pahikkala, T. (eds.) FinTAL 2006. LNCS (LNAI), vol. 4139, pp. 99–109. Springer, Heidelberg (2006). https://doi.org/10.1007/11816508_12
15. Kralj, J., Robnik-Sikonja, M., Lavrac, N.: NETSDM: semantic data mining with network analysis. J. Mach. Learn. Res. **20**(32), 1–50 (2019)
16. Loper, E., Bird, S.: Nltk: the natural language toolkit. arXiv cs/0205028 (2002)
17. McKinney, W.: pandas: a foundational python library for data analysis and statistics. Python High Perform. Sci. Comput. **14** (2011)
18. Mota, N.B., Vasconcelos, N.A., Lemos, N., Pieretti, A.C., Kinouchi, O., Cecchi, G.A., Copelli, M., Ribeiro, S.: Speech graphs provide a quantitative measure of thought disorder in psychosis. PLoS ONE **7**(4), e34928 (2012)
19. Nivre, J., et al.: Universal dependencies v1: a multilingual treebank collection. In: Proceedings of the 10th International Conference on Language Resources and Evaluation (LREC 2016), pp. 1659–1666 (2016)
20. Port, A., et al.: Persistent topology of syntax. Math. Comput. Sci. **12**(1), 33–50 (2018)
21. Rama, T., Kolachina, P.: How good are typological distances for determining genealogical relationships among languages? In: Proceedings of COLING 2012: Posters, pp. 975–984 (2012)
22. Rosvall, M., Axelsson, D., Bergstrom, C.T.: The map equation. Eur. Phys. J. Spec. Top. **178**(1), 13–23 (2009)
23. Škrlj, B., Kralj, J., Lavrač, N.: Py3plex: a library for scalable multilayer network analysis and visualization. In: Aiello, L.M., Cherifi, C., Cherifi, H., Lambiotte, R., Lió, P., Rocha, L.M. (eds.) COMPLEX NETWORKS 2018. SCI, vol. 812, pp. 757–768. Springer, Cham (2019). https://doi.org/10.1007/978-3-030-05411-3_60
24. Smoot, M.E., Ono, K., Ruscheinski, J., Wang, P.L., Ideker, T.: Cytoscape 2.8: new features for data integration and network visualization. Bioinformatics **27**(3), 431–432 (2010)
25. Tiedemann, J.: Parallel data, tools and interfaces in opus. In: LREC, vol. 2012, pp. 2214–2218 (2012)
26. Trask, R.L. (ed.): Dictionary of Historical and Comparative Linguistics. Edinburgh University Press, Edinburgh (2000)
27. Zhang, D., Yin, J., Zhu, X., Zhang, C.: Network representation learning: a survey. IEEE Trans. Big Data (2018)

Speech Analysis and Synthesis

An Incremental System for Voice Pathology Detection Combining Possibilistic SVM and HMM

Rimah Amami$^{(\boxtimes)}$, Rim Amami, and Hassan Ahmad Eleraky

Deanship of Preparatory Year and Supporting Studies, Imam Abdulrahman Bin
Faisal University, Dammam, Kingdom of Saudi Arabia
{raamami,rabamami,haeleraky}@iau.edu.sa

Abstract. The voice pathology detection using automatic classification
systems is a useful way to diagnose voice diseases. In this paper, we
propose a novel tool to detect voice pathology based on an incremental
possibilistic SVM-HMM method which can be applied to serval practical
applications using non-stationary or a very large-scale data in purpose to
reduce the memory issues faced during the storage of the kernel matrix.
The proposed system includes the steps of using SVM to incrementally
compute possibilitic probabilities and then they will be used by HMM
in order to detect voice pathologies. We evaluated the proposed method
on the task of the detection of voice pathologies using voices samples
from the Massachusetts Eye and Ear Infirmary Voice and Speech Laboratory (MEEI) database. According to the detection rates obtained by
our system, the performance sounds robust, efficient and speed applied
to a task of voices pathology detection.

Keywords: Possibilistic degree · Incremental learning · HMM ·
SVM · Voice pathology

1 Introduction

In the last decades, there has been a remarkable advance in the automatic systems dealing with the voice pathology diagnostic. However, the discrimination
between pathological and normal voices is still a complex field of research in
speech classification. The aim of this paper is to help the diagnosis of pathological voices among normal voices. Currently, the traditional way to detect voice
pathology is to visit a specialist who examines the vocal folds of the patient
using endoscopic tools. This process is considered time consuming, complex and
expensive. Thus, this area of science has attracted a lot of attention in purpose
to develop an accurate automatic device able to help the speech specialists for
early diagnosing voices pathologies. In this work, we propose an automatic system for the detection of pathological voices combining incremental possibilistic
SVM and HMM.

© Springer Nature Switzerland AG 2019
C. Martín-Vide et al. (Eds.): SLSP 2019, LNAI 11816, pp. 127–138, 2019.
https://doi.org/10.1007/978-3-030-31372-2_11

Hidden Markov Model (HMM) [17] is a statistical model which consists of a finite number of unknown states. Each of those states is associated with a respective probability distribution. HMM are considered as probabilistic framework which able to model a time series of any observations. HMM are successfully used for classification tasks in particular in bioinformatics and speech processing.

In the past 20 years, Support Vector Machines (SVM) technique acquired an important place in solving classification and regression problems since they provide a valuable learning design that generalize accurately by handling high dimensional data [5,10]. SVM were first introduced by Vapnik as an approximate implementation of the Structural Risk Minimization (SRM) induction principle [5,7].

Various studies using HMM and SVM have been proposed for voice pathology detection and classification; Dibazar et al. [14], propose to investigate HMM performance on a task of the detection of pathological voices using 5 pathologies. They suggest HMM approach using MFCC (Mel frequency cepstral coefficients) achieved a classification accuracy of 70%. The authors in [15] present a method based on HMM which classifies speeches into normal class and pathological class. The performance of this system for detection of vocal fold pathology is equal to 94%. In the study of [16], HMM was applied in order to classify voices of the database composed by 11 normal voices and 11 pathological voices. The proposed system obtains accuracy rate of 100% for pathological voices and 98% for normal voices.

Pend et al. [11] propose to combine PCA (Principal component analysis) to SVM using 27 features in order to classify the normal and pathological voice. Four classifiers were evaluated in [12] based on voice pathology problem. The Support vector machines achieved the best performance. In [13], the authors developed an incremental method combining density clustering and Support Vector Machines for voice pathology detection. This proposed method achieved a performance equal to 92%.

The main idea of SVM was to seek for a model with the optimal generalization performance while building the solution to the minimization problem of SRM through a quadratic programming optimization [6].

For the given data points (x_i, y_i) where $i = 1, .., n$ and n is the number of the data, SVM learn a classifier $f(x) = w^T x + b$ where the hyperplane that optimally separates the data is the one that minimises:

$$\frac{1}{2}\|w^{ij}\|^2 + C\sum_{i=1}^{n}\xi^{ij} \tag{1}$$

Where C is a regularization term and ξ is a positive slack variable. Subject to the inequality constraints:

$$y_i[w^T.x_i + b] \geq 1 - xi^i; i = 1, 2, ...n \tag{2}$$

On the other hand, the solution to the optimization quadratic programming problem can be cast to the Lagrange's function and we obtain the following dual objective function:

$$L_d = \max_{\alpha_i} \sum_{i=1}^{n} \alpha_i - \sum_{i=1}^{n} \sum_{j=1}^{n} \alpha_i \alpha_j y_i y_j K(x_i, x_j). \tag{3}$$

where $K(x_i, x_j)$ is the kernel of data x_i and x_j and the coefficients α_i are the lagrange multipliers and are computed for each sample of the data set.

In the context of speech classification area, the major problems of detection systems using batch methods can be resumed in two facts; the time-varying of speech samples and the amount of available data for learning stage. Hence, the online or incremental techniques provide a valuable solution in applications that handle speech data.

Given these facts, the main contribution of this paper is the application of an incremental possibilistic SVM technique combined to HMM to the problem of voice pathology detection. The implementation of the proposed system begin with the parameters extraction of samples and then proceed with the application of the proposed method to classify the voices samples.

This paper is organized as follows: in Sect. 2, the proposed system of voice pathology detection is presented and discussed. In Sect. 3, the features extraction step is described. In Sect. 4, experiments conditions and results are presented and evaluated. In Sect. 5, the conclusion and the perspectives of this work are illustrated.

2   The Proposed Incremental Possibilistic SVM-HMM System

In this paper, we consider the output of the incremental possibilistic Support Vector Machines (SVMs) as probabilities used into the HMM-based decoder and in particular used in the computation of HMM's likelihood (see Fig. 1). The combination of incremental possibilistic SVM and HMM seems interesting and a robust solution since the SVM lack the ability to model time series. Hence, the probabilities outputted from possibilistic SVM are used by the HMM in order to provide their state-dependent likelihoods as follows:

$$P(x|q_i) \propto \sum_{k=1}^{K} c_{ik} \cdot \frac{Pr(k|x)}{Pr(k)} \tag{4}$$

where for a given feature vector x, the posterior probability of the class k are given by $Pr(k|x)$ and $Pr(k)$ is the a-priori probability of class k. c_{ik} are the mixture weights for each HMM state.

The proposed method is shown in the Fig. 1. It must be pointed out that for our voice database, the Mel Frequency Cepstral Coefficients (MFCC) are extracted for each voice sample. Furthermore, in this work, we use the incremental learning which behaves exactly like an online learning by introducing repeatedly a new data at the current classifier. In other words, each step of the incremental learning of SVM consists of adding a new sample to the solution and retiring the old samples while keeping their Support Vectors (SV) which describes the learned decision boundary. Indeed, the training samples needed for the next step of the incremental learning process are obtain by incorporating the new incoming sample and the SV of the previous samples.

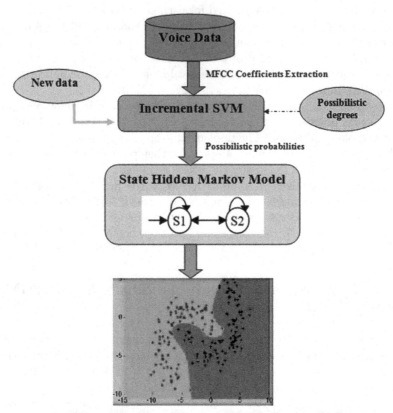

Classification (Normal/Pathology voice)

Fig. 1. The block diagram of the proposed method

The key idea of incremental SVM is to keep the Karush–Kuhn–Tucker (KKT) conditions satisfies while retiring old samples and adding a new one to the solution. Recalling that the KKT conditions are:

$$g_t = -1 + K_t, : \alpha + \mu y_i \begin{cases} \geq 0, if \alpha_t = 0 \\ = 0, if 0 < \alpha_t < C \\ \leq 0, if \alpha_t = C \end{cases} \frac{\delta W}{\delta \mu} = y^T \alpha = 0 \qquad (5)$$

The following algorithm summarizes the incremental SVM steps to learn new incoming samples (x_c, y_c). It consists to construct a classifier h^c from the classifier h^{c-1} [1].

Algorithm 1. Incremental SVM

Initialize lagrangian multiplier α_c to zero

If $g_c > 0$, terminates and set h^{c-1} as new classifier h^cs

If $g_c \leq 0$, apply the largest possible increment α_c in order to obtain one of those conditions:

(a) $g_c = 0$: Add the new sample to margin set S, then update the present Jacobian inverse $\Re$ and terminate.

(b) $g_c = C$: Add the new sample to error set E, and terminate.

(c) Elements of D^n migrate across S, E and $\Re$: update membership of elements and if S changes, then update $\Re$.

and repeat as necessary.

To our knowledge, efficient application using Support vector machines (SVM) based incremental learning in the field of voice pathology detection has not been reported in the last years. Thus, we propose the incremental SVM learning in the context of voice pathology detection based on the possibilistic degrees combined to HMM.

2.1 Possibilistic SVM

SVM was first introduced to solve the problems of pattern classification. In recent years, SVM have demonstrated robustness and have been successfully used to various practical applications. However, in many real-world applications, the performance of SVM would be seriously affected by the nature of available data i.e. data may be accompanied by noise. The voice pathology detection applications are often considered as a very complicated and delicate problems since the voices samples are non-stationary signals with a high amount of variation in the way how and by whom the sample is pronounced.

Let us note that the speech which is generally produced on a short-time scale, includes non-stationary parts due to the physiological system of the speaker which defines the amplitude and frequency modulation.

Furthermore, the behavior of SVM depends mostly on the training data and the optimal hyperplane is identified mainly from the support vectors. Thus, the variation in the voice sample may lead SVM to misclassify the data set [4].

In the literature, various solutions were proposed to solve this kind of problems such as weighted SVM, adaptive SVM and central SVM. In this paper, we propose a possibilitic SVM based on a geometric distance to improve the performance of the conventional SVM on a task of voice pathology detection. The main idea is to assign different possibilistic degrees to the different voice samples while SVM is computing class posterior probabilities. Those degrees calculates an euclidean distance between the point and the center of each class. As a result, the membership degree of the sample x_i near of the center of the class y_i is more important than the degrees of the points far from the center. We use the euclidian distance algorithm to generate the possibilitic degrees [2].

The formulation of the proposed possibilitic SVM is defined in three steps (see Fig. 2):

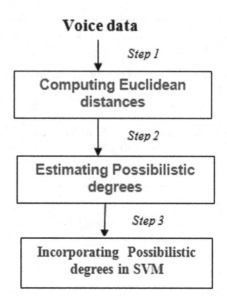

Fig. 2. The process of possibilistic SVM

As shown in the Fig. 2, the first step consists of computing the Euclidean distance between the center of the different classes y_k and the data x_i to be detected. Then, the possibilitic degree is evaluated which measure the degree that the data x_i belong to the class y_i. The final step of the formulation of the possibilitic SVM consists in incorporated into SVM those degrees in order to help the HMM based decoder to classify the voice pathologies.

Euclidean Distance. The Euclidean distance is computed between X_i and the center of the class CY_i where $i \in (1, \ldots, k)$. We suppose that it exists a possibility that the data X_i belongs to one of the classes Y_i. The lowest measured value of the Euclidean distance given by $d(CY_i, X_i)$ is assigned to the nearest data X_i to the class Y_i and the highest computed value is associated with the farthest class to the data X_i.

Possibilistic Degrees. The possibilitic degrees noted $m_i(X)$ measure the membership degree of every voice X_i of our data set to a given class Y_i. Those degrees are computed as follows:

$$m_i(X) := 1/d(C_i, X_i) \tag{6}$$

Where C_i is the center of the ith class and d is the Euclidean distance previously calculated.

The Fig. 3 shows the possibilitic degrees generated by step 2 of a given class (where two samples are misclassified from the Class 1). As we can see, the degrees of training voices samples closer to the center of the class 1 are much larger and the samples farthest from the center are much smaller.

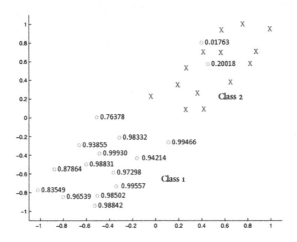

Fig. 3. An example of possibilitic degrees generated by the possibilitic SVM

Formulation of Possibilistic SVM. The purpose of incorporating possibilistic degrees is to limit the restrictions when the data have a larger degree into a given class.

Hence, with the formulation of possibilistic SVM for non-separable data, all the training data set must satisfy the following constraints:

$$
\mathbf{m(x)}(w^{ij})^T \phi(x_t) + b^{ij} \geq 1 - \xi_t^{ij}, \text{if } y_t = i
$$
$$
\mathbf{m(x)}(w^{ij})^T \phi(x_t) + b^{ij} \leq 1 - \xi_t^{ij}, \text{if } y_t = j
$$
$$
\xi_t^{ij} \geq 0 \tag{7}
$$

with $m(x)$ is the possibilitic degree of the sample x.

We optimize, also, the formulation of the possibilitic SVM in order to obtain a new dual representation including the possibilistic degrees $m(x)$:

$$
L_d = \max_{\alpha_i} \sum_{i=1}^{m} \alpha_i - \sum_{i=1}^{m}\sum_{j=1}^{m} m(x_i)m(x_j)\alpha_i\alpha_j y_i y_j \Phi(x_i)\Phi(x_j). \tag{8}
$$

In the new formulation the of possibilistic SVM, the decision function is given by:

$$
\sum_{i=1}^{m} \mathbf{m(x)}\alpha_i y_i \Phi(x_i) + b \tag{9}
$$

2.2 Incremental Possibilistic SVM

In this paper, the training sample x_i represents the vector MFCC features of voices files coming from the MEEI database. In the supervised learning, the label y_i represents the class to which belong the sample x. The Figure below shows the process of the proposed incremental possibilistic SVM.

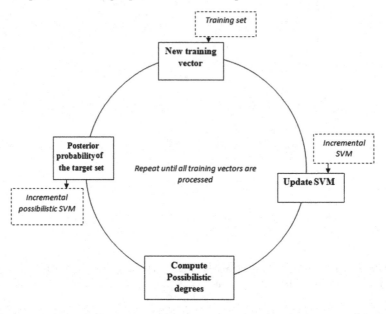

Fig. 4. Incremental possibilitic SVM

As seen in the Fig. 4, the incremental possibilitic SVM get, first, a new training vector from the data X. Then, the existing SVM is updated to add the new training sample. Before computing the probabilities, a possibilitic degree is calculated for the given data and incorporated into SVM formulation. This process will be repeated until all posteriors probabilities for training samples are computed.

2.3 Hidden Markov Model

Hidden Markov Model (HMM) is considered as statistical model to estimate the probability of a set of observations based on the sequence of hidden state transitions. The use of HMM for speech recognition has become popular for the last decade thanks to its the inherent statistical framework. HMM are simple networks that can generate speech using a sequence of states for each model and modeling the short-term spectra associated with each state. The following equation shows a state transition probability distribution, a_{ij}:

$$a_{ij} = Pq_{t+1} = j/qt = i, 1 \prec i, j \prec N_n \sum a_{ij} = 1; 1 \prec i, j \prec N_n \qquad (10)$$

where N is number of states in given model and q_t is the current state.

3   Feature Extraction

The feature extraction is the first step in a recognition system whose scheme is summarized in the Fig. 1. In this study, Mel-frequency cepstral coefficients (MFCCs) features [3] are extracted. Those coefficients are a very well-known extractor that allows to select significant features which are able to be used in several pattern detection problems.

The differential (Delta) and acceleration coefficients (Delta-Delta) were also calculated and used. Furthermore, the frame energy is appended to each feature vector. In the MEEI database, the pathological speakers have different voice disorders like traumatic, organic and psychogenic problems. The MEEI database contains 53 healthy samples and 724 samples with voice disorders. The speech samples were recorded in a controlled environment with a rate of 25 kHz or 50 kHz and 16 bits of resolution. The samples of healthy voices have duration of 3 s, and pathological voices samples have duration of 1 s. In this work, we set the duration of each frame to 20 ms and the *Hamming window* was used to extract the speech frames. For our experiments, the voice samples set consists of 53 normal voice samples and 139 for pathological (Keratosis/Vocal Poly/Adductor) voice samples.

The speakers have similar age, gender and different voice pathologies. The following table describes the MEEI database used in this work (Table 1):

Table 1. Normal and pathological speakers from the MEEI database

Disorder	Male	Female
Non-pathological speakers		
	21	32
Pathological speakers		
Paralysis	35	36
Keratosis	15	11
Vocal Polyp	12	8
Adductor	3	19

4   Experimental Results

The voice pathology detection performance is evaluated by four algorithms; standard SVM, standard HMM, batch possibilistic SVM and the proposed incremental possibilistic SVM-HMM. The results of our work will be presented for a classification into five classes normal/Keratosis/Vocal Poly/Adductor for all voices. For SVM method, we have to set several parameters such as the kernel width γ and the regularization parameter C which is the regularization parameter. Hence, we used the optimum values of $\gamma = \frac{1}{K}$ and $C = 10$ found in a grid search using a cross-validation. RBF is selected as the kernel function. The

Table 2. Comparison of EER (%), Efficiency (DCF(%)), Sensitivity (%) and Specificity (%) for the different voice pathology detection systems and the proposed incremental possibilistic SVM-HMM system

System	Disorder	EER	DCF	Sensibility	Specificity
Standard SVM	Normal	07.98 ± 01.33	91.18 ± 0.17	86.75	86.91
	Paralysis	09.12 ± 1.04	92.33 ± 0.49	86.65	87.08
	Keratosis	08.45 ± 01.10	90.35 ± 0.69	86.15	86.81
	Vocal Polyp	09.14 ± 0.62	92.02 ± 0.57	86.39	87.02
	Adductor	8.80 ± 0.51	92.13 ± 0.77	87.51	86.64
Standard HMM	Normal	09.38 ± 0.47	89.86 ± 1.03	87.83	89.15
	Paralysis	09.14 ± 0.82	91.20 ± 0.95	87.28	88.39
	Keratosis	08.75 ± 01.11	92.31 ± 0.67	88.05	87.81
	Vocal Polyp	09.19 ± 01.02	91.80 ± 0.53	86.92	87.37
	Adductor	8.99 ± 0.63	92.13 ± 0.74	86.75	86.99
Batch possibilistic SVM	Normal	05.84 ± 0.36	93.87 ± 1.55	91.82	91.12
	Paralysis	7.91 ± 0.47	92.28 ± 01.07	90.25	89.75
	Keratosis	05.18 ± 0.72	91.69 ± 01.39	91.45	89.91
	Vocal Polyp	07.24 ± 01.22	92.33 ± 01.17	91.21	90.19
	Adductor	06.92 ± 01.15	92.00 ± 0.82	90.05	89.68
Incremental possibilistic SVM-HMM	Normal	02.11 ± 0.57	98.81 ± 0.88	98.95	99.27
	Paralysis	02.11 ± 0.57	99.22 ± 0.88	98.95	99.27
	Keratosis	02.36 ± 0.48	98.72 ± 1.02	97.92	98.20
	Vocal Polyp	02.07 ± 1.10	99.57 ± 0.91	99.08	98.97
	Adductor	01.48 ± 1.58	99.12 ± 0.73	98.55	98.77

choice to use RBF (Gaussian) Kernel was made after a study done on our data with different kernel functions such Linear, Polynomial, and Sigmoid.

The voices samples are subdivided into portions for training (70%) and testing (30%) steps. In order to investigate the performance of our voice pathology detection system, we consider four measures: Error Equal rate (EER), the performance accuracy (DCF), Sensitivity and Specificity.

The results, given in Table 2, show that the detection system based on the hybrid incremental possibilistic SVM-HMM yield the best results in this study. Obviously, the proposed system using the Incremental Possibilistic SVM-HMM and MFCC coefficients with their first and second derivatives outperforms the standard HMM, the standard SVM and the batch possibilistic SVM with an obtained accuracy equal to 99%.

The voice pathologies detection using the standard HMM give the worst results within a rate of 90%. Moreover, the detection system using the possibilistic SVM give a decent rate of 93%. Table 2 presents the performance of the proposed incremental hybrid method compared with the standard methods SVM, HMM and the batch possibilistic SVM method. The results obtained in this study for voice pathology detection are very encouraging. As a future work, we suggest to investigate different multi-pathologies detectors and also, to improves the incremental hybrid classifiers in order to determine the degree of voice pathology.

Furthermore, the Table 3 presents a comparison of the proposed hybrid incremental method with other recent methods from the state-of-art for the voice pathology detection problem using the MEEI datasets in similar experimental conditions.

Table 3. Comparison of the performance of our proposed incremental method possibilistic SVM-HMM and different methods in the state-of-art

System	Accuracy (%)
Patil et al. [9]	97%
Dibazar et al. [8]	97%
Amara et al. [18]	96%
Zulfiqar et al. [19]	93%
Proposed method	**99%**

The following table shows that the proposed incremental possibilistic SVM-HMM method improves the robustness of the voice pathology detection system and achieved the highest accuracy compared to several existing methods in the state-of-art.

5   Conclusion

Standard SVM and standard HMM works correctly in a batch setting where the algorithm has a fixed collection of samples and uses them to construct a hypothesis, which is used, thereafter, for detection and classification tasks without further modification. This paper proposes to combine an incremental possibilitic SVM to HMM for voice pathology detection task based an online setting. In the proposed method, we incorporate possibilistic degrees to the class posterior probabilities computing by SVM. Then, to improve the detection decision, we have given those possibilistic probabilities to HMM-based decoder in order to detect normal voices among pathological voices. The experimental results on the normal/pathology voices from MEEI database suggest that the proposed method gives high accuracies compared with several methods in the literature in detection task.

References

1. Cauwenberghs, G., Poggio, T.: Incremental and decremental support vector machine learning. In: Advances in Neural Information Processing Systems (NIPS), vol. 13, pp. 409–415 (2001)
2. Smets, P.: The combination of evidence in the transferable belief model. IEEE Trans. Pattern Anal. Mach. Intell. **12**, 447–458 (1990)

3. Davis, S.B., Mermelstein, P.: Comparison of parametric representations for mono-syllabic word recognition in continuously spoken sentences. IEEE Trans. Acoust. Speech Signal Process. **28**, 357–366 (1980)
4. Amami, R., Ben Ayed, D., Ellouze, N.: Incorporating belief function in SVM for phoneme recognition. In: Polycarpou, M., de Carvalho, A.C.P.L.F., Pan, J.-S., Woźniak, M., Quintian, H., Corchado, E. (eds.) HAIS 2014. LNCS (LNAI), vol. 8480, pp. 191–199. Springer, Cham (2014). https://doi.org/10.1007/978-3-319-07617-1_17
5. Vapnik, V.: The Nature of Statistical Learning Theory. Springer, New York (1995). https://doi.org/10.1007/978-1-4757-2440-0
6. Hofmann, M.: Support Vector Machines-kernels and the Kernel Trick. An elaboration for the Hauptseminar, Reading Club: Support Vector Machines (2006)
7. Cortes, C., Vapnik, V.: Support-vector networks. Mach. Learn. **20**, 273–297 (1995)
8. Alireza Dibazar, A., Narayanan, S., Berger, T.W.: Feature analysis for automatic detection of pathological speech. In: Conference of the Engineering in Medicine and Biology, vol. 1 (2002)
9. Patil, H.A., Baljekar, P.N.: Classification of normal and pathological voices using TEO phase and Mel cepstral features. In: International Conference on Signal Processing and Communications (SPCOM), pp. 01–05 (2012)
10. Amami, Rimah, Ben Ayed, Dorra: Robust noisy speech recognition using deep neural support vector machines. In: De La Prieta, Fernando, Omatu, Sigeru, Fernández-Caballero, Antonio (eds.) DCAI 2018. AISC, vol. 800, pp. 300–307. Springer, Cham (2019). https://doi.org/10.1007/978-3-319-94649-8_36
11. Pend, C., Xu, Q.J., Wan, B.K., Chen, W.X.: Pathological voice classification based on features dimension optimization, Transactions of Tianjin University, vol. 13, no. 6 (2007)
12. Hariharan, M., Polat, K., Sindhu, R., Yaacob, S.: A hybrid expert system approach for telemonitoring of vocal fold pathology. Appl. Soft Comput. **13**, 4148–4161 (2013)
13. Amami, R., Smiti, A.: An incremental method combining density clustering and Support Vector Machines for voice pathology detection. Comput. Electr. Eng. **57**, 257–265 (2016)
14. Dibazar, A.A., Berger, T.W., Narayanan, S.S.: Pathological voice assessment. In: IEEE 28th Engineering in Medicine and Biology Society, New York, USA, pp. 1669–1673, August 2006
15. Majidnezhad, V., Kheidorov, I.: A HMM-based method for vocal fold pathology diagnosis. IJCSI Int. J. Comput. Sci. Issues **9**(6) (2012). No 2
16. Maragos, P., Kaiser, J., Quatieri, T.: On separating amplitude from frequency modulations using energy operators. In: Proceedings of International Conference on Acoustics, Speech, and Signal Processing ICASSP, vol. 2, pp. 1–4 (1992)
17. Rabiner, L.R.: A tutorial on Hidden Markov Models and selected applications in speech recognition (1989)
18. Amara, F., Fezari, M., Bourouba, H.: An improved GMM-SVM system based on distance metric for voice pathology detection. Appl. Math. Inf. Sci. J. **10**(3), 1061–1070 (2016)
19. Zulfiqar, A., Irraivan, E., Mansour, A., Ghulam, M.: Automatic voice pathology detection with running speech by using estimation of auditory spectrum and cepstral coefficients based on the all-pole model. J. Voice **30**(6), 757.e7–757.e19 (2016)

External Attention LSTM Models for Cognitive Load Classification from Speech

Ascensión Gallardo-Antolín[1]([⊠]) [iD] and Juan M. Montero[2] [iD]

[1] Department of Signal Theory and Communications, Universidad Carlos III de Madrid, Avda. de la Universidad, 30, 28911 Leganés, Madrid, Spain
gallardo@tsc.uc3m.es
[2] Speech Technology Group, ETSIT, Universidad Politécnica de Madrid, Avda. de la Complutense, 30, 28040 Madrid, Spain
juancho@die.upm.es

Abstract. Cognitive Load (CL) refers to the amount of mental demand that a given task imposes on an individual's cognitive system and it can affect his/her productivity in very high load situations. In this paper, we propose an automatic system capable of classifying the CL level of a speaker by analyzing his/her voice. We focus on the use of Long Short-Term Memory (LSTM) networks with different weighted pooling strategies, such as mean-pooling, max-pooling, last-pooling and a logistic regression attention model. In addition, as an alternative to the previous methods, we propose a novel attention mechanism, called external attention model, that uses external cues, such as log-energy and fundamental frequency, for weighting the contribution of each LSTM temporal frame, overcoming the need of a large amount of data for training the attentional model. Experiments show that the LSTM-based system with external attention model outperforms significantly the baseline system based on Support Vector Machines (SVM) and the LSTM-based systems with the conventional weighed pooling schemes and with the logistic regression attention model.

Keywords: Computational Paralinguistics · Cognitive load · Speech · LSTM · Weigthed pooling · Attention model

1 Introduction

Cognitive Load (CL) refers to the amount of mental demand that a given task imposes on a subject's cognitive system and it is usually associated to the working memory that refers to the capacity of holding short-term information in the brain [8]. As overload situations can affect negatively the individual's performance, the automatic detection of the cognitive load levels has many applications in real scenarios such as drivers' or pilots' monitoring.

The work leading to these results has been partly supported by Spanish Government grants TEC2017-84395-P and TEC2017-84593-C2-1-R.

© Springer Nature Switzerland AG 2019
C. Martín-Vide et al. (Eds.): SLSP 2019, LNAI 11816, pp. 139–150, 2019.
https://doi.org/10.1007/978-3-030-31372-2_12

Speech-based CL detection systems are particularly interesting since they are non-intrusive and speech can be easily recorded in real applications. In fact, in 2014, an international challenge (Cognitive Load Sub-Challenge inside the INTERSPEECH 2014 Computational Paralinguistics Challenge) was organized with the aim of studying the best acoustic features and classifiers for this task [23]. Following this line of research, this work focuses on the design of an automatic system for CL level classification from speech.

Different features have been proposed for this task, as spectral-related parameters such as, Mel-Frequency Cepstral Coefficients (MFCC) [13,23], spectral centroid, spectral flux [23], and prosodic cues (intensity, pitch, silence duration, etc.) [2,24]. For the classifier module itself, Gaussian Mixture Models (GMM) [13] and Support Vector Machines (SVM) [23,24] are the most common choices.

However, in the last years, the application of deep learning models to speech-related tasks, such as Automatic Speech Recognition (ASR) [21,22], Language Recognition (LR) [27] or Speech Emotion Recognition (SER) [10,11,19] has allowed to increase the performance drastically. As a consequence, nowadays, Deep Neural Networks (DNN) have become the state of the art in this kind of systems. Among all the architectures proposed in the literature for speech-related tasks, Convolutional Neural Networks (CNN) [21], Long Short-Term Memory (LSTM) networks [7] and their combination are the most commonly used. On the one hand, CNNs exhibit the capability of learning optimal speech representations. On the other hand, LSTMs are capable to perform temporal modeling, so they are very suitable for dealing with sequences as it is the case of speech signals.

The so-called attention modeling is a new line of research, complementary to CNs and LSTMs, that tries to learn the structure of the temporal sequences aiming at determining the relevance of each frame to the task under consideration. Attention models have been successfully proposed for ASR [4], machine translation [17] or SER [10,11,19].

In this paper, we propose to adopt the previous findings to cognitive load level classification from speech. As this task has many similarities to SER, our work is mainly based on previous research on emotion classification from speech, especially, [10] and [19]. In particular, we focus on the use of LSTMs in combination with different weighted pooling strategies for CL classification, and we propose an external attention model that tries to take advantage of the benefits offered by attentional schemes, overcoming the need of a large amount of data for their training. Note that one of the main challenges of this kind of tasks is the lack of training data due to the difficulty of collecting and annotating recordings with the appropriate characteristics.

The remainder of this paper is organized as follows: Sect. 2 describes the fundamentals of LSTM with weighted pooling networks, Sect. 3 covers the different weighting schemes used in this work, together to our proposed external attention weighting method. Our results are presented in Sect. 4, followed by some conclusions of the research in Sect. 5.

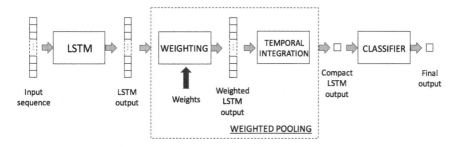

Fig. 1. General scheme of an LSTM with weighted pooling architecture. For simplicity, it is assumed that the LSTM layer is composed by only one LSTM cell.

2   LSTM with Weighted Pooling Networks

Long Short-Term Memory networks are a special kind of Recurrent Neural Networks (RNNs) that have the ability to store information from the past in the so-called memory blocks [7], in such a way that they are capable of learning long-term dependencies, overcoming the vanishing gradient problem. Therefore, LSTM outputs depend on the present and previous inputs, and, for this reason, they are very suitable for modeling temporal sequences, as speech.

The sequence-to-sequence learning carried out by LSTMs can be thought as a transformation of an input sequence of length T, $x = \{x_1, ..., x_T\}$ into an output sequence $y = \{y_1, ..., y_T\}$ of the same length, assuming that the classification process is easier in the y-space than in the x-space. However, as in the case of SER, CL classification can be seen as a many-to-one sequence-to-sequence learning problem [10]. Specifically, the input is a sequence of acoustic vectors and the final output must be the predicted CL level for the whole utterance (one single value). For this reason, it is advisable to include an intermediate stage in order to generate a more compact representation of the temporal LSTM output sequence that, in turn, will be the input to the classifier itself [10,11]. A most common option is the so-called Weighted Pooling (WP) module [19], as shown in Fig. 1. It consists of two different steps: weighting and temporal integration.

A desirable characteristic of WP is the ability for retaining the relevant information regarding the considered task while discarding the non-significant one. This issue can be addressed in the first step, where a weight α_t is computed and assigned to each temporal LSTM output y_t, following a certain criterion. For the CL task, it is reasonable to expect that not all the frames within an utterance reflect the subject's CL state with the same intensity, and therefore, larger weights should be assigned to frames containing significant cues about the speaker's CL, whereas smaller weights should be set to neutral or not relevant frames to the task. Different weighting schemes are discussed in Sect. 3.

In the second step, temporal aggregation, the weighted elements of the LSTM output sequence are somehow combined over time for producing a summarized representation of the information contained in it. For doing this, the most common choice is to perform a simple aggregation operation as follows,

$$z = \sum_{t=1}^{T} \alpha_t y_t \tag{1}$$

where $y = \{y_1, y_2, ..., y_T\}$ is the LSTM output sequence, $\alpha = \{\alpha_1, \alpha_2, ..., \alpha_T\}$ is the weight vector and z is the final utterance-level representation.

Note that it is possible to find a parallelism between this method and the temporal feature integration technique that is part of many parameterization modules in conventional hand-crafted feature-based systems, and whose aim is to obtain segment- or utterance-level representations of sequences of short-time features. Temporal integration has been successfully used in different speech/audio-related tasks, such as SER [5] or acoustic event classification [15,16]. Well-known methods comprise the computation of the statistics (mean, standard deviation, skewness, ...) of short-time acoustic vectors over longer time scales or their filtering [16]. Although out of the scope of this paper, weighted pooling could be performed by applying any of these techniques instead of the simple aggregation operation in Eq. (1).

3   Weighting Schemes

Several weighting schemes have been proposed in the literature. They can be classified into three categories: fixed, local attention and external attention weights.

3.1   Fixed Weights

This is the most simplistic alternative in which the same weights are used across all the utterances. The most used variants are the following:

- **Mean-pooling.** In this case, it is assumed that all the LSTM frames are equally important and, therefore, the weights α are set to,

$$\alpha_t = \frac{1}{T}, \quad \forall t \tag{2}$$

- **Max-pooling.** Here, it is assumed that the whole LSTM output sequence is optimally represented by its maximum, so the weights follow this expression,

$$\alpha_t = \begin{cases} 1, & y_t = \max\{y\} \\ 0, & \text{otherwise} \end{cases} \tag{3}$$

- **Last-pooling.** As in LSTM networks every output relies on previous and present inputs, it can be expected that the last outputs are the most reliable ones since for their computation, the LSTM uses to some extent information from the whole utterance [27]. This is equivalent to take into account only the last M frames of the LSTM output, according to the following expression,

$$\alpha_t = \begin{cases} \frac{1}{M}, & T - M < t \leq T \\ 0, & \text{otherwise} \end{cases} \tag{4}$$

3.2 Local Attention Weights

The aim of this approach is to focus on the frames of the utterance that convey more information about the classification task, therefore, a different weight is assigned to each temporal frame. Although when enough training data is available, it is possible to design more complex attention models, as those described in [10,11], in this work, we adopt the strategy proposed in [19] where the weights are computed as a simple logistic regression as follows,

$$\alpha_t = \frac{\exp(u^T y_t)}{\sum_{t=1}^{t=T} \exp(u^T y_t)} \tag{5}$$

where u and y are the attention parameters and the LSTM output, respectively. Both, the attention parameters and the LSTM outputs, are obtained in the whole training process of the system.

3.3 External Attention Weights

As mentioned before, the lack of training data prevents the use of complex attention models. Our hypothesis is that, in these cases, the attention model is not going to be properly trained and therefore, it should be more effective to use attention weights derived from external cues.

Previous studies about speech production under cognitive load conditions have shown that the level of CL may affect speech by producing changes in the prosody with respect to the neutral voice. In fact, variations in intensity (energy) [12,14], fundamental frequency (F_0) [2,12,14] and duration [2,14] are correlated to the speaker's cognitive load. We propose to incorporate the information contained in these prosody-related parameters in the weighted pooling scheme of the LSTM network.

Specifically, we consider the energy (actually, we use the log-energy) and F_0 as external attention signals $e_{ATT}(t)$ with the assumption that frames with high energies and F_0 values are more likely to present a strong content about the subject's CL level. The weights of the external attention model are computed from these signals. For doing this, firstly, e_{ATT} is normalized at utterance-level in the range [0, 1] yielding to a normalized signal $\bar{e}_{ATT}$, and secondly, the weights are obtained as the result of the softmax transformation applied to the normalized attention signal as follows,

$$\alpha_t = \frac{\exp(\bar{e}_{ATT}(t))}{\sum_{t=1}^{t=T} \exp(\bar{e}_{ATT}(t))} \tag{6}$$

This last operation guarantees that the sum of the weights across all the frames of the utterance is one.

4   Experiments and Results

4.1   Database and Baseline System

To the best of our knowledge, nowadays, there are a few speech databases containing utterances pronounced in different CL conditions by a significant number of speakers and conveniently labeled. One of the databases fulfilling these requirements is the "Cognitive Load with Speech and EGG" (CSLE) database [23, 26] that we have adopted for our experiments. It has been used in the Cognitive Load Sub-Challenge inside the INTERSPEECH 2014 COMPARE [23] whose main objective was the assessment of different speech features and classifiers for the prediction of subjects' cognitive load from their voice characteristics.

The CSLE database contains speech from 26 Australian English speakers recorded at 16 kHz by using a close-talk microphone while performing a set of tasks designed for inducing different levels of cognitive load (low, medium and high, denoted as *L1*, *L2* and *L3*, respectively). As in the challenge, in this paper, we have considered the following three tasks:

- *Reading Sentence (RS)*. In this case, speakers were asked to read a set of short sentences and recall an isolated letter between them. The degree of cognitive load was objectively assigned according to the number of read sentences before remembering the letter. Each speaker pronounced 75 utterances with a duration of 4 s on average, yielding a total of 1950 speech files.
- *Stroop Time Pressure (STP)*. It is based on the well-known Stroop test [25] where speakers were required to indicate the color of a set of printed words that, in turn, are names of colors. In medium and high load tasks, there was a mismatch between color names and color fonts. In addition, in the case of high load conditions, there was a time constraint for finishing the task. It contains 234 utterances (9 per speaker) with an mean duration of 17 s.
- *Stroop Dual (SD)*. It is similar to the previous task, but in this case, speakers had to execute another simultaneous task (tone counting) in the high load scenario. In total, for this task, 234 utterances (9 per speaker) with an average duration of 21 s were recorded.

The challenge organizers provided a partition of the database into training+development and testing subsets, where it was guaranteed that speakers belong to only one of these subsets (speaker independence). The number of speakers is 18 and 8 in the training+development and testing subsets, respectively. Table 1 shows the details about the database composition.

The baseline system is the one provided by the challenge organizers whose details can be found in [23]. In summary, it uses the standard parameterization adopted in the last Computational Paralinguistics Challenges (6373 characteristics), obtained with the open-source openSMILE feature extractor [6]. The classifier is a linear kernel SVM implemented by using the WEKA toolkit [9].

Following the challenge recommendations, each task was considered separately. This way, for both, the baseline and the LSTM-based models, an independent system per task has been trained with its specific training+development data and evaluated with the corresponding testing data.

Table 1. Composition of the CSLE database. For each task, the number of utterances per subset and cognitive load level are indicated.

Task	Number of utterances				
	Subset		L1	L2	L3
Reading Sentence	Train+Dev	1350	378	378	594
	Test	600	168	168	264
Stroop Time Pressure	Train+Dev	162	54	54	54
	Test	72	24	24	24
Stroop Dual	Train+Dev	162	54	54	54
	Test	72	24	24	24
Total	Train+Dev	1674	486	486	702
	Test	744	216	216	312

4.2 LSTM-Based Systems Configuration

Figure 2 shows the LSTM architectures with the three main weighting schemes evaluated in this work. In particular, Fig. 2(a) represents the fixed weight approach and its three variants: last-pooling, max-pooling and mean-pooling, Fig. 2(b) shows the system with logistic regression attention weights and Fig. 2(c) depicts our proposal, the LSTM system with external attention model. All systems were implemented with the Tensorflow [1] and Keras [3] packages.

In all cases, the same input acoustic features were used. The feature set consists of $n_B = 64$ log-Mel filterbank energies (log-Mels), computed every 10 ms using a Hamming window of 32 ms long and a mel-scaled filterbank composed of $n_B = 64$ filters by using the Librosa Python toolkit [18]. After feature extraction, mean and standard deviation normalization are applied at utterance-level yielding to a set of normalized log-Mels sequences x_I with T x 64 dimensions, where T is the number of frames of each utterance.

In all architectures, the length of the LSTM input sequences is set to $L = 1024$ for the RS task and $L = 2048$ for the STP and SD tasks, which corresponds to approximately 10 s and 20 s, respectively. Shorter utterances are padded with zeros by using a Masking layer, in such a way that these masked values are not used in further computations. Longer utterances are cut (this is only necessary in a few cases in the SD task). The output sequence of the Masking layer is denoted as x and its dimensions are L x 64.

This sequence is passed through an LSTM recurrent layer with $n_L = 128$ memory cells and 25% dropout to avoid over-fitting in the training process. The LSTM output, denoted as y, is a sequence of size L x 128. Next, the information contained in y is summarized by using the considered weighting scheme with weights α, yielding to a 128-dimensional vector, z. The length of the weight vector α is L. However, note that when $T < L$, $\alpha_t = 0$, $T < t \leq L$. The vector z is the input of a dense layer with $n_C = 3$ nodes (as the classes of our system are the 3 CL levels, L_1, L_2 and L_3) with softmax activation producing

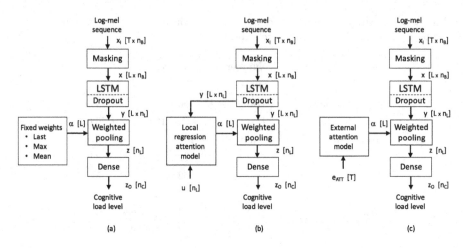

Fig. 2. Different LSTM-based architectures for cognitive load classification. (a) Fixed weights (last-pooling, max-pooling, mean-pooling); (b) Local logistic regression attention model; (c) External attention model. In brackets, the dimension of each variable, where T, L, n_B, n_L and n_C, stand for the number of frames of the input signal, the length of the LSTM input/output sequence, the number of mel filters, the number of LSTM units and the number of classes (CL levels), respectively.

a 3-dimensional output, z_O, representing the probabilities of each class. Finally, the class with the highest probability is assigned to the utterance.

In all cases, the LSTM models were trained using stochastic gradient descent and the Adam method with an initial learning rate of 0.001. We used a batch size of 32 and a maximum number of 60 epochs.

In the logistic regression attention model, the attention parameter vector u has a dimension of $n_L = 128$. All its components were initialized to $1/L$ and then refined during the training stage of the whole system.

In the external attention model, e_{ATT} denotes the external attention signal from which the weights α are derived. In this work, we have considered two alternatives for e_{ATT}. In the first case, it corresponds to the fundamental frequency F_0 of the speech signal computed every 10 ms using a Hamming window of 32 ms long and constraining the maximum F_0 to 500 Hz. In the second case, e_{ATT} is the log-energy of the speech signal extracted every 10 ms using a Hamming window of 32 ms long. Both, F_0 and log-energy were computed with the Librosa Python toolkit [18].

4.3 Results

This Subsection contains the experiments carried out in order to assess the performance of the proposed LSTM-based systems. As the number of instances for each class (CL levels) is unbalanced, results are given in terms of the Unweighted Average Recall (UAR) that is computed as the unweighted mean of the class-specific recalls.

Table 2 contains the results achieved for the baseline system and different LSTM architectures for the three tasks under consideration, *Reading Sentence*, *Stroop Time Pressure* and *Stroop Dual*. The column "Average" refers to the micro-average across the tasks. In the case of the LSTM-based systems, each experiment was run 10 times and therefore, results in Table 2 are the average UAR across the 10 subexperiments and the respective standard deviation.

LSTM corresponds to the conventional approach where no weighted pooling is applied and only the last frame of the LSTM output is passed through the following dense softmax layer. In the *LSTM+VAD* alternative, a Voice Activity Detector (VAD) is applied to the raw speech signals before the feature extraction in order to remove the silence/noise frames. As can be observed, the use of a VAD produces a decrease in performance. This suggests that silence pauses convey important information for discriminating between different CL levels, as they are related to the rhythm, elocution speed and disfluencies that can be heavily affected by the speaker's CL state. This result corroborates the observations about the effects of CL on speech production mentioned in, for example, [20].

The fixed weighting schemes evaluated are *Last-pooling* (in this case, the last $M = 200$ frames of the LSTM output were picked and averaged), *Max-pooling* and *Mean-pooling*. All these strategies outperform the conventional LSTM showing that not only the last frame contains relevant information for the task. Among these approaches, *Mean-pooling* achieves the best performance, and therefore, it seems better not to completely discard LSTM frames.

The *Logistic Regression Attention* method outperforms the previous ones, although its results overlap with *Mean-pooling* in the *RS* task and in the average across the three tasks. Nevertheless, it is clear that focusing on frames conveying more CL characteristics can help to improve the performance of the system.

Our proposal, the two *External Attention* approaches, produces the best results for all the tasks in comparison to the rest of LSTM-based systems. Comparing both approaches, using the log-energy as external attention signal slightly outperforms the F_0 alternative. Any case, these results support our hypothesis that the log-energy and F_0 could be used for establishing to some extent the relative importance of the frames for the CL level classification task.

Figure 3 depicts the weights used in the weighted pooling stage of the *Logistic Regression Attention* (top) and the *External Attention* strategy with log-energy (bottom). Contrary to the observations made in [19], in our case, the regression attention weights are very uniform and closely resemble the mean-pooling weights. This justifies the fact that the results achieved by *Mean-pooling* and *Logistic Regression Attention* are rather similar. We hypothesize that one possible reason for this behaviour is the lack of data for adequately training both, the attention and the LSTM model. However, the weights derived from the log-energy in the *External Attention* approach presents a large degree of variation, suggesting that the log-energy becomes a good approximation of the amount of cognitive load content of a speech frame when no enough data is available for training more sophisticated attention models. On average, *External Attention Energy* achieves 9.61% relative error reduction with respect to *Mean-pooling* and 6.85 % with respect to *Logistic Regression Attention*.

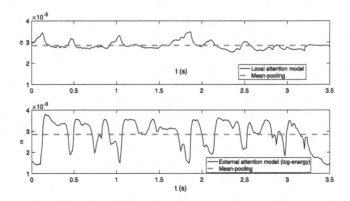

Fig. 3. Attention weights for one utterance belonging to the *Reading Sentence* task. Top: Weights obtained with the local regression attention strategy. Bottom: Weights derived from the log-energy used in the external attention approach.

Regarding the comparison of the LSTM-based systems to the baseline, it can be observed that *Logistic Regression Attention, External Attention Energy* and *External Attention F_0* clearly outperforms the SVM-based system for the *RS* task and on average across the three tasks. For the *STP* and *SD* tasks, these systems perform similarly, but these results are not very reliable as the number of test files in both cases is rather small (72 utterances). In summary, *External Attention Energy* achieves a relative error reduction with respect to the baseline of 11.04% an 9.64% for the *RS* task and on average, respectively.

Table 2. Unweighted Average Recalls (UARs) [%] for the baseline system and different LSTM-based classifiers for the Reading Sentence (RS), Stroop Time Pressure (STP) and Stroop Dual (SD) tasks and on Average.

System	RS	STP	SD	Average
SVM [23]	61.50	66.70	56.90	61.60
LSTM	48.87 ± 1.36	55.42 ± 1.02	45.83 ± 4.09	49.61 ± 1.33
LSTM+VAD	45.34 ± 1.79	54.01 ± 2.02	46.60 ± 4.06	46.36 ± 1.51
LSTM Last-Pooling	52.42 ± 1.53	59.57 ± 2.81	46.60 ± 4.11	52.67 ± 1.30
LSTM Max-Pooling	59.87 ± 1.28	53.48 ± 0.98	41.95 ± 1.83	57.54 ± 1.18
LSTM Mean-Pooling	62.99 ± 0.82	60.69 ± 0.67	50.00 ± 2.07	61.61 ± 1.01
LSTM Logistic Regression Attention	63.58 ± 0.48	63.47 ± 0.67	54.59 ± 0.67	62.75 ± 0.59
LSTM External Attention F0	65.24 ± 0.95	64.68 ± 0.52	56.35 ± 1.38	64.32 ± 0.83
LSTM External Attention Energy	65.75 ± 0.44	65.97 ± 0.76	59.20 ± 1.03	65.30 ± 0.70

5   Conclusions and Future Work

In this paper, we have developed an automatic system capable of classifying the cognitive load level of a speaker by analyzing his/her voice, based on LSTM models with different weighted pooling strategies. We have evaluated and compared the performance of mean-pooling, max-pooling, last-pooling and a logistic regression attention model. In addition, we have proposed a novel attention mechanism, called external attention model, that uses external cues, such as log-energy and fundamental frequency, for weighting the contribution of each LSTM temporal frame and that it is suitable in situations with scarce training data, as in this case. Experiments have shown that our proposal achieves, on average, a relative error reduction of 9.64% and 6.85% with respect to the baseline SVM and the LSTM with logistic regression attention systems, respectively.

For future work, we plan to extend our research in two directions: to explore different data augmentation techniques for increasing the amount of data for training the LSTM-based system and to study the use of other external cues.

Acknowledgments. We would like to thank Prof. J. Epps for kindly providing the CSLE dataset and Prof. B. Schuller and the rest of the ComParE 2014 organizers for kindly providing the dataset partition and the baseline system.

References

1. Abadi, M., et al.: TensorFlow: large-scale machine learning on heterogeneous systems. Software (2015). tensorflow.org
2. Boril, H., Sadjadi, O., Kleinschmidt, T., Hansen, J.: Analysis and detection of cognitive load and frustration in drivers speech. In: Proceedings of INTERSPEECH 2010, pp. 502–505 (2010)
3. Chollet, F., et al.: Keras: the python deep learning library. Software (2015). https://github.com/fchollet/keras
4. Chorowski, J., Bahdanau, D., Serdyuk, D., Cho, K., Bengio, Y.: Attention-based models for speech recognition. In: Proceedings of NIPS 2015, pp. 577–585 (2015)
5. Eyben, F., Huber, B., Marchi, E., Schuller, D., Schuller, B.: Real-time robust recognition of speakers' emotions and characteristics on mobile platforms. In: Proceedings of ACII 2015, pp. 778–780 (2015)
6. Eyben, F., Weninger, F., Groß, F., Schuller, B.: Recent developments in openSMILE, the munich open-source multimedia feature extractor. In: Proceedings of MM 2013, pp. 835–838 (2013)
7. Gers, F.A., Schraudolph, N.N., Schmidhuber, J.: Learning precise timing with LSTM recurrent networks. J. Mach. Learn. Res. **3**, 115–143 (2003)
8. van Gog, T., Paas, F.: Cognitive load measurement. In: Seel, N.M. (ed.) Encyclopedia of the Sciences of Learning, pp. 599–601. Springer, Boston (2012). https://doi.org/10.1007/978-1-4419-1428-6
9. Hall, M., Frank, E., Holmes, G., Pfahringer, B., Reutemann, P., Witten, I.: The WEKA data mining software: an update. SIGKDD Explor. **11**, 10–18 (2009)
10. Huang, C., Narayanan, S.: Attention assisted discovery of sub-utterance structure in speech emotion recognition. In: Proceedings of INTERSPEECH 2016, pp. 1387–1391 (2016)

11. Huang, C., Narayanan, S.: Deep convolutional recurrent neural network with attention mechanism for robust speech emotion recognition. In: Proceedings of ICME 2017, pp. 583–588 (2017)
12. Huttunen, K., Keränen, H., Väyrynen, E., Pääkkönen, R., Leino, T.: Effect of cognitive load on speech prosody in aviation: evidence from military simulator flights. Appl. Ergon. **42**(2), 348–357 (2011)
13. Kua, J.M.K., Sethu, V., Le, P., Ambikairajah, E.: The UNSW submission to INTERSPEECH 2014 compare cognitive load challenge. In: Proceedings of INTERSPEECH 2014, pp. 746–750 (2014)
14. Lively, S.E., Pisoni, D.B., Summers, W.V., Bernacki, R.H.: Effects of cognitive workload on speech production: acoustic analyses and perceptual consequences. J. Acoust. Soc. Am. **93**(5), 2962–2973 (1993)
15. Ludeña-Choez, J., Gallardo-Antolín, A.: Feature extraction based on the high-pass filtering of audio signals for acoustic event classification. Comput. Speech Lang. **30**(1), 32–42 (2015)
16. Ludeña-Choez, J., Gallardo-Antolín, A.: Acoustic event classification using spectral band selection and non-negative matrix factorization-based features. Expert. Syst. Appl. **46**(1), 77–86 (2016)
17. Luong, M.T., Pham, H., Manning, C.D.: Effective approaches to attention-based neural machine translation. arXiv preprint arXiv:1508.04025 (2015)
18. McFee, B., et al.: Librosa: audio and music signal analysis in python. In: Proceedings of SCIPY 2015, pp. 18–25 (2015)
19. Mirsamadi, S., Barsoum, E., Zhang, C.: Automatic speech emotion recognition using recurrent neural networks with local attention. In: Proceedings of ICASSP 2017, pp. 2227–2231 (2017)
20. Müller, C., Großmann-Hutter, B., Jameson, A., Rummer, R., Wittig, F.: Recognizing time pressure and cognitive load on the basis of speech: an experimental study. In: Bauer, M., Gmytrasiewicz, P.J., Vassileva, J. (eds.) UM 2001. LNCS (LNAI), vol. 2109, pp. 24–33. Springer, Heidelberg (2001). https://doi.org/10.1007/3-540-44566-8_3
21. Qian, Y., Bi, M., Tan, T., Yu, K.: Very deep convolutional neural networks for noise robust speech recognition. IEEE/ACM Trans. Audio Speech Lang. Process. **24**(12), 2263–2276 (2016)
22. Rao, K., Peng, F., Sak, H., Beaufays, F.: Grapheme-to-phoneme conversion using long short-term memory recurrent neural networks. In: Proceedings of ICASSP 2015, pp. 4225–4229 (2015)
23. Schuller, B., et al.: The INTERSPEECH 2014 computational paralinguistics challenge: cognitive & physical load. In: Proceedings of INTERSPEECH 2014 (2014)
24. van Segbroeck, M., Travadi, R., Vaz, C., Kim, J., Black, M.P., Potamianos, A., Narayanan, S.S.: Classification of cognitive load from speech using an i-vector framework. In: Proceedings of INTERSPEECH 2014, pp. 751–755 (2014)
25. Stroop, J.R.: Studies of interference in serial verbal reactions. J. Exp. Psychol. **18**(6), 643 (1935)
26. Yap, T.F.: Speech production under cognitive load: effects and classification. Ph.D. dissertation, The University of New South Wales, Sydney, Australia (2012)
27. Zazo, R., Lozano-Díez, A., González-Domínguez, J., Toledano, D.T., González-Rodríguez, J.: Language identification in short utterances using long short-term memory (LSTM) recurrent neural networks. PLoS ONE **11**(1), e0146917 (2016)

A Speech Test Set of Practice Business Presentations with Additional Relevant Texts

Dominik Macháček[(✉)] , Jonáš Kratochvíl , Tereza Vojtěchová ,
and Ondřej Bojar

Charles University, Faculty of Mathematics and Physics, Institute of Formal and
Applied Linguistics, Malostranské náměstí 25, 118 00 Prague, Czech Republic
{machacek,jkratochvil,vojtechova,bojar}@ufal.mff.cuni.cz

Abstract. We present a test corpus of audio recordings and transcriptions of presentations of students' enterprises together with their slides and web-pages. The corpus is intended for evaluation of automatic speech recognition (ASR) systems, especially in conditions where the prior availability of in-domain vocabulary and named entities is benefitable. The corpus consists of 39 presentations in English, each up to 90 s long. The speakers are high school students from European countries with English as their second language. We benchmark three baseline ASR systems on the corpus and show their imperfection.

Keywords: Speech recognition · ASR evaluation · Speech corpus · Non-native English

1 Introduction

Nowadays, English is being widely used as lingua franca for communication between people without common first language (denoted as L1). Europe is populated by dozens of nations with various and unique languages. In need for cooperation or interaction, English is often used as a universal first foreign language (or, in other words, the second language a human learns, L2) even between neighboring nations with closely related national languages, e.g. Czech and Polish. At the same time, many people are still not capable of using English and are dependent on translation services, which in turn often rely on human experts. We see an opportunity to boost availability, speed and language coverage of skilled professional translators and interpreters with the help of machines.

In spoken communication, such as during business conferences, the translation relies on speech comprehension. In Europe there are as many varieties of L2 English as there are European languages because many speakers have an accent derived from their L1. Current commonly used corpora for training the ASR systems are often based on audio recordings of English L1 speakers [6,11], which may not be optimal for ASR of European L2 English. Furthermore, the outputs

© Springer Nature Switzerland AG 2019
C. Martín-Vide et al. (Eds.): SLSP 2019, LNAI 11816, pp. 151–161, 2019.
https://doi.org/10.1007/978-3-030-31372-2_13

of ASR systems to date heavily depend on domain coverage of training data and they could be considerably improved by domain adaptation techniques. Also, the pronunciation of named entities from primarily non-English speaking areas usually differs significantly between English L1 and L2 speakers. Big corpora of L1 speakers often do not cover these differences and named entities are a big source of ASR errors and misunderstandings.

In certain situations, it is possible to prepare the ASR or spoken language translation (SLT) system for the specifics of a given talks and speakers. This is due to the fact that the sessions such as conferences and meetings are often planned ahead of time and additional relevant materials such as accompanying presentations to the talks or relevant websites are available.

With this in mind, we have created a corpus consisting of practice presentations of student fictional firms. The corpus contains audio recordings, transcriptions and additional relevant texts (presentation slides and web pages) of the participants. The audio recordings cover English L2 speakers with eight European L1s (cs, sk, it, de, es, ro, hu, nl, fi). Some of the practise firms' web pages are in English, some of them in local languages. Our corpus is suitable for evaluation of ASR systems, both in settings with and without additional materials provided ahead of time.

In Sect. 2, we describe the methodology that was used to collect the corpus data. In Sect. 3, we describe the corpus and its possible applications for the ASR systems. In Sect. 4, we present evaluation on three distinct English ASR systems. We summarize related works in Sect. 5 and conclude in Sect. 6.

2   Methodology

In this section, we explain the methodology we followed when creating the corpus. We collected the data at an international trade fair of student firms (see Sect. 2.1), during a competition of business presentations (Sect. 2.2). We motivated the speakers to transcribe their own speech presentations by introducing the Clearest voice competition for valuable prizes (Sect. 2.3). Additionally, we collected documents related to the student firms (Sect. 2.4). Throughout the corpus creation, we adhered to ethical standards (Sect. 2.5).

2.1   Background of Data: Student Firms and Trade Fair

"Student firms" are mock companies established for the practice of running a real company. The participants who run the companies are high-school students, mainly from economically-oriented schools or departments. The firms meet at trade fairs, where they practise promoting their fictional goods or services, issuing invoices for mock trades, and bookkeeping. They also compete in aforementioned tasks and are evaluated based on various criteria by field professionals. The best firms advance into higher rounds of trade fairs, from regional rounds through national into international.

We collected the data at an international trade fair held recently in the Czech Republic. The firms involved in our data collection were from 7 European countries. See Table 1 for a summary.

The trade fair organizers provided us the firms' presentation slides, which were used by students during the fair. In many cases, we were able to find their web pages and included them into the corpus. See Sect. 3.3 for more details.

Table 1. Number of student firms included in corpus and their countries of origin.

Country	Firms
Czech Republic	18
Italy	8
Romania	4
Slovakia	3
Austria	2
Spain	2
Belgium	2
Total	39

2.2 Presentation Competition

One of the activities during the fair, in which students could participate, was a competition of mock presentations of their businesses. The subject of the competition was to promote the firm to a random stranger in an elevator. The maximal allowed duration of the presentation was 90 s and no additional materials were allowed to be shown. The participants had to use English and either one or two students were allowed to give the presentation. A professional three-member committee was evaluating the content considering various aspects of the presentation. The selected competition winners were awarded prices for their performances.

We equipped the speakers with headset microphones to ensure the best possible quality of recordings. Despite of that, there was loud background noise that leaked to the recordings. On the one hand, this adds an extra obstacle for ASR, but on the other hand, the recordings thus represent a real environment where humans interact.

2.3 Manual Transcriptions

In order to obtain manual transcriptions of all the recordings, we asked the participants to transcribe their speech, given only their own recording. To motivate the students, we presented the task as an additional competition for valuable prizes. The objective of this competition was to find out who has the "clearest voice" for ASR. We processed the recordings with English ASR systems, evaluated them and awarded the students based on their respective ASR recognition scores. See Sect. 4 for more details.

The quality of the transcription was one of the major factors of the competition (together with clarity of speech) because the students had no access to any ASR outputs and had to assume that anything could be recognized correctly. We therefore believe that the students had a strong incentive to provide as accurate transcripts as possible. Furthermore, we reviewed all the transcriptions and edited them to include the missing parts, normalize punctuation and correct the misspellings, but for authenticity, we preserved the original grammar and vocabulary, even when it was not considered as standard English (e.g. *massageses* as a plural of *massage*, or *botel*, pronounced as *bottle*, as a term for a *hotel on a boat*).

2.4   Additional Resources

As mentioned above, the participants of this trade fair competed in various disciplines, which included also the preparation of slides and web pages for the fictional companies and their products.

Thanks to our close collaboration with the main organizer of the event, we were able to obtain additional materials, where available. While none of these additional materials were directly used in the presentation competition, they were closely linked to the mock companies and their activity subject. More details on the obtained and processed collection are available in Sect. 3.3.

We are confident that the students did their best when preparing these materials, motivated by the various competitions. For the purposes of ASR adaptation, the practical usability and overall quality of these materials highly differ from company to company. The relevant topics and named entities for each company are nevertheless mentioned in the corresponding materials.

2.5   Ethical Standards

During the competition and subsequent data evaluation we did comply with the ethical standards, which are in Europe given by General Data Protection Regulation (GDPR). Before the competition has started, all the participants gave us their consent to use and release collected data for research purposes, except of their names and any other personal data. Therefore, we removed the real names of students from the recordings, transcriptions and additional materials, and their photographs from the slides and downloaded web pages.

3   Corpus

The main motivation for collecting the corpus was to test our current ASR models and to gather data for further improvement of their robustness. We believe that the audio recordings contained in the corpus can be beneficial for anyone who wants to deploy their ASR models in real world applications. We also believe that the model performance on these data is a good approximation of its general accuracy in noisy environment.

3.1 Audio Recordings and Transcriptions

The corpus consists of 39 recordings of presentations of fictional student firms. The content of the audio recording corpus is summarized in Table 2 and the native languages of the speakers in Table 3.

Recordings contain different types of background noise including live music, announcements by organisers of the fair at main stage, conversations in different languages and noise produced by attendees of the presentations.

Table 2. Audio and content of the corpus.

	Single speaker	Two speakers	Total
Number of recording	17	22	39
Total audio duration	24 m 20 s	24 m 8 s	58 m 28 s
Transcription words	2891	3722	6613
Distinct speakers	17	44	61

Table 3. Native languages of the speakers in corpus.

Language	cs	de	it	es	ro	sk	hu	nl	fi	Total
Single speakers	9	-	-	1	3	1	1	-	-	17
Two speakers	18	4	16	2	-	-	-	3	1	44

3.2 Topics

The mock firms involved in the corpus represent a large variety of small or medium-sized companies. We summarize their business fields in Table 4. The most common are travel agencies followed by various food or beverage producers. Each firm is unique, focusing on a very specific segment of the market. Most of the firms fictionally operate only in their local areas.

Table 4. Business categories of student firms included in the corpus.

Business category	Firms
Travel agencies	7
Food and beverage producers	4
Beauty and health	3
Clothes and shoes	3
Household equipment	3
Online promotion	2
Accessories	2
Logistics	2
Others	13
Total	39

3.3   Additional Resources

We collected additional resources of 36 student firms participating in corpus creation. We are including either their presentation slides, web page or both. The numbers and types of resources are described in Table 5. In total, the additional resources contain 97000 of words, with a total vocabulary size of 15000.

Table 5. Types of additional materials and number of firms providing them.

Slides	Web	Firms
✓	✓	20
✓	✗	12
✗	✓	4
✗	✗	3

Table 6. Languages of presentation materials

Lang.	cs	en	de	it	es	ro	sk	cs/en	ro/en	sk/en	it/en/es/de	Total
Slides	14	15	-	-	-	-	1	1	-	1	-	32
Web	14	2	2	2	1	1	-	-	1	-	1	23

In order to protect the privacy of participants, we remove their real names and photographs, however, we preserve all facts that are related to the companies themselves. These include real or fictious email addresses, phone numbers, websites and locations.

The resources included in the corpus come in three distinct formats: the original (either Microsoft Office presentation format, or original web content format such as HTML or pictures), XLIFF format generated by MateCat Filters tool,[1] which is an XML-based format preserving the original structure of the document and may be useful for translation of the content, advanced information extraction tools, proper sentence segmentation or word-sense disambiguation. We also provide a plaintext format, which we created from XLIFF simply by extracting textual data from the documents. We included plaintexts because they may be convenient for the corpus users, and originals and XLIFF files, because they contain the complete information about the original structure of documents.

3.4   Additional Resources by Languages

The slides are either in Czech, Slovak or English. The web pages are mostly in national languages. Two of them are in multiple parallel language variants and two are in English only. Despite this fact, we believe they can still be valuable

---

[1] http://filters.matecat.com/.

resource for ASR or SLT improvement with English as a source. We believe
that the named entities or specific in-domain vocabulary of the spoken presentation, which could otherwise be left unrecognized, may be inferred from these
documents even automatically.

We provide the language counts of presentations and web pages in Table 6.
We note that there is one company in the corpus whose presenter's L1 was
Hungarian, their slides were English and web page in Romanian.

All the documents in the corpus are marked with language tags.

Table 7. WER of JRTk, Kaldi BBC and Google model scores on all recordings in the
corpus (right) and on the recordings on which all the systems produced some output
(left). WER of 100% indicates that no output was provided.

	Recognized by all			All recordings		
	Google	Kaldi B.	JRTk	Google	Kaldi B.	JRTk
Mean	73.59	87.55	45.21	89.32	87.47	45.63
Min	20.90	83.96	25.00	20.90	83.96	25.00
Max	98.31	91.03	74.08	100.00	91.03	99.58
Median	87.50	87.59	43.41	100.00	87.04	46.31
Stddev	27.87	2.29	15.28	21.82	1.92	15.23

4   Evaluation of ASR

In order to document the state of the art of ASR, we evaluated three ASR
systems on the corpus.

4.1   The ASR Systems

We consider three different ASR systems:

Janus Recognition Toolkit (JRTk) [9] featuring the IBIS single pass
decoder [15]. Its acoustic model was trained on TED talks [6] and Broadcast
News [4]. This system was designed to recognize lecture talks from IWSLT 2017
workshop [17].

Google Cloud Speech-to-Text[2] with English (United States) language
option.

Kaldi [12] based model trained on data from Multi-Genre Broadcast Challenge [2], on 1600 h of broadcast audio from BBC TV and several hundred million
words of subtitle text for language modeling. This model is thus suitable mainly
for native British English speakers.

We also tried Microsoft Cloud ASR but it failed for all our recordings.

---

[2] https://cloud.google.com/speech-to-text/.

4.2   Evaluation Metric

We use the standard word error rate (WER) metric, which is the minimum number of text insertions, deletions and substitutions needed to transfer one document to another, normalized by the total number of words in the document. As customary in ASR development, we disregard letter case and punctuation for this evaluation. We took the transcriptions obtained from the participants as the ground truth against which the automatic speech recognition outputs were evaluated.

4.3   Results

The descriptive statistics of respective word error rate scores are listed in Table 7 and visualized in Fig. 1. Note that the lower WER, the better recognition.

As already discussed in Sect. 3, the audio files contain a significant amount of background noise. Due to this fact, Google returned an empty output in some cases, resulting in the WER of 100%. In order to account this, we selected only the recordings on which all the systems had less than 100% WER, and measured a second set of descriptive statistics on this subset.

By manually inspecting the recordings on which the systems had the highest error rate we observed that the ASR difficulties could have been caused by a very strong accent of the speaker, or by the fact that the microphone was not in the appropriate distance from the mouth, or that the speaker did not articulate

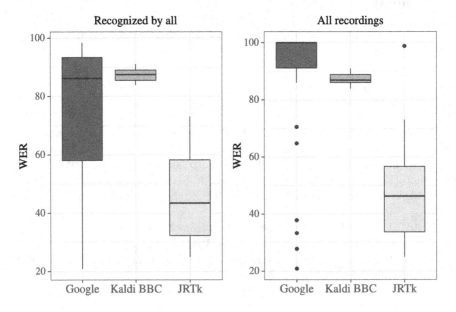

Fig. 1. Boxplot showing the word error rate scores of Google, Kaldi BBC and JRTk models on all recordings (right) and on a subset where all the systems produced some non-empty output (left).

clearly. Also, the background conditions such as a music band playing or students entering the presentation room may have affected the recognition quality.

5   Related Works

Tests sets for ASR are usually released together with speech corpora [5,6,11]. Our corpus is unique in a way that it contains L2 English, similarly as [20], but in our corpus there is a large variety of speakers, European L1s and realistic background noise conditions. Also, to our best knowledge, there is not any other speech corpus with additional in-domain resources.

Robustness to Noise: There are some corpora intended for noisy speech recognition: [1,7,14,18]. In [10], the authors show that model trained on a large dataset of distorted data with background noise is able to generalize much better than domain-specific models. Similar conclusions were derived in [8], where the authors experimented with random sampling of noise and intentionally corrupting the training data.

Non-native Speech: Adaptation for non-native speech in low-resource scenarios was studied by [19], who proposed interpolation of acoustic models or polyphone decision tree specialization. This can be incorporated into statistical ASR systems. For hybrid HMM-DNN (Hidden Markov models and deep neural networks) models, data selection methods can be used. In [16], combination of L2 out-of-domain read speech and L2 in-domain spontaneous speech led to the highest improvements, as opposed to using L1 speech.

Domain Adaptation: For purely neural LF-MMI (Lattice-free maximum mutual information) models [13], multi-task learning with large out-of-domain data as a first task and in-domain data as a second task, or various approaches of transfer learning can be beneficial [3].

6   Conclusion

We presented a small English speech corpus (only about 1 hour in total) intended as a test set for challenging speech recognition conditions: 61 distinct speakers, none of which were native speakers of English, a diverse set of vocabulary domains and noisy background.

   We have demonstrated that current ASR systems have severe difficulties in processing the test set, with WER ranging from 40 to 100% on individual audio recordings. The test set is equipped with additional text materials which can serve as evaluation of domain adaptation.

   The corpus is publicly released and available under the following link: http:// hdl.handle.net/11234/1-3023.

Acknowledgements. This research was supported in parts by the grants H2020-ICT-2018-2-825460 (ELITR) of the European Union and 19-26934X (NEUREM3) of Czech Science Foundation.

We are grateful to the organization team of the fictional student firms fair, who allowed us to conduct the competition during the event. We are also grateful to the students, who presented their firm and transcribed their audio recordings. Last but not least we are thankful to the team in Karlsruhe Institute of Technology and to the PerVoice team, who helped us overcome the technical difficulties that we have encountered.

References

1. Abdulaziz, A., Kepuska, V.: Noisy TIMIT speech LDC2017S04. In: Linguistic Data Consortium (LDC). Linguistic Data Consortium (LDC), University of Pennsylvania (2017)
2. Bell, P., et al.: The MGB challenge: evaluating multi-genre broadcast media recognition. In: Proceedings of the ASRU (2015)
3. Ghahremani, P., Manohar, V., Hadian, H., Povey, D., Khudanpur, S.: Investigation of transfer learning for ASR using LF-MMI trained neural networks. In: 2017 IEEE Automatic Speech Recognition and Understanding Workshop (ASRU), pp. 279–286, December 2017
4. Graff, D.: The 1996 broadcast news speech and language-model corpus. In: Proceedings of the 1997 DARPA Speech Recognition Workshop, pp. 11–14 (1996)
5. Gretter, R.: Euronews: a multilingual speech corpus for ASR. In: Proceedings of the Ninth International Conference on Language Resources and Evaluation (LREC 2014), pp. 2635–2638. European Language Resources Association (ELRA), Reykjavik, Iceland, May 2014. http://www.lrec-conf.org/proceedings/lrec2014/pdf/695_Paper.pdf
6. Hernandez, F., Nguyen, V., Ghannay, S., Tomashenko, N.A., Estève, Y.: TED-LIUM 3: twice as much data and corpus repartition for experiments on speaker adaptation. CoRR abs/1805.04699 (2018). http://arxiv.org/abs/1805.04699
7. Hu, Y., Loizou, P.: Subjective comparison of speech enhancement algorithms. In: Proceedings of ICASSP, vol. 1, June 2006
8. Kim, C., et al.: Generation of large-scale simulated utterances in virtual rooms to train deep-neural networks for far-field speech recognition in Google Home. In: Proceedings of INTERSPEECH, August 2017
9. Lavie, A., Waibel, A., Levin, L., Gates, D., Zeppenfeld, T., Zhan, P.: JANUS III: speech-to-speech translation in multiple languages. In: Proceedings of ICASSP 1997, January 1997
10. Narayanan, A., et al.: Toward Domain-Invariant Speech Recognition via Large Scale Training. In: 2018 IEEE Spoken Language Technology Workshop, SLT 2018, Athens, Greece, December 18–21, 2018, pp. 441–447 (2018). https://doi.org/10.1109/SLT.2018.8639610
11. Panayotov, V., Chen, G., Povey, D., Khudanpur, S.: Librispeech: an ASR corpus based on public domain audio books. In: 2015 IEEE International Conference on Acoustics, Speech and Signal Processing (ICASSP), pp. 5206–5210, April 2015
12. Povey, D., et al.: The kaldi speech recognition toolkit. In: IEEE 2011 Workshop on Automatic Speech Recognition and Understanding. IEEE Signal Processing Society, December 2011. iEEE Catalog No.: CFP11SRW-USB

13. Povey, D., et al.: Purely sequence-trained neural networks for ASR based on lattice-free MMI. In: Interspeech 2016, pp. 2751–2755 (2016). https://doi.org/10.21437/Interspeech.2016-595
14. Schmidt-Nielsen, A., et al.: Speech in Noisy Environments (SPINE) training audio LDC2000S87. In: Linguistic Data Consortium (LDC). Linguistic Data Consortium (LDC), University of Pennsylvania (2000)
15. Soltau, H., Metze, F., Fugen, C., Waibel, A.: A one-pass decoder based on polymorphic linguistic context assignment. In: IEEE Workshop on Automatic Speech Recognition and Understanding. ASRU 2001, pp. 214–217, December 2001
16. Tchistiakova, S.: Acoustic Models for Second Language Learners. master thesis, Universität des Saarlandes, Università degli studi di Trento (2018)
17. Nguyen, T.-S., Müller, M., Sperber, S., Zenkel, T., Stüker, S., Waibel, A.: The 2017 KIT IWSLT speech-to-text systems for English and German. In: The International Workshop on Spoken Language Translation (IWSLT), Tokyo, Japan, 14–15 December 2017
18. Vincent, E., Barker, J., Watanabe, S., Le Roux, J., Nesta, F., Matassoni, M.: The second 'CHiME' speech separation and recognition challenge: an overview of challenge systems and outcomes. In: 2013 IEEE Workshop on Automatic Speech Recognition and Understanding, pp. 162–167, December 2013
19. Wang, Z., Schultz, T., Waibel, A.: Comparison of acoustic model adaptation techniques on non-native speech. In: ICASSP, IEEE International Conference on Acoustics, Speech and Signal Processing - Proceedings, vol. 1, May 2003
20. Zhao, G., et al.: L2-ARCTIC: a non-native english speech corpus. In: Proceedings of the Interspeech 2018, pp. 2783–2787 (2018). https://doi.org/10.21437/Interspeech.2018-1110

Investigating the Relation Between Voice Corpus Design and Hybrid Synthesis Under Reduction Constraint

Meysam Shamsi[(✉)], Damien Lolive, Nelly Barbot, and Jonathan Chevelu

Univ Rennes, CNRS, IRISA, Lannion, France
{meysam.shamsi,damien.lolive,nelly.barbot,jonathan.chevelu}@irisa.fr

Abstract. Hybrid TTS systems generally try to optimise their cost function with the voice provided to generate the best signal. The voice is based on a speech corpus usually designed for a specific purpose. In this paper, we consider that the voice creation is realized through a corpus design step under reduction constraints. During this stage, a recording script is crafted to be optimal for the target TTS engine and its purpose. In this paper, we investigate the impact of sharing information between the corpus design step and the hybrid TTS optimisation step.

We start from a reduced voice optimized for a unit selection system using a CNN-based model. This baseline is compared to a hybrid TTS system that uses, as its target cost, a linguistic embedding built for the recording script design step. This approach is also compared to a standard hybrid TTS system trained only on the voice and so that does not have information about the corpus design process.

Objective measures and perceptual evaluations show how the integration of the corpus design embedding as target cost outperforms a classical hard-coded target cost. However, the feed-forward DNN acoustic model from the standard hybrid TTS system remains the best. This emphasizes the importance of acoustic information in the TTS target cost, which is not directly available before the voice recording.

Keywords: Hybrid speech synthesis · Corpus reduction · Linguistic and Phonological embeddings

1 Introduction

Nowadays, there are two main strategies for Text-To-Speech (TTS) synthesis. The first one is based on unit selection [1] and the second one is the Statistical Parametric Speech Synthesis (SPSS) [2,3]. The basic idea of unit selection-based TTS is to choose and concatenate a sequence of units from a natural speech corpus. The selected units should have linguistic features as close as possible to the target ones, associated to the text to vocalize, and the concatenations of consecutive unit signals should minimize differences in their joins. SPSS uses a vocoder and is known for the smoothness of its generated signals and

© Springer Nature Switzerland AG 2019
C. Martín-Vide et al. (Eds.): SLSP 2019, LNAI 11816, pp. 162–173, 2019.
https://doi.org/10.1007/978-3-030-31372-2_14

its flexibility. Conversely, unit selection based TTS systems provide more natural-sounding signals than SPSS [2,4].

The advantages and disadvantages of these TTS systems naturally led to the design of hybrid systems. The combination of both systems usually involves statistical models trained on the voice to predict parameters of an ideal generated speech and to guide a unit selection decoder that concatenates real signal segments extracted from the voice. Recent studies and the last Blizzard challenges have revealed good achievements of hybrid systems (see for instance [4–6]). In the last years, deep learning models such as Deep Neural Networks (DNNs) and Recurrent Neural Networks (RNNs) have been successfully used as acoustic models in hybrid systems, replacing HMMs, like in [7]. The main challenge in designing acoustic models is that the linguistic sequence does not have the same length as the acoustic sequence. For instance, in [8], a one to many approach is followed to deal with this problem. A LSTM-based auto-encoder is employed and permits to generate a sequence of acoustic frames representative of the input phoneme. As another example, in [9], each candidate phone unit is converted into a fix-length unit vector, called *Unit2Vec*, and DNNs are used as target and concatenation cost functions.

In order to manage the variable sequence length problem, a similar process has been applied in [10], a feed forward DNN for a one to one approach models phoneme frames, based on frame position, and the euclidean distance in the embedding space is used as the TTS target cost function. This approach also provides better results than an expert target cost.

In all cases, hybrid TTS systems are trained on a speech corpus independently of how it has been built. It may not lead to a significant difference when the voice is large enough, offering a good internal acoustic diversity. But, when the size of the voice is constrained in some ways, as in industrial applications which often need a high quality recorded voice, the adequacy between the voice and the TTS engine may impact the quality of the generated signals [11–13].

The cost, e.g. in terms of annotation time or recording time, to build a TTS voice for a professional usage is correlated to the length of the recording script. Hence, creating a voice under cost constraints requires to craft carefully the sentences to guarantee a good TTS quality in the end. To design such a script, a usual method is the selection of a subset of sentences as short and linguistically rich as possible from a large text corpus. This approach can be formalized as an optimisation problem in a discrete space [14]. The properties that the linguistic and phonological content of this subset has to achieve can stem from TTS engine needs or from the considered application independently or not of the TTS system. For instance, in [18,19], the phonological distribution in the script has to be close to a target one: natural, uniform or representative of a given domain. Conversely, the constraints and the nature of attributes to cover can be specific to the TTS engine, like in [17] where the phonological attributes used for the target cost function are covered, or in [20,21] where the internal descriptors of a SPSS system are considered, or, also in [19] where a pruning is done to remove units that are least used by the unit selection TTS system.

The resolution of this set covering problem for TTS corpus design has been widely studied in past studies [14–17].

Whereas the unit selection approach can support a small well-adapted voice corpus, the learning processes in hybrid systems are greedy in terms of voice data. Therefore, one may ask how to address and improve the use of hybrid TTS systems in a context of parsimonious voice building. In this paper, we investigate how the information from the voice creation process can be useful to help a hybrid TTS engine. To avoid disruption in the experiments, it focuses only on the inclusion of the information as the target cost of the hybrid TTS system. Using a unique voice, built from a simulated and controlled corpus design process, three variants of the same system are compared. The first one is based on an expert target cost function as in classical unit selection framework, whereas the target cost function of the second one is trained on linguistic, phonological and acoustic contents of the voice. This second approach illustrates a usual hybrid TTS system, as described in [10]. At last, the third approach uses a target cost function whose definition takes into account the voice creation process. The proposed method relies on the partition and covering of the embedding space used to design the recording script. Since this embedding is learnt before the recording phase, only linguistic and phonological features are required. Using objective and foremost perceptual evaluations, the experiments help to understand relations between corpus design and hybrid TTS.

This paper is organized as follows. First, Sect. 2 introduces the experimental framework. Especially, it explains how the corpus design is simulated and presents the resources used for training. Since all compared systems use the same voice, a voice creation process under size constraint is described in Sect. 3. This process is compared with the standard set-covering approach as a preliminary experiment. Section 4 details the different systems considered and especially differences between the hybrid ones. Evaluations and results are given in Sect. 5 before an overall discussion in Sect. 6.

2   Experimental Framework

In order to carry out the experiments presented in this paper and take into account the assumption of a recording phase, we have avoided the constraint of this recording work by reducing an already recorded and annotated corpus as in [11,25]. We have chosen an audio-book read by a professional speaker as initial corpus, thus limiting the bias inherent to the recording phase (speaker experience, recording conditions, etc.).

From this book, a randomly selected continuous part T has been taken away as a test set and the other part, denoted F, is named the full corpus in the remainder.

The voice creation step is simulated by the selection of a sentence set S from F, based on linguistic and phonological features only; the voice corresponds to the set of the signals associated to S. The objective is to find the best set S to synthesize the entire book, and the voice quality is evaluated using the subset T.

To illustrate the recording time constraint, $\mathcal{S}$ may be not longer than a given ratio of $\mathcal{F}$ in number of phoneme instances, and for the presented experiments, this ratio has been set to 10% of $\mathcal{F}$.

The initial corpus, i.e. the entire audio-book, contains 3,339 utterances of a French expressive audio-book spoken by a male speaker. The overall length of the speech corpus is 10 h44. More information on the annotation process can be found in [23]. $\mathcal{F}$ is composed of 3,005 utterances and 362,126 phoneme instances. The test set $\mathcal{T}$ contains the 334 other sentences from the initial corpus.

For all experiments, synthesis is done by the IRISA TTS unit selection system [22]. It can also be used as a hybrid TTS system like in [10].

3   Voice Construction and Preliminary Experiment

As explained in Sect. 1, several approaches can be used for corpus design under size constraint. The corpus design method used to build the voice that feeds all evaluated systems should be carefully selected so that the comparisons are fair. It needs to be usable in a hybrid TTS context and also leads to good performances with a unit-selection system. Among the methods optimized for a specific TTS engine (as in [17]) and others based on distributional information about the target domain (as in [19]), the latter seems preferable. This is particularly true since the corpus used here is in a consistent domain (a full audio-book as explained in 2). At last, distributional information can be well modeled by Neural Networks and can then be integrated into a hybrid TTS workflow. This section details the proposed corpus design method used to create the voice in further experiments. Moreover, in a preliminary experiment, this method will be compared to standard approaches to ensure its relevance.

The way to select sentences for the voice is accomplished as follows. Utterances from $\mathcal{F}$ are used to train an auto-encoder based on a multi-layer Convolution Neural Network (CNN) as illustrated in Fig. 1. The activation function is tanh and the loss function is the Mean Squared Error (MSE). The input vectors are composed of 296 components of categorical and numerical types automatically computed. The categorical attributes represent information about quinphonemes, syllables, articulatory features, and Part Of Speech for the current, previous and following words. These features are converted to a one-hot vector. The numerical features take into account information such as the phoneme position inside the word or utterance. These numerical features are normalized so that all the entries of the linguistic vector are in the range $[0, 1]$. The linguistic content of each input utterance is then represented by the sequence of linguistic feature vectors associated to the phonemes that compose it.

By taking the encoder part as the embedding model, each utterance of $\mathcal{F}$ is transformed into a sequence of vectors in the embedding space and its associated average vector is chosen to characterize this utterance. This results in a fixed-length vector whose size is the number of features (30) in the embedding space for each utterance. A K-Means algorithm is then applied to partition the set of average vectors represented $\mathcal{F}$. From each cluster, the utterance whose the

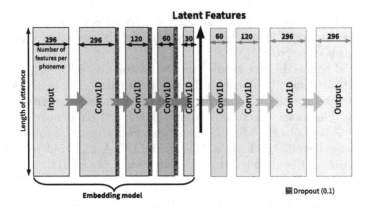

Fig. 1. Deep convolutional auto-encoder used to train linguistic and phonological embeddings.

Table 1. Objective evaluation of the proposed script design strategy using the TTS global cost.

Corpus design method	Average TTS global cost	95% confidence interval
Random	1.77	± 0.01
Set covering Greedy	1.75	± 0.02
CNN+KMeans	**1.60**	± 0.02

average vector is the closest to the center is selected and add to $\mathcal{S}$. This subset $\mathcal{S}$ is thus built to represent the linguistic richness of F by covering all its clusters, with the length about 10% from that of $\mathcal{F}$. The natural speech signals associated to elements of $\mathcal{F}$ are used as the TTS voice corpus of the experiments described below.

In order to assess the achievements of this script design method and its derived voice, a second voice with an identical length is built using a classic set covering strategy [17]. For this, the features used are diphones with the same linguistic as for the CNN. The utterances of $\mathcal{T}$ are then vocalized using the two voices respectively but the same TTS system, namely the IRISA system based on an expert target cost function. Generated outputs are objectively evaluated using the TTS global cost (a linear combination of target and concatenation costs) and also compared using a perceptual assessment. Besides, as baseline, for each utterance of $\mathcal{T}$, the average TTS global cost stemming from the use of 10 randomly selected voices is added. As for the perceptual evaluation, it is conducted in the form of an AB test with 17 listeners. From the 334 samples of $\mathcal{T}$, 100 samples are evaluated at least 6 times. Results are summarized in Table 1 and Fig. 2.

Whereas the TTS global cost mean provided by the standard set covering is close to the one resulting from the random selection method, the CNN-Kmeans

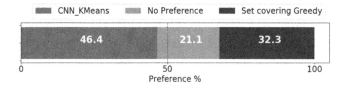

Fig. 2. Perceptual evaluation of the proposed script design strategy.

based corpus design method gives a significantly lower TTS global cost mean. This latter approach is also preferred during the listening test.

For the next experiments, we then keep this voice and the associated creation process based on the combination of CNN and KMeans algorithm. They will be used with different TTS engine configurations to investigate the relation between voice creation and hybrid synthesis.

4   TTS Systems Under Comparison

The objective of the paper is answering to this question: *Is it helpful to use the same phoneme representation in the corpus design step and in the TTS target cost?*

To do so, three methods for calculating the TTS target cost are compared. An expert target cost function which is a weighted sum of linguistic features is used as the baseline. The two other methods are based on embedded representations at phone level. The first one uses the same embedding for the corpus design step and the target cost function while the third one uses a specific embedding for the target cost function taking into account acoustics. The target cost is the euclidean distance in the embedding space between the candidate phone and the target one.

In the following, these three systems are described and then compared.

4.1   Expert-Based Target Cost (*Exp*)

In this method, the system used is a state of the art unit selection system. The target cost is defined as a weighted sum of linguistic features and has since been improved over the years [10]. The concatenation cost is the same as in [22], defined as a sum of euclidean distances on acoustic features between consecutive units.

4.2   Same Embedding for Corpus Design and TTS (*CNN*)

The second method replaces the expert target cost function by a cost function relying on the phoneme level embedding created during the corpus design step. Consequently, we propose here to use the same embedding model and phoneme representation for both corpus design and TTS target cost. The *CNN* auto-encoder described in Sect. 3 represents the linguistic information of phoneme by a vector of latent features. The TTS target cost is the euclidean distance in the embedding space between the candidate and target units. The *CNN* model is trained at the utterance level with $\mathcal{F}$ corpus and uses only linguistic information. One of the assets of this model is having contextual information of phonemes at the utterance level which could help a better representation in the embedding space.

4.3 Different Embeddings for Corpus Design and TTS (*MLP*)

The third method uses an embedding model specific to the target cost function using both linguistic and acoustic information. According to the proposition in [10], a feed-forward DNN is trained to predict the acoustic information at frame level for each input phoneme vector. The timing features are concatenated to embedding features in order to help prediction of the corresponding acoustic features. As in the previous system, the target cost function corresponds to the euclidean distance in the embedding space.

The learning data is the linguistic and acoustic information corresponding to phoneme/frame of the voice corpus $\mathcal{S}$. The timing features are the phoneme duration in seconds and the relative position of the corresponding frame inside the phoneme. The acoustic features consist of a 60 dimension Mel-Frequency Cepstral Coefficients (MFCC) vector, and the log of fundamental frequency F_0. The acoustic features are centered and reduced (unit variance). The frame length is 10 ms.

After training, the encoder part that transforms linguistic vector of phonemes into embedding space is detached and used as the embedding model.

4.4 Systems Differences

Table 2 summarizes and highlights the differences of the two embedding models described above and Fig. 3 displays the three approaches compared in this study.

Table 2. Embedding models comparison for both hybrid systems.

Method	CNN	MLP
Training data	Full corpus ($\mathcal{F}$)	Voice corpus ($\mathcal{S}$)
Input	Linguistic	Linguistic+Timing
Output	Linguistic	Acoustic
Training Level	Utterances (Sequence of phonemes)	Frames of signals

It is important to notice that the *MLP* model benefits from acoustics while the *CNN* model is only learnt with linguistic data. Also, both models learn, by

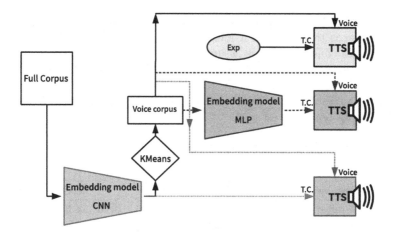

Fig. 3. TTS systems considered, namely Exp, MLP and CNN from top to bottom. The only difference come from the target cost (T.C.) computation.

construction, an embedding at the phoneme level, even if the *MLP* model is trained at the frame level (see [10]).

Besides, the *CNN* model is trained on the full corpus $\mathcal{F}$ and not only on the voice corpus $\mathcal{S}$ to maximize the quantity of data used for learning. The learning data is samples at the utterance level for the *CNN* model whereas the *MLP* one considers samples at the frame level. Hence, the *MLP* has much more data for training. It would not have been efficient to train the *CNN* model just with 300 samples from the $\mathcal{S}$ corpus.

Considering all this, we want to see if the consistency of embeddings between the corpus design step and the synthesis step helps to improve synthesis. However, the use of an acoustic model, with the *MLP* model, might not be completely fair. To be complete, further experiments are planned to try to inject acoustics in the corpus design step.

5   Experiments and Results

In the following subsections, we report the objective and perceptual evaluation results for the three methods.

5.1   Objective Evaluation

Since for the three methods, the target cost functions measure distances in three different (embedding or not) spaces, it is not possible to compare their outputs based on TTS costs. However, the same script is used as the test set and the *Concatenation rate* is then more appropriate to compare TTS performances. For each test utterance, this statistic is the number of concatenations in synthetic signal divided by the total number of possible concatenations. As for this measure, the lower is the better as it means more consecutive units from the same utterance. Less concatenation is assumed to result in higher quality. This measurement is computed for the test part $(\mathcal{T})$ and the rest of full corpus $(\mathcal{F} - \mathcal{S})$. It helps to find how methods can be generalized to other scripts than $\mathcal{F}$.

As shown in Table 3, the *CNN* method has better statistics than *Exp* method and *MLP* beats both for test part.

Table 3. Concatenation rate (%) results; confidence interval are calculated by using boot strap method with alpha $= 0.05$.

Measures/Methods	*Exp*	*CNN*	*MLP*
Rest of full corpus $(\mathcal{F} - \mathcal{S})$	56.63 ± 0.16	$\mathbf{54.36 \pm 0.16}$	$\mathbf{54.34 \pm 0.15}$
Test part $(\mathcal{T})$	56.64 ± 0.52	56.24 ± 0.51	$\mathbf{53.98 \pm 0.50}$

5.2   Perceptual Evaluation

In [10], the use of an acoustic model for the derivation of target cost has proved to be superior to an expert-based model. So two AB listening tests have been prepared to compare the synthetic quality of systems. The first one is between the *Exp* method and the *CNN* method and the other one is between the *CNN* and the *MLP* method. According to the protocol proposed for perceptual evaluation in [24], each AB test is composed of the 100 samples extracted from $\mathcal{T}$ with the highest DTW on MCep features. The samples are shorter than 7 s. The listeners have been asked to compare 40 pairs in terms of overall quality. The results are reported on Fig. 4.

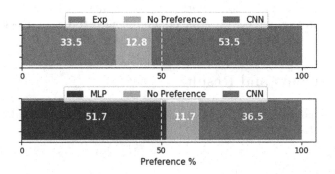

Fig. 4. Listening test results.

There are 14 listeners who have participated to the first test and 10 listeners as for the second test. Each pair of samples in the first test has been compared at least 5 times and in the second test at least 4 times. The result of the first test shows that the *CNN* based embedding as input of target cost can generate synthetic signals with significantly higher quality than the expert target cost. The second test indicates the preference of listeners for *MLP* model, which takes advantage of linguistic and acoustic information, rather than *CNN* model.

6 Conclusion

In this paper, we have investigated the relation between the corpus design process and a hybrid TTS. The TTS voice corpus has been selected based on an embedding model which uses the phonological information of the full corpus. This embedding model can be applied instead of the expert TTS cost or an acoustic model of phonemes. It has then be used to build a hybrid system by computing the target cost function as the euclidean distance between units in the embedding space.

In the first step, we have presented a phoneme embedding model which is basically the encoder part of a *CNN* auto-encoder. The transformation of utterances in the embedding space is followed by the KMeans algorithm to select a subset of full corpus in order to compose a voice corpus. Our preliminary experiment has shown that this method could achieve perceptually higher quality of synthetic signals than a voice designed by a classical set covering method.

The proposed *CNN* model has been applied to provide a phoneme embedding in hybrid TTS instead of an acoustic model (*MLP*) trained on the selected voice corpus. The perceptual test has shown that although the *CNN* model has better performances than expert-based target cost TTS, the *MLP* model has been preferred to the *CNN* model.

The *CNN* may be tuned or changed to improve performances. However, these results seem to emphasize the importance of acoustic information in any phone-embedding process for TTS tasks. The *CNN* model has been used for both corpus design and hybrid TTS, it is learnt on the full corpus, and takes into account more contextual information by the use of utterances as training samples (instead of frames). On the other side, the *MLP* model profits from acoustic information besides the linguistic one. Consequently, in future works, the use of an acoustic model as the embedding model for corpus reduction or corpus design should be investigated.

Acknowledgements. This study has been realized under the ANR (French National Research Agency) project SynPaFlex ANR-15-CE23-0015 and also funded by the Région Bretagne and the Conseil Départmental des Côtes d'armor.

References

1. Hunt, A., Black, A.: Unit selection in a concatenative speech synthesis system using a large speech database. ICASSP **1**, 373–376 (1996)
2. Zen, H., Tokuda, K., Black, A.: Statistical parametric speech synthesis. Speech Commun. **51**(11), 1039–1064 (2009)

3. Zen, H., Senior, A., Schuster, M.: Statistical parametric speech synthesis using deep neural networks. In: ICASSP, pp. 7962–7966 (2013)
4. King, S., Wihlborg, L., Guo, W.: The Blizzard Challenge 2017. In: Blizzard Challenge workshop (2017)
5. Fan, Y., Qian, Y., Xie, F., Soong, F.: TTS synthesis with bidirectional LSTM based recurrent neural networks, In: Interspeech, pp. 1964–1968 (2014)
6. King, S., Crumlish, J., Martin, A., Wihlborg, L.: The Blizzard Challenge 2018. In: Blizzard Challenge Workshop (2018)
7. Merritt, T., Clark, R., Wu, Z., Yamagishi, J., King, S.: Deep neural network-guided unit selection synthesis. In: ICASSP, pp. 5145–5149 (2016)
8. Wan, V., Agiomyrgiannakis, Y., Silen, H., Vit, J.: Google's next-generation real-time unit-selection synthesizer using sequence-to-sequence LSTM-based auto-encoders. In: Interspeech, pp. 1143–1147 (2017)
9. Zhou, X., Ling, Z., Zhou, Z., Dai, L.: Learning and modeling unit embeddings for improving HMM-based unit selection speech synthesis. In: Interspeech, pp. 2509–2513 (2018)
10. Perquin, A., Lecorvé, G., Lolive, D., Amsaleg, L.: Phone-level embeddings for unit selection speech synthesis. In: Dutoit, T., Martín-Vide, C., Pironkov, G. (eds.) SLSP 2018. LNCS (LNAI), vol. 11171, pp. 21–31. Springer, Cham (2018). https://doi.org/10.1007/978-3-030-00810-9_3
11. Chevelu, J., Lolive, D.: Do not build your TTS training corpus randomly. In: EUSIPCO, pp. 350–354 (2015)
12. Szklanny, K., Koszuta, S.: Implementation and verification of speech database for unit selection speech synthesis. In: Federated Conference on Computer Science and Information Systems (FedCSIS), pp. 1262–1267 (2017)
13. Nose, T., Arao, Y., Kobayashi, T., Sugiura, K., Shiga, Y., Ito, A.: Entropy-based sentence selection for speech synthesis using phonetic and prosodic contexts. In: Interspeech, pp. 3491–3495 (2015)
14. François, H., Boëffard, O.: Design of an optimal continuous speech database for text-to-speech synthesis considered as a set covering problem. In: Interspeech, pp. 829–832 (2001)
15. Cadic, D., D'Alessandro, C.: Towards optimal TTS corpora. In: LREC, pp. 99–104 (2010)
16. Isogai, M., Mizuno, H., Mano, K.: Recording script design for corpus-based TTS system based on coverage of various phonetic elements. In: ICASSP, pp. 301–304 (2005)
17. Barbot, N., Boëffard, O., Chevelu, J., Delhay, A.: Large linguistic corpus reduction with SCP algorithms. Computat. Linguist. 41(3), 355–383 (2015)
18. Krul, A., Damnati, G., Yvon, F., Moudenc, T.: Corpus design based on the kullback-leibler divergence for text-to-speech synthesis application. In: ICSLP, pp. 2030–2033 (2006)
19. Krul, A., Damnati, G., Yvon, F., Boidin, C., Moudenc, T.: Approaches for adaptive database reduction for text-to-speech synthesis. In: Interspeech, pp. 2881–2884 (2007)
20. Cooper, E., Chang, A., Levitan, Y., Hirschberg, J.: Data selection and adaptation for naturalness in HMM-based speech synthesis. In: Interspeech, pp. 357–361 (2016)
21. Nose, T., Arao, Y., Kobayashi, T., Sugiura, K., Shiga, Y.: Sentence selection based on extended entropy using phonetic and prosodic contexts for statistical parametric speech synthesis. IEEE/ACM Trans. Audio, Speech, Lang. Process. 25(5), 1107–1116 (2017)

22. Alain, P., Barbot, N., Chevelu, J., Lecorvé G., Simon, C., Tahon, M.: The IRISA text-to-speech system for the blizzard challenge 2017. In: Blizzard Challenge Workshop (2017)
23. Boeffard, O., Charonnat, L., Le Maguer, S., Lolive, D., Vidal, G.: Towards fully automatic annotation of audio books for TTS. In: LREC, pp. 975–980 (2012)
24. Chevelu, J., Lolive, D., Le Maguer, S., Guennec, D.: How to compare TTS systems: a new subjective evaluation methodology focused on differences. In: Interspeech (2015)
25. Lambert, T., Braunschweiler, N., Buchholz, S.: How (not) to select your voice corpus: random selection vs. phonologically balanced. In: SSW6, pp. 264–269 (2007)

Speech Recognition

An Amharic Syllable-Based Speech Corpus for Continuous Speech Recognition

Nirayo Hailu Gebreegziabher[(✉)] ⓘ and Andreas Nürnberger ⓘ

Fakultät für Informatik, Data and Knowledge Engineering Group,
Otto von Guericke Universität Magdeburg, Universitätsplatz 2, 39106
Magdeburg, Germany
{nirayo.hailugebreegziabher,
andreas.nuernberger}@ovgu.de

Abstract. Speech recognition systems play an important role in solving problems such as spoken content retrieval. Thus, we are interested in the task of speech recognition for low-resource languages, such as Amharic. The main challenges in solving Amharic speech recognition are the limited availability of corpora and complex morphological nature of the language. This paper presents a new corpus for the low-resource Amharic language which is suitable for training and evaluation of speech recognition systems. The corpus prepared contains 90 h of speech data with word and syllable-based annotation. Moreover, the use of syllable units for acoustic and language model in comparison with a morpheme-based model is presented. Syllable-based triphone speech recognition system provides a lower word error rate of 16.82% on the subset of the dataset. Moreover, syllable-based hybrid deep neural network with hidden Markov model provides a 14.36% word error rate.

Keywords: Speech recognition · Corpus · Neural and hidden Markov model · Syllable units

1 Introduction

With the increasing amount of spoken data being stored, shared, and processed nowadays, there is a need for systems performing automatic speech recognition, audio indexing, and search on audio streams. Hence, researchers are interested in the task of speech recognition and retrieving data from spoken contents, such as for Amharic. The domain of spoken contents includes broadcast news, oral historic archives, online lectures, meeting dialogues, and call-center conversations [14]. There are numerous amount of research that has been done on speech recognition [6, 7, 19–21]. However, performing speech recognition on low-resource languages raises some of the major research challenges in the area. There should be an open research with publicly available datasets and methodologies to speed up the progress in the field and to make speech recognition systems available for wider use.

There are efforts made to develop both morpheme-based [8, 9, 11] and syllable-based [7, 9] speech recognition systems for Amharic. However, all published works used only 20 h of training data [10]. In this paper, an effort has been made to collect

C. Martín-Vide et al. (Eds.): SLSP 2019, LNAI 11816, pp. 177–187, 2019.
https://doi.org/10.1007/978-3-030-31372-2_15

more Amharic speech and text corpora to make it publicly available for researchers in the field. We have also demonstrated the advantage of using Amharic syllable units instead of other units like morpheme.

The remainder of the paper is organized as follows: Sect. 2 describes the Amharic language. Section 3 discusses the corpus preparation. In Sect. 4, Amharic speech recognition components acoustic, language, and pronunciation models are described. Section 5 presents the experiments on the corpus. The last section, Sect. 6, provides discussion, conclusions, and highlights of the future work.

2 The Amharic Language

Amharic is the official language spoken in Ethiopia. It is a Semitic language of the Afro Asiatic Language group that is related to Hebrew and Arabic. There are more than 25 million users according to Ethnologue[1]. The language has its own writing system. As it is true in other languages, Amharic has its own phonetic and phonological character-istics. Amharic orthography, also known as ፊደል (fidəl), represents a consonant-vowel sequence, which is modified for the vowel.

There are seven vowels in Amharic namely, እ[ə], ኡ[u], ኢ[i], አ[a], ኤ[e], እ[ɨ], ኦ[o] [4, 5] (see Table 1). The language has 33 basic characters with each having seven forms for each consonant-vowel combination (33 × 7 = 231) with additional charac-ters there are 276 distinct orthography.

Table 1. Amharic vowels category

	Front	Central	Back
High	ኢ[i]	እ [ɨ]	ኡ [u]
Mid	ኤ[e]	እ [ə]	ኦ [o]
Low		አ [a]	

To create a complete inventory of Amharic sounds there are a set of thirty-eight phones, seven vowels, and thirty-one consonants [5]. The consonants are classified as stops, fricatives, nasals, liquids, and semi-vowels. Table 2 shows the first three of the Amharic phone inventory.

---

[1] https://www.ethnologue.com/language/amh (last accessed on 30.11.2018).

Table 2. A few Amharic orthographic inventories

	ə	u	i	a	e	ɨ	o
h	ሀ	ሁ	ሂ	ሃ	ሄ	ህ	ሆ
l	ለ	ሉ	ሊ	ላ	ሌ	ል	ሎ
m	መ	ሙ	ሚ	ማ	ሜ	ም	ሞ

2.1 Amharic Morphology

Amharic inflectional morphology exhibits addition of prefixes, suffixes, and modifications of root words. A single Amharic word could give hundreds of morphologically inflected different form of words. This morphological richness of the language increases the size of lexicons in speech recognition. The use of morphemes as a sub-word unit for Amharic speech recognition system is shown on [8, 9, 11] however, there is problem of out of vocabulary (OOV) morphemes. It is practically difficult to use the Amharic rule-based morphological analyzer like HornMorpho[2] for speech recognition purpose. Therefore, researches usually use Morfessor [25] to automatically segment words into morphemes. The tool allows supervised, semi-supervised, and unsupervised training statistical approaches. In this paper, unsupervised training method has been used to prepare morpheme-based corpus. The comparison of morpheme and syllable-based speech recognition model is presented in Sect. 5.

2.2 Amharic Syllabification

Syllable is a unit of sound composed of a central peak of sonority (usually a vowel (V)), and the consonants (C) which cluster around this central peak. Syllabification is the task of segmenting words whether spoken or written into syllables. Technically, the basic elements of syllables are Onset (the first phone in the sequence) and Rhyme (the remaining sequence of phones), which includes nucleus (central peak of sonority) and Coda (the remaining consonants other than the onset) [26]. A syllable can be described by a series of grammars such as consonant-vowel-consonant (CVC) sequence or onset, nucleus & coda (ONC).

Amharic is a syllabic language in which every orthography represents consonant-vowel assimilation. However, not all syllables in Amharic follow the CV sequence represented by the graphemes. Instead, Amharic syllables may follow various patterns, such as V, VC, CV, and CVC, including possible consonant clusters and gemination. Moreover, Amharic orthography did not show epenthetic vowel & geminated consonants that make it challenging to perform syllabification simply following the templates.

A novel syllabification algorithm for Amharic has been shown in [12]. In the paper, acoustic evidence, Amharic syllable template (V, VC, CV, VCC, CVC and CVCC [26]) and the well-known linguistic syllabification implementation principles namely,

---

[2] https://github.com/hltdi/HornMorpho.

maximum onset and sonority hierarchy principles, have been used to develop a rule-based syllabification algorithm. The algorithm considered gemination and the irregular nature of Amharic epenthesis vowel (ɨ). In this paper, the algorithm has been re-implemented in python with minor improvements. The algorithm is used to prepare syllable-based text corpus for the experiments.

3 Speech and Text Corpus Preparation

Speech recognition research in major languages such as English, German and Chinese has been conducted since 1950s. However, for low-resource languages such as Amharic, there are only a few attempts as it is mentioned in Sect. 1. There are only 20 h of speech data available [10] for the language, which is very less data to develop a better speech recognizer system. It is also challenging to develop Amharic speech-to-speech translation and spoken content retrieval systems [13, 15].

Collecting and preparing a very large speech corpus suited for the development of speech recognizer is costly and labor-intensive task. In this paper, we have prepared approximately 90 h of speech corpus from audiobooks and radio show archives with word and syllable-based transcription. The corpus is merged with the existing dataset and partitioned into training and evaluation set which is made publicly available[3]. An effort has been made to better estimate the number of speakers and age range in the audiobooks and radio show subsets, since we could not found such details.

There are two alternatives in preparing speech corpus. The first alternative is collecting text corpus and ask the native speakers of the language to read the text while recording. The other alternative is finding a variety of prerecorded and transcribed speech and preprocess it for the development of speech recognizer. In this paper, the second alternative is used. However, very few audiobooks and transcribed speech found, which limited the size of the corpus prepared. We have also used publicly available radio program archives. Table 3 provides a summary of all subsets in the corpus.

Table 3. Amharic speech corpus subset summary

Subset	Hours	Gender		Age	#Sentences	#Tokens
		Male	Female			
Existing	20	70	54	18–40	11234	109125
Audiobooks	81	40	–	18–40	22026	339342
Radio	9	30	20	18–50	2780	50208

3.1 Audio Segmentation

For segmenting the audiobooks and the radio show archives, Audacity[4] open source tools have been used. The segmentation process was semi-automatic. Since most of

---

[3] http://www.findke.ovgu.de/findke/en/Research/Data+Sets/Amharic+Speech+Corpus.html.

[4] https://www.audacityteam.org/.

speech recognition toolkits expect relatively shorter utterances, the average length of the segments is made 14 s. To align the text and spoken sentence command line tools and manual effort has been made. The preprocessing step includes fine-tuning such as removing non-speech contents, removing long silences, and correcting the audio samples using audio processing tools. The sampling frequency for each subset is normalized to 16 kHz with sample size of 16 bits, 256 kbs bitrate with mono channel.

Finally, the corpus is merged with the existing 20 h of data which contains varieties of speakers based on gender, age and dialects. The summary of the dataset could be found in [10].

3.2   Text Preprocessing

After aligning the text with the speech, numbers are converted into equivalent Amharic text as it is spoken in the recordings. Punctuation marks, foreign words, special characters, and symbols have been eliminated, abbreviations are also expanded manually. For some preprocessing tasks, simple python script has been used. For the language model (LM) preparation, CACO the 1.39 million (M) Amharic sentence from [24] has been used. The corpus is merged with our domain-specific text for speech recognition task, which makes it 1.4 M sentences. The text is converted into morphemes for morpheme-base LM using Morfessor 2.0 and into syllables using syllabification algorithm mentioned in Sect. 2.2 for syllable-based LM.

4   Amharic Speech Recognition System

To solve the general speech recognition problem there are three basic modeling approaches, namely, Hidden Markov model (HMM) [1, 16], hybrid Deep Neural Network with HMM model (DNN-HMM) and end-to-end [17, 19, 20] or all neural model. HMM-based automatic speech recognition is a very popular and successful one [16], nevertheless more recently deep neural network (DNN) is becoming state-of-the-art [21, 22].

Since Amharic suffer from lack of standard dataset it is not feasible to go for all neural model. However, in this paper, we have demonstrated the development of a syllable-based DNN-HMM model on the subset of the corpus. In this work, we have tried to balance the benefit of DNN, the use of less dataset and the advantage of getting n-best recognition result. Getting more than one best result is beneficial especially for indexing in spoken content retrieval [2, 3, 15, 18].

In HMM-based ASR, the aim of the system is to find the most likely sentence $W = \{w_1, w_2, w_3, .., w_t\}$ (word sequence) as it is shown in the Eq. (1) which transcribes the speech audio $O = \{o_1, o_2, o_3, \ldots, o_t\}$ (acoustic observation).

$$W = \underset{W}{argmax}\ P(W|O) = \underset{W}{argmax}\ P(O|W)P(W) \tag{1}$$

Given the phone set, lexicon and the audio files the HMM generates the probability of pronunciation and particular observation sequence given a state sequence which is also referred to as Acoustic model. In the training phase, all the models including the language model are represented as a weighted finite state transducer (WFST) and they become composed to form one large WFST graph.

4.1 Acoustic Modeling

Acoustic model, $P(O|W)$, represents the relationship between an audio signal and the phonemes or other linguistic units that make up speech. The model learned from a set of audio recordings and their corresponding transcripts [1]. A simple 3-state HMM with its transition probabilities a_{ij} and output probabilities $b_i(o_t)$ is illustrated in Fig. 1.

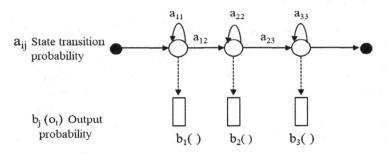

Fig. 1. A simple left-to-right 3-state HMM

Each states capture the beginning, central and ending parts of a phone. In order to capture the articulation effects, triphone models are preferred to context-independent phone models. A mixture of multivariate Gaussian probability distribution functions represented the emission probabilities. The parameters of Gaussian distributions estimated using the Baum-Welch algorithm [16]. In the decoding phase, the dynamic programming Viterbi algorithm is used to get the most probable speech unit sequence (syllable, morpheme or word) sequence from the graph generated.

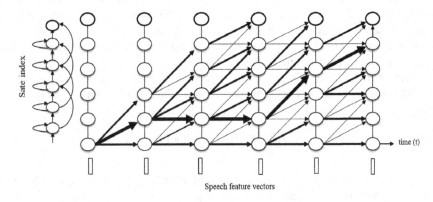

Fig. 2. A Viterbi algorithm for speech unit recognition

As shown in Fig. 2, this algorithm can be seen as finding the best path through a matrix where the vertical dimension represents the states of the HMM and the horizontal dimension represents the frames of speech (i.e. time). Each small circles in the picture represents the probability of observing that frame at that time and each arrow between circles corresponds to a transition probability. Instead of summing over all possible state sequences, we just consider the most likely path which can be achieved by changing the summation to a maximization in the recursion. The score for state j, given the input at time t is computed using Eq. (2).

$$P_j(t) = \max_i \left[P_i(t-1) a_{ij} b_j(t) \right] \tag{2}$$

The paths are grown from left-to-right column-by-column. We need to keep track of the states that make up this path by keeping a sequence of back-pointers. At time t, each partial path is known for all states i, finally we backtrack to find the state sequence of the most probable path. An interesting detail of the application of HMM in speech recognition can be found in [1, 16].

4.2 Language Model

Language model, P(W), it is a probabilistic model used to guide the search algorithm (predict next word given history). It assigns a probability to a sequence of tokens to be finally recognized. The most common modeling approach is the N-gram model $P(w_N | w_1, w_2, \ldots, w_{N-1})$ but recurrent neural network (RNN) is also used as a modeling approach [23]. In this paper, two language models are prepared using a subset of the CACO text corpus mentioned in Sect. 3.2. The first language model is a 1.4 M morpheme-base 5-gram LM and the other is 73k syllable-based 5-gram LM.

4.3 Pronunciation Model

Pronunciation model (lexicon model), P(W|L), forms the bridge between the acoustic and language models [1]. Prior knowledge of language mapping between words and the acoustic units (phoneme is most common). Two different lexicons are prepared for our experiment. The first one is prepared by selecting the most frequent 51k morphemes from the morpheme-based language model text. The second syllable-based lexicon is prepared in the same way by selecting only 16.7k unique syllables from the syllable-based language model corpus.

5 Experiments

The acoustic features extracted for our experiments consist of 13 dimensional Mel Frequency Cepstral Coefficient (MFCC), with their first- and second-order derivatives. A window size of 25 ms with an overlap of 10 ms has been used in the estimation of the MFCCs. The acoustic models have been trained and tested using Kaldi[5], one of the most widely used open source speech recognition toolkit.

All the language models mentioned in Sect. 4.2 are generated using KenLM[6] statistical language modeling toolkit. The language models are smoothed with modified Kneser-Ney smoothing technique.

5.1 Morpheme-Based System

For the morpheme-based system monophone and triphone models have been experimented in the Kaldi toolkit. The pronunciation dictionary consists of 51k most frequent morphemes described in Sect. 4.3 is used. Moreover, the pronunciation dictionary has been prepared as explained in Sect. 2.1.

In all the models, a 3-state left-to-right HMM topology is used. The monophone model is a context-independent HMM model which does not consider the neighboring phones in the acoustic modeling. The alignment from the monophone model is used as input to the triphone HMMs. Unsurprisingly, the monophone model has worse performance than both the triphone and hybrid DNN-HMM model. Table 4. shows summary of morpheme error rate for each models.

Table 4. Morpheme-based model system performance

Model	Morpheme error rate (MER) %
GMM-HMM monophone	70.97
GMM-HMM triphone	56.36
DNN-HMM triphone	44.62

5.2 Syllable-Based System

As it has been indicated in Sect. 2.2, Amharic has six syllable templates [26]. However, researchers have been considering only the CV syllable template [7, 8]. Moreover, epenthesis vowel and gemination are not handled in those research works [9]. In this paper, all the six syllable templates, as well as epenthesis vowel has been realized using the Amharic syllabification algorithm.

The experimental setup for syllable-based system is the same as morpheme-based system explained in Sect. 5.1, except the lexicon and language model is prepared from syllable units. The lexicon contains only 16.7k syllables and the language model is prepared using 73k syllable-based sentences, which is a small subset of the text corpus. A hybrid DNN-HMM model is experimented with similar setup used in the morpheme-based model.

In the DNN-HMM model, the GMM-HMM alignment from the triphone model is passed into a simple feedforward network (vanilla network with tanh nonlinearities

---

[5] https://github.com/kaldi-asr/kaldi.git.

[6] https://kheafield.com/code/kenlm/.

adapted from Kaldi script). The network architecture has been built with only 300 hidden layer dimension, 3 hidden layers, minibach size of 128, with initial learning rate 0.04 and final learning rate 0.004. The model is trained for 15×2 plus extra 2×5 epochs which is 40 epochs in total.

The syllable-based system performed better in all the models even with fewer data in the language model. Moreover, the OOV using syllable units is only three, which is extremely low compared with the morpheme-based system. Table 5 shows a summary of syllable error rate for each model.

Table 5. Syllable-based models system performance

Model	Syllable error rate (SER) %
GMM-HMM monophone	38.00
GMM-HMM triphone	16.82
DNN-HMM triphone	14.36

All the model performance shown in Tables 4 and 5 gained using the subset (20 h) of data to compare the performance of the morpheme-based model and syllable-based model.

6   Discussion and Conclusions

In this paper, new Amharic speech corpus is presented and made available for public access. The dataset is semi-automatically segmented and aligned with word and syllable-based transcript in order to make it suitable for speech recognition and spoken content retrieval tasks. Moreover, syllable-based speech recognition and language models are also introduced. Morpheme and syllable-based models are trained using the existing and the newly prepared corpus. The syllable-based models showed a better result compared with all the morpheme-based models even with language model prepared from a relatively small corpus. The syllable-based system showed a negligible amount of OOV syllables compared with the morpheme-based system. The size of the vocabulary required to prepare the pronunciation dictionary is also noticeably reduced when syllable units are used. The DNN-HMM model showed a better result in all the models even though a simple network with less number of epochs and hidden layers are used. The system provides n-best results in the form of a lattice which makes it a good starting point for tasks like lattice indexing for spoken content retrieval which is planned to be evaluated in future work. As a future work we have also planned to go for all neural model using all the subsets of the dataset.

Acknowledgments. The authors would like to thank the DAAD and MoSHE for funding this research work and DW for allowing us to use Amharic radio program audio from their online archive.

References

1. Gales, M., Steve, Y.: The application of hidden Markov models in speech recognition. Found. Trends® Signal Process. **1**(3), 195–304 (2008)
2. Chelba, C., Timothy, H., Murat, S.: Retrieval and browsing of spoken content. IEEE Signal Process. Mag. **25**(3), 39–49 (2008)
3. Larson, M., Stefan, E.: Using syllable-based indexing features and language models to improve German spoken document retrieval. In: Eighth European Conference on Speech Communication and Technology (2003)
4. Getahun, A.: ዘመናዊ የአማርኛ ሰዋስው በቀላል አቀራረብ (Modern Amharic Grammar in a simple approach), Addis Ababa (2008)
5. Baye, Y.: አጭርና ቀላል የአማርኛ ሰዋስው (Short and simple Amharic Grammar). Addis Ababa (2008)
6. Solomon, T.: Automatic speech recognition for Amharic. Ph.D. thesis (2006). http://www.sub.unihamburg.de/opus/volltexte/2006/2981/pdf/thesis.pdf
7. Solomon, T., Wolfgang, M.: Syllable-based speech recognition for Amharic. In: Proceedings of the 2007 Workshop on Computational Approaches to Semitic Languages: Common Issues and Resources. Association for Computational Linguistics (2007)
8. Martha, Y., Solomon, T., Wolfgang, M.: Morpheme-based automatic speech recognition for a morphologically rich language-Amharic. In: Spoken Languages Technologies for Under-Resourced Languages (2010)
9. Martha, Y., Solomon, T., Laurent, B.: Using different acoustic, lexical and language modeling units for ASR of an under-resourced language–Amharic. Speech Commun. **56**, 181–194 (2014)
10. Solomon, T., Wolfgang, M., Bairu, T.: An Amharic speech corpus for large vocabulary continuous speech recognition. In: 9th European Conference on Speech Communication and Technology (2005)
11. Michael, M., Laurent, B., Million, M.: Amharic speech recognition for speech translation. Atelier Traitement Automatique des Langues Africaines (TALAF). JEP-TALN (2016)
12. Nirayo, H., Sebsibe, H.: Modeling improved syllabification algorithm for Amharic. In: Proceedings of the International Conference on Management of Emergent Digital EcoSystems. ACM (2012)
13. Chelba, C., Timothy, H., Ramabhadran, B., Saraçlar, M.: Speech retrieval. Spoken language understanding: systems for extracting semantic information from speech (2011)
14. Lee, L., et al.: Spoken content retrieval: beyond cascading speech recognition with text retrieval. IEEE/ACM Trans. Audio Speech Lang. Process. (TASLP) **23**(9), 1389–1420 (2015)
15. Larson, M., Gareth, J.: Spoken content retrieval: a survey of techniques and technologies. Found. Trends® Inf. Retr. **5**(4–5), 235–422 (2012)
16. Huang, X., et al.: Spoken Language Processing: A Guide to Theory, Algorithm, And System Development, vol. 95. Prentice Hall PTR, Upper Saddle River (2001)
17. Hinton, G., et al.: Deep neural networks for acoustic modeling in speech recognition: the shared views of four research groups. IEEE Signal Process. Mag. **29**(6), 82–97 (2012)
18. Can, D., Murat, S.: Lattice indexing for spoken term detection. IEEE Trans. Audio Speech Lang. Process. **19**(8), 2338–2347 (2011)
19. Amodei, D., et al.: Deep speech 2: end-to-end speech recognition in English and Mandarin. In: International Conference on Machine Learning (2016)
20. Bahdanau, D., Chorowski, J., Serdyuk, D., Bengio, Y., et al.: End-to-end attention-based large vocabulary speech recognition. In: ICASSP, pp. 4945–4949. IEEE (2016)

21. Chan, W., Jaitly, N., Le, Q., Vinyals, O.: Listen, attend and spell: a neural network for large vocabulary conversational speech recognition. In: ICASSP, pp. 4960–4964. IEEE (2016)
22. Kim, S., Seltzer, M. L.: Towards language-universal end-to-end speech recognition. In: 2018 IEEE International Conference on Acoustics, Speech and Signal Processing (ICASSP), pp. 4914–4918. IEEE (2018)
23. Mikolov, T., et al.: Recurrent neural network based language model. In: 11th Annual Conference of the International Speech Communication Association (2010)
24. Andargachew, M.G., Binyam, E.S., Michael, G., Andreas, N.: Contemporary Amharic corpus: automatically morpho-syntactically tagged Amharic corpus. In: Proceedings of the First Workshop on Linguistic Resources for Natural Language Processing, pp. 65–70 (2018)
25. Sami, V., Peter, S., Stig-Arne, G., Mikko, K.: Morfessor 2.0: python implementation and extensions for Morfessor Baseline. Aalto University publication series SCIENCE + TECHNOLOGY, 25/2013. Aalto University, Helsinki (2013)
26. Mulugeta, S.: The syllable structure and syllabification in Amharic. Masters of philosophy in general linguistic thesis. Department of Linguistics, Trondheim, Norway (2001)

Building an ASR Corpus Based on Bulgarian Parliament Speeches

Diana Geneva, Georgi Shopov, and Stoyan Mihov[✉]

IICT - BAS, 2, Acad. G. Bonchev Street, 1113 Sofia, Bulgaria
{dageneva,gshopov,stoyan}@lml.bas.bg

Abstract. This paper presents the methodology we applied for building a new corpus of Bulgarian speech suitable for training and evaluating modern speech recognition systems. The Bulgarian Parliament ASR
(BG-PARLAMA) corpus is derived from the recordings of the plenary
sessions of the Bulgarian Parliament. The manually transcribed texts and
the audio data of the speeches are processed automatically to build an
aligned and segmented corpus. NLP tools and resources for Bulgarian
are utilized for the language specific tasks. The resulting corpus consists of 249 hours of speech from 572 speakers and is freely available
for academic use. First experiments with an ASR system trained on the
BG-PARLAMA corpus have been conducted showing word error rate
of around 7% on parliament speeches from unseen speakers using time-
delay deep neural network (TD-DNN) architecture. The BG-PARLAMA
corpus is to our knowledge the largest speech corpus currently available
for Bulgarian.

Keywords: Speech corpus · Automatic speech recognition ·
Low resource language

1 Introduction

Acoustic data acquisition for low resource languages like Bulgarian is an important and challenging task. Only few speech resources are available for Bulgarian
– e.g. GLOBALPHONE[1] and BulPhonC[2]. As far as we know the largest currently available Bulgarian ASR speech corpus is BulPhonC [5]. This corpus consists of less than 40 hours of read speech, which makes it rather insufficient for
the purpose of training modern DNN and RNN acoustic models. On the other
hand, recent developments in speech and language technology made it possible
to significantly reduce the manual work required for building a speech corpus
from transcribed audio content. Also, the increase in the amount of multimedia
content on the Internet in the recent years makes it possible to automatically
collect data. The English ASR corpus LibriSpeech [9] has been derived from thousands of public domain audio books. Unfortunately, the availability of Bulgarian

---

[1] ISLRN:250-105-856-478-2 http://www.islrn.org/resources/250-105-856-478-2/.
[2] ISLRN:755-406-235-455-4 http://www.islrn.org/resources/755-406-235-455-4/.

© Springer Nature Switzerland AG 2019
C. Martín-Vide et al. (Eds.): SLSP 2019, LNAI 11816, pp. 188–197, 2019.
https://doi.org/10.1007/978-3-030-31372-2_16

audio content with transcriptions on the Internet is very limited. One of the few sources of such content is the Bulgarian Parliament, Narodno sabranie, where all speeches from the plenary sessions are transcribed manually and recorded on video. The texts and videos are published on the parliament's web page[3]. Other projects for building speech corpora from transcribed audio of parliament speeches are presented in [12] for Catalan and [6] for Icelandic.

This paper describes the work and methodology used for building the BG-PARLAMA corpus, which is a speech data set based on the speeches of the Bulgarian Parliament members. The corpus is freely available for academic use[4]. Section 2 describes the raw data available from the Bulgarian Parliament. In Sect. 3 we describe the process we used to build the language models for the ASR system used for audio alignment. Section 4 presents the ASR based alignment procedure for long audio that we used in the creation of this corpus. Section 5 describes the procedure for the selection and segmentation of the content and the structure of the corpus. Finally, in Sect. 6 we present and compare experimental results for ASR trained on the new data set.

2   Data Source

The website of the Bulgarian Parliament provides mp4 video files for all plenary sessions from 2010 up to now. The speeches are recorded using stationary directed microphones on the parliament's platform. The distance between the microphone and the speaker varies, which causes differences in the speech level between the speakers. Occasionally background noise occurs in the recordings. The format of the audio stream in the video files is 44100 Hz mono compressed with the AAC codec at 75 kb/s. In some files from 2013–2015 the audio signal is corrupted by electric hum. There are around 120 sessions per year. Every session is split into parts by session breaks. Every part of a session is stored in a separate video file. For our corpus we have downloaded all videos from 2010 up to the end of June 2018 with a total duration of 4839 h.

The manual transcriptions available in textual format go back to year 1991. Older transcriptions are provided only as scanned PDF files. For every session of the parliament exactly one text file with transcriptions is given. In most cases the session breaks are marked in the transcriptions. We downloaded all texts from 1991, consisting of 94 million words.

The corpus preparation procedure had to overcome the following specifics:

1. The correspondence between the speech and its manual transcription is not exact. For example unintended repetitions in the speech are not present in the transcription. Also, some grammatical or lexical mistakes in the speech are corrected in the transcription. On some occasions the texts are modified by changing the word order and inserting or deleting words in order to increase its clarity.

---

[3] https://www.parliament.bg.
[4] http://lml.bas.bg/BG-PARLAMA.

2. The transcriptions make use of digits to express number, date, time and currency expressions. Common abbreviations are used as well.
3. In the transcriptions the name of the speaker together with the name of his party is written in free text in front of each speech. The free text of the names occasionally contains variations or mistakes in the spelling. There are also cases where a member of the parliament switches parties. This complicates the unique identification of the speaker.
4. There are additional annotations in the text files (e.g. indication of what is happening in the room) which are not represented in the speech.

3   Building Language Models for ASR

3.1   Text Corpus

All transcribed texts are first normalized by converting them to lower-case and then are tokenized. The verbalization (expansion) of vocabulary items which differ in verbal and written form (such as dates, numbers and abbreviations) is done next. Since we have no ground truth for verbalization, we reuse the finite-state rewrite rules developed for the Text-to-Speech system [3]. Those rules make use of contextual information to resolve ambiguities caused by inflections in the Bulgarian language. For example, the verbalization of a number written with digits is ambiguous because it could be verbalized either to ordinal or cardinal number and should agree with the preceding or following noun on number and gender. In Bulgarian if a sequence of digits is preceded by the word "article" it has to be verbalized as a masculine ordinal number; if it is preceded by "chapter" it has to be verbalized as a feminine ordinal number and if followed by "meters" it has to be verbalized as a cardinal number etc. We can express these rules with regular expressions such as:

$$\text{Digits-to-OrdinalMasculine} / \text{"член"} \_$$
$$\text{Digits-to-OrdinalFeminine} / \text{"глава"} \_$$
$$\text{Digits-to-Cardinal} / \_ \text{"метра"}$$

The verbalization rules are further compiled into finite-state transducers and applied to the texts. Generally, verbalization is an ambiguous task and therefore the process introduces verbalization errors. Nevertheless, since we later require perfect alignment, texts which include verbalization errors shall hardly be included in the dataset. Finally, all punctuation marks are removed from the texts.

3.2   Dictionary

The lexicon that we use is based on the lexicon used in [7] consisting of 440 K entries. This lexicon has been further extended to cover 99% of the vocabulary found in the text corpus by adding the phonetizations of the most frequent unknown words and acronyms from the transcribed texts. To achieve this we manually added accent marks and then generated the phonetizations automatically using phonetization rules from the Bulgarian TTS system [3].

3.3   Language Models

We used the SRILM tool [2] to train two 3-gram (modified) Kneser-Ney smoothed language models on the verbalized texts. The first model is trained on all verbalized texts (94 million words) while the second one is trained only on the texts corresponding to the available video files (32 million words). The smaller language model is used only for the ASR system in the alignment phase (see Sects. 4 and 5). Perplexities of both language models were measured on verbalized texts from parliament sessions not included in the training texts. The perplexity of the smaller language model is 59.52 with (and 85.48 without) end-of-sentence tokens and the perplexity of the larger is 59.48 with (and 85.42 without) end-of-sentence tokens.

4   Segmentation into Speeches

The training of acoustic models for ASR systems requires as input relatively short utterances (around ten seconds) with the corresponding nearly perfectly matching reference texts. To achieve this, we have to align the audio, extracted from the available video files, with the verbalized texts and split it into small segments. This is done in two phases. The first phase, described in this section, aims to split the audio and transcription files for a whole plenary session into separate audio and corresponding text files for each speech. In the second phase (see Sect. 5) short audio and text segments with nearly perfect match are extracted.

4.1   Splitting Transcriptions into Session Parts

Initially, the original video files are processed with the ffmpeg[5] tool in order to extract the audio stream in 16KHz PCM wav file format. As mentioned above, for each session of the parliament there are several audio files but only one text transcription. We use the marked breaks to split the transcribed text to match the audio parts. Ideally, this would give us equal number of audio and text files. However, occasionally there are wrongly indicated breaks in the transcriptions and missing or corrupted audio files which leads to differences in the number of audio and text files. We manually corrected the inconsistencies by splitting or merging text files or deleting the corresponding transcriptions when the audio file was unavailable. Nevertheless, occasional inaccuracies remain in the alignment of the session audio and text parts.

4.2   Speaker Identification

We had to identify the speaker for each speech. As explained, the name and the party of the parliament member, which we call speaker label, is given in free text in front of each speech in the transcription file. Unfortunately, for speeches of the same speaker this label may be different. This is caused by

---

[5] http://www.ffmpeg.org.

members of the parliament switching parties over the years, middle or family names being included or omitted, spelling variations and mistakes and others. In total we extracted 4033 different speaker labels. To deal with the problem we chose for each parliament member a unique identifier – 1113 speaker identifiers in total. Then we manually assigned to each label the corresponding identifier. The identifiers also reflect the gender of the speaker. This procedure was aided by automatically grouping the labels using Levenshtein similarity.

Using the speaker identifiers, the transcribed texts were split into separate speech transcriptions. For each such transcription the speaker, the session, and the session part it came from are known.

4.3 ASR for Alignment

Next, we aim to align the verbalized speech transcriptions with the corresponding audio. For this purpose we use an ASR system for Bulgarian trained on the BulPhonC corpus [5]. For the speech alignment we apply the Kaldi ASR toolkit [10]. We train a triphone hidden Markov model using Gaussian mixture models, applying LDA and MLLT, on the acoustic data from the BulPhonC corpus (see Sect. 6 for more details). Then we align each of the transcribed speeches of the speaker with the recognized text produced by the Kaldi ASR decoder using the BulPhonC acoustic model and the smaller language model (see Subsect. 3.3).

4.4 Audio Segmentation

In the segmentation stage we search for the best match between each speech transcription and an arbitrary region of the ASR recognized text from the corresponding session part. This is done by using the dynamic programming framework [11] with symbol level Levenshtein distance. If the best match is above a chosen threshold (in this case 30% letter error rate) we search for matches of the given speech transcription in the ASR text of the other parts of that session. In this way we compensate for the inaccuracies in the alignment of the session audio and text parts. Using the timestamps for each word from the ASR output we extract the audio segment that corresponds to the matched region in the speech transcription.

As a result we obtain for each speaker a set of text-audio pairs for each of his or her speeches. Those pairs may contain inaccuracies (up to 30% letter error rate) between the audio and the transcription.

5 Audio Splitting and Selection

The corpus is divided into three parts: training, development and test. The training part has to satisfy the following constraints:

- (near) perfect match between the audio and the transcription;
- sufficient speech duration for each speaker and no big imbalances in the speech duration between the speakers;
- audio split into relatively short segments.

Table 1. Data subsets in the BG-PARLAMA corpus

Subset	Duration	per-spk minutes	Male spks	Female spks	Total spks
Train	248 h 47 m	26	422	150	572
Dev	1 h 10 m	6	6	5	11
Test	1 h 8 m	5	9	5	14

On the other hand, the development and test parts should provide:

- unbiased selection of texts and speakers;
- no speaker overlap between training, development and test parts;
- audio split on sentence boundaries.

5.1 ASR for Selection

For the purpose of segment selection we again used the ASR system for Bulgarian trained on the BulPhonC corpus. In this case, since the speeches are already aligned to the speakers, we were able to improve the recognition quality by using speaker adaptation (SAT) with feature-space Maximum Likelihood Linear Regression (fMLLR). All other parameters are as described in Subsect. 4.3.

5.2 Alignment Procedure and Extraction of Corpus Candidates

The alignment procedure also uses symbol level Levenshtein distance. We align the text recognized by the ASR with the corresponding speech transcription. From the best alignment we extract the subsequences with exact matches (with no corrections). Those subsequences are further trimmed in order to start and finish with silence of at least 0.1 s. We use the timestamps in the decoder output to split the audio of the speech into the selected subsequences with perfect match. As a result of the phase we obtain a set of corpus candidates consisting of 1.6 million extracted short audio segments with the corresponding text excerpts with perfect match of a total duration of 1979 h.

N.B. The perfect match between the ASR recognized text and the transcription does not necessarily imply that the audio always matches the transcription. On some occasions there still remain some inaccuracies.

5.3 Selection of Training Corpus

The Training Corpus is obtained from the corpus candidates by:

1. discarding all audio segments with transcriptions shorter than 20 symbols (avoiding very short utterances);
2. discarding all speakers with total duration of audio with perfect match less than 10 min;
3. limiting the audio for each of the remaining speakers to 30 min[6].

---

[6] In our setup training with audio for individual speakers limited to 60 min did not improve the ASR accuracy.

Finally, in the training corpus we obtained 148607 speech segments from 572 speakers of a total duration of 249 hours (see Table 1).

Table 2. Bulgarian ASR phonetic system

A	мер<u>а</u>к	l	лек, лампа	g	<u>г</u>рад
a	м<u>а</u>зе, кед<u>ъ</u>р	m	<u>м</u>ама	d	<u>д</u>ар
e	т<u>е</u>л, п<u>е</u>ро	n	<u>н</u>ар	v	<u>ж</u>ар
i	б<u>и</u>к, п<u>и</u>рон	p	<u>п</u>ек	z	<u>з</u>ар
O	к<u>о</u>нче	r	<u>р</u>ъка	k	<u>к</u>ана
o	б<u>о</u>рба, кич<u>у</u>р	s	<u>с</u>ин	4	<u>ч</u>ар
U	Т<u>у</u>нис	t	<u>т</u>их	6	<u>ш</u>ах
Y	кат<u>ъ</u>р	f	<u>ф</u>ар	j	кра<u>й</u>, Кол<u>ь</u>о
b	<u>б</u>аба	h	<u>х</u>ол	0	чорба<u>дж</u>ия
w	<u>В</u>арна	c	<u>ц</u>ар	9	го<u>дз</u>ила

5.4   Selection of Development and Test Corpus

The speakers for the development dataset are chosen among the speakers not included in the training corpus. First, we extracted the speakers with total audio duration in the corpus candidates between 6 and 10 min. We obtained in total 69 speakers (21 female and 48 male speakers) with corpus candidates of total duration of 9 h. However the corpus candidates are not suitable for development and test because their ASR output matches perfectly the corresponding transcription. Because of that we extracted the raw speeches for those speakers. A random selection of those speeches was further manually split into sentences for inclusion in the development corpus. Finally, we prepared for the development corpus 173 segments from 11 speakers of total duration of 70 min. In order to obtain a correct language model the selected speeches are removed from the text corpus.

The dataset for the test corpus was prepared in a similar way. Instead of using the corpus candidates, we selected new speakers and their speeches from newer plenary sessions. The training and development datasets make use of the records of the plenary sessions up to 30 June 2018 whereas for the testing corpus we use the data from plenary sessions from 1 July to 30 September 2018. In this way we ensured that the speakers and the texts in the test dataset have no overlap with the train and development datasets. The alignment and splitting of the segments was done manually as well. As result we produced a test corpus consisting of 219 segments from 14 speakers with a duration of 68 min.

Table 1 summarizes the quantitative information for the three datasets in the BG-PARLAMA corpus.

6   ASR Experiments and Results

We present the ASR results using HMM-GMM and DNN based acoustic models trained on the BulPhonC and the BG-PARLAMA datasets. All ASR experiments are performed with the Kaldi Toolkit [10] using the dictionary and language model described in Sect. 3 and the phonetic system presented in [5, 7] (see Table 2). This phonetic system achieved better accuracy compared to the Bulgarian IPA phonetic system. We train three acoustic models:

Table 3. ASR word error rate of the acoustic models trained on different datasets

Acoustic model		BG-PARLAMA dev	BG-PARLAMA test
BulPhonC	GMM	36.95%	33.43%
	GMM+SAT	25.06%	21.39%
	DNN	44.54%	34.25%
BG-PARLAMA	GMM	16.55%	16.06%
	GMM+SAT	13.32%	12.27%
	DNN	7.45%	6.80%

1. GMM – a triphone hidden Markov model using Gaussian mixture models, applying LDA and MLLT [4] (used in Subsect. 4.3);
2. GMM+SAT – a triphone hidden Markov model with the addition of speaker adaptation with fMLLR [1, 8] (used in Subsect. 5.1);
3. DNN – deep neural networks with p-norm non-linearities [13].

We used the same recipes and parameters for the models as in LibriSpeech [9].

We tested the ASR accuracy on the BG-PARLAMA development and the BG-PARLAMA test sets. Table 3 summarizes the recognition results. As shown in the table the DNN model trained on the BulPhonC dataset performs worse than the corresponding GMM and GMM+SAT models. We speculate that this is due to overfitting caused by insufficient training data. Trained on the BG-PARLAMA dataset, however, the DNN models delivered significantly better results. This could be explained with the much larger size of the audio data. The best ASR result on the test data is achieved by the DNN model trained on the BG-PARLAMA dataset – 6.80%.

We performed a preliminary analysis of the recognition errors and identified the following two main sources of errors:

– mismatches between the speeches and their transcriptions;
– errors in the verbalizations caused by the inherent ambiguity of this task.

We plan to address those issues in future versions of the corpus.

7   Conclusion

We described the compilation procedure for the BG-PARLAMA corpus, applying automatic alignment and segmentation of speeches from the Bulgarian Parliament. As result we produce the largest currently available corpus of Bulgarian speech suitable for training modern ASR systems. The presented experiments show that the larger size of the resulting corpus (249 h) outweighs the audio mismatches caused by the automatic alignment procedure. The ASR system trained on the new BG-PARLAMA corpus achieves an error rate of around 7%. The corpus is freely available for academic use.

In the future we plan to experiment with varying the model's parameters in order to further improve the accuracy. We also plan to explore the recognition accuracy using the BG-PARLAMA corpus on other domains of Bulgarian speech and language models.

Acknowledgments. The research presented in this paper is partially funded by the Bulgarian Ministry of Education and Science via grant DO1-200/2018 'Electronic healthcare in Bulgaria' (e-Zdrave). We acknowledge the provided access to the e-infrastructure of the Centre for Advanced Computing and Data Processing, with the financial support by the Grant No BG05M2OP001-1.001-0003, financed by the Science and Education for Smart Growth Operational Program (2014–2020) and co-financed by the European Union through the European structural and Investment funds.

References

1. Anastasakos, T., McDonough, J., Schwartz, R., Makhoul, J.: A compact model for speaker-adaptive training. In: Proceeding of Fourth International Conference on Spoken Language Processing. ICSLP 1996. vol. 2, pp. 1137–1140 (Oct 1996).https://doi.org/10.1109/ICSLP.1996.607807
2. Andreas Stolcke: SRILM - an extensible language modeling toolkit. In: INTER-SPEECH (2002)
3. Andreeva, M., Marinov, I., Mihov, S.: SpeechLab 2.0: A high-quality text-to-speech system for bulgarian. In: Proceedings of the RANLP International Conference 2005. pp. 52–58 (September 2005)
4. Gopinath, R.A.: Maximum likelihood modeling with Gaussian distributions for classification. In: Proceedings of the 1998 IEEE International Conference on Acoustics, Speech and Signal Processing, ICASSP 1998 (Cat. No.98CH36181). vol. 2, pp. 661–664 (May 1998). https://doi.org/10.1109/ICASSP.1998.675351
5. Hateva, N., Mitankin, P., Mihov, S.: BulPhonC: bulgarian speech corpus for the development of ASR technology. In: Proceedings of the Tenth International Conference on Language Resources and Evaluation LREC 2016, Portorož, Slovenia, May 23–28, 2016. pp. 771–774 (2016). http://www.lrec-conf.org/proceedings/lrec2016/summaries/478.html
6. Helgadóttir, I.R., Kjaran, R., Nikulásdóttir, A.B., Guonason, J.: Building an ASR corpus using Althingi's parliamentary speeches. In: Proceedings INTER-SPEECH. pp. 2163–2167 (2017). 10.21437/Interspeech.2017-903. http://dx.doi.org/10.21437/Interspeech.2017-903

7. Mitankin, P., Mihov, S., Tinchev, T.: Large vocabulary continuous speech recognition for Bulgarian. Proc. RANLP **2009**, 246–250 (2009)
8. Gales, M.J.F., Woodland, P.C.: Mean and variance adaptation within the MLLR framework. Comput. Speech Lang. **10**(4), 249–264 (1996). https://doi.org/10.1006/csla.1996.0013.   http://www.sciencedirect.com/science/article/pii/S0885230896900133
9. Panayotov, V., Chen, G., Povey, D., Khudanpur, S.: Librispeech: an ASR corpus based on public domain audio books. In: 2015 IEEE International Conference on Acoustics, Speech and Signal Processing (ICASSP). pp. 5206–5210 (April 2015). https://doi.org/10.1109/ICASSP.2015.7178964
10. Povey, D., et al.: The Kaldi speech recognition toolkit. In: IEEE 2011 Workshop on Automatic Speech Recognition and Understanding. IEEE Signal Processing Society (Dec 2011), IEEE Catalog No.: CFP11SRW-USB
11. Wagner, R.A., Fischer, M.J.: The string-to-string correction problem. J. ACM **21**(1), 168–173 (1974). https://doi.org/10.1145/321796.321811. http://doi.acm.org/10.1145/321796.321811
12. Anguera Miró, X., Luque, J., Gracia, C.: Audio-to-text alignment for speech recognition with very limited resources. In: INTERSPEECH. pp. 1405–1409 (2014)
13. Zhang, X., Trmal, J., Povey, D., Khudanpur, S.: Improving deep neural network acoustic models using generalized maxout networks. In: 2014 IEEE International Conference on Acoustics, Speech and Signal Processing (ICASSP). pp. 215–219 (May 2014). https://doi.org/10.1109/ICASSP.2014.6853589

A Study on Online Source Extraction in the Presence of Changing Speaker Positions

Jens Heitkaemper[1(✉)], Thomas Fehér[2], Michael Freitag[2],
and Reinhold Haeb-Umbach[1]

[1] Department of Communications Engineering,
Paderborn University, Pohlweg 47-49, 33098 Paderborn, Germany
{heitkaemper,haeb}@nt.upb.de
[2] voice INTER connect GmbH, Ammonstr. 35, 01067 Dresden, Germany
{thomas.feher,michael.freitag}@voiceinterconnect.de

Abstract. Multi-talker speech and moving speakers still pose a significant challenge to automatic speech recognition systems. Assuming an enrollment utterance of the target speakeris available, the so-called SpeakerBeam concept has been recently proposed to extract the target speaker from a speech mixture. If multi-channel input is available, spatial properties of the speaker can be exploited to support the source extraction. In this contribution we investigate different approaches to exploit such spatial information. In particular, we are interested in the question, how useful this information is if the target speaker changes his/her position. To this end, we present a SpeakerBeam-based source extraction network that is adapted to work on moving speakers by recursively updating the beamformer coefficients. Experimental results are presented on two data sets, one with artificially created room impulse responses, and one with real room impulse responses and noise recorded in a conference room. Interestingly, spatial features turn out to be advantageous even if the speaker position changes.

Keywords: Robust speech recognition and Multi-channel speech enhancement · Speaker adaptation · Conference scenario

1 Introduction

In recent years, robust multi-channel Automatic Speech Recognition (ASR) has been a major focus of research which led to large improvements in transcription accuracy [1]. These gains are mainly due to the development of novel neural network (NN) architectures [2,3] and the combination of neural network (NN)s with well-known speech enhancement techniques like statistical beamforming [4,5] and dereverberation [6]. However, realistic application environments often still present a challenge to Automatic Speech Recognition (ASR) systems because of overlapped speech and moving speakers [7].

© Springer Nature Switzerland AG 2019
C. Martín-Vide et al. (Eds.): SLSP 2019, LNAI 11816, pp. 198–209, 2019.
https://doi.org/10.1007/978-3-030-31372-2_17

Recently, several promising approaches for source separation [8–10] and source extraction [11–14] in the presence of multiple simultaneously active speakers were presented. This contribution focuses on source extraction, where one is interested in only one of the speakers in a mixture.

Different techniques have been proposed to identify the target speaker. In the so-called SpeakerBeam (SB) approach, the target speaker is identified by an enrollment, also called adaptation utterance (AU), which the speaker has to provide in advance and from which his/her spectral characteristics are obtained [11,13]. This information is then used to guide a neural network for mask estimation to focus on the target speaker.

The desired speaker can also be identified by the speaker's position as in [14], where a neural network uses oracle information of the target speaker location to focus on a specific source, assuming the speaker does not move. In [12] a beamforming vector is estimated on a keyword preceding the user's command. While this setting may be appropriate for operating a digital home assistant, in many other application scenarios, such as a meeting, it would be very inconvenient if utterances had to start with a keyword to identify and locate the target speaker. Additionally, a fixed beamformer estimated on a AU or a keyword cannot capture changes in the speaker position or noise statistics.

In this contribution we are concerned with the extraction of a target speaker from multi-talker speech. We would like to take advantage of the spatial diversity present in the speech mixture while facing the problem that the spatial characteristics of the target speaker may change. To be specific, we allow speakers to change their position from one utterance to the next. The proposed system is based on the SpeakerBeam concept developed in [11], which we extend to a block-online source extraction system. We assume that an AU has been recorded for each speaker in advance, when no competing speakers are present. This AU is used to estimate a beamforming vector, which is applied to the AU itself to improve the extraction of the speaker embedding vector, which captures the target speaker's spectral characteristics. It is further used to enhance the distorted input signal of the neural network. Thereby, emphasizing all signal components originating from the position of the target speaker during the AU. To cope with subsequent changing speaker positions, the beamformer coefficients are recursively updated.

Spatial features have proven very effective in enhancing the performance of neural network supported acoustic beamforming [15–17]. It is, however, unclear, to which extent they are also useful if speaker positions change. We therefore test the effectiveness of those features by comparing results for stationary speakers and speaker position changes between utterances. It will be shown that spatial features computed on the speech mixtures remain to be effective.

The paper is structured as follows: In Sect. 2 a short overview over the system is presented, where Sect. 2.1 focuses on the beamforming vector estimation and Sect. 2.2 explains the neural network structure used for mask estimation. In Sect. 3 the systems are evaluated on a database presented in Sect. 3.1. Final conclusions are drawn in Sect. 4.

2   System Overview

We assume a multi-channel signal captured by D microphones. In the short-time Fourier transform (STFT) domain the overlapped speech $\mathbf{Y}$ and the adaptation utterance $\mathbf{A}$ can be expressed as

$$\mathbf{Y}(t,f) = \mathbf{X}_i(t,f) + \sum_{j \neq i} \mathbf{X}_j(t,f) + \mathbf{N}(t,f) \tag{1}$$

$$\mathbf{A}(t,f) = U(t,f) + \mathbf{N}(t,f). \tag{2}$$

Here, $\mathbf{Y}(t,f)$, $\mathbf{N}(t,f)$ and $\mathbf{X}_k(t,f)$ are the STFT coefficient vectors of the speech mixture, of the noise and of the k-th source image at the microphones. $\mathbf{A}(t,f)$ represents the distorted and $U(t,f)$ the clean AU. The time and frequency indices t and f will be dropped wherever possible without sacrificing clarity.

2.1   Beamforming

Speech enhancement is done using the well known Minimum Variance Distortion-less Response (MVDR) beamformer, which minimizes the noise power without introducing distortions on signals originating from a target direction, by optimizing the cost function [18]:

$$\mathbf{F}_{\mathrm{MVDR}} = \underset{\mathbf{F}}{\mathrm{argmin}}\; \mathbf{F}^{\mathsf{H}} \boldsymbol{\Phi}_{\mathbf{NN}} \mathbf{F} \quad \text{s.t.} \quad \mathbf{F}^{\mathsf{H}} \tilde{\mathbf{H}} = 1, \tag{3}$$

where $\tilde{\mathbf{H}} = [1, ..., \tilde{H}_D]^{\mathsf{T}}$ is the target speaker acoustic transfer function (ATF) normalized to a reference microphone, which is called relative transfer function (RTF), and $\boldsymbol{\Phi}_{\mathbf{NN}}$ is the noise spatial correlation matrix (SCM).

We employ the solution of the MVDR cost function in the form presented in [19]:

$$\mathbf{F}_{\mathrm{MVDR}} = \frac{\boldsymbol{\Phi}_{\mathbf{NN}}^{-1} \boldsymbol{\Phi}_{\mathbf{XX}}}{\mathrm{tr}\left\{ \boldsymbol{\Phi}_{\mathbf{NN}}^{-1} \boldsymbol{\Phi}_{\mathbf{XX}} \right\}} \mathbf{u}, \tag{4}$$

where $\mathbf{u}$ is a unit vector pointing to the reference microphone, $\mathrm{tr}\{\cdot\}$ is the trace operator and $\boldsymbol{\Phi}_{\mathbf{XX}}$ is the target speech SCM. Here, the target speech SCM is forced to follow the rank-1 approximation [20] by using:

$$\tilde{\boldsymbol{\Phi}}_{\mathbf{XX}} = \mathbf{a}\mathbf{a}^H \cdot \mathrm{tr}\{\boldsymbol{\Phi}_{\mathbf{XX}}\}/\mathrm{tr}\{\mathbf{a}\mathbf{a}^H\} \tag{5}$$

with $\mathbf{a} = \boldsymbol{\Phi}_{\mathbf{NN}} \mathcal{P}\left\{ \boldsymbol{\Phi}_{\mathbf{NN}}^{-1} \boldsymbol{\Phi}_{\mathbf{XX}} \right\}$ and $\mathcal{P}\{\cdot\}$ as the principal component of the matrix given in parentheses. Both the noise and target speaker SCMs are estimated using speech and noise masks M_ν, where $\nu \in [\mathbf{X}, \mathbf{N}]$. In case of block-wise estimation a recursive update of the SCM is applied [21]:

$$\boldsymbol{\Phi}_{\nu\nu}(nN) = \beta_\nu \boldsymbol{\Phi}_{\nu\nu}((n-1)N) + (1 - \beta_\nu)\hat{\tilde{\boldsymbol{\Phi}}}_{\nu\nu}(nN), \tag{6}$$

with n as the block-index, β_ν as the forgetting factor and

$$\hat{\boldsymbol{\Phi}}_{\nu\nu}(nN) = \frac{1}{\sum_{l=0}^{N-1} M_\nu(nN - l)} \sum_{l=0}^{N-1} M_\nu(nN - l)\mathbf{Y}(nN - l)\mathbf{Y}^H(nN - l). \quad (7)$$

In the offline (batch) case, $\boldsymbol{\Phi}_{\nu\nu}(nN)$ is estimated on the whole utterance, i.e., $\beta_\nu = 0$ and N is set to the number of frames in the utterance.

Equation (6) requires an initialization. The noise SCM is initialized either by assuming white noise and thereby a diagonal matrix or by estimating the SCM of diffuse noise:

$$\boldsymbol{\Phi}_{\mathrm{diff}}(f) = \varphi_{\mathbf{N}} \cdot \mathrm{sinc}\left(2\pi f \cdot \frac{F_{\max}}{F} \cdot \mathbf{d}/c\right), \quad (8)$$

where $\mathbf{d}$ is the matrix of distances between the microphones, c is the velocity of sound, $F_{\max}$ the Nyquist frequency, F the number of frequency bins, and $\varphi_{\mathbf{N}}$ is the noise power.

The target speech SCM may either be initialized using the RTF of the speaker position and the rank-1 approximation $\tilde{\boldsymbol{\Phi}}_{\mathbf{XX}} = \varphi_{\mathbf{X}}\tilde{\mathbf{H}}\tilde{\mathbf{H}}^H$ with $\varphi_{\mathbf{X}}$ as the speech power, or using the SCM of the AU.

For comparison purposes, a second speech enhancement method is employed using non-adaptive beamforming. A set of MVDR beamforming coefficient vectors is precomputed, assuming concentrated sources at fixed, predefined positions and a diffuse noise field, as described in [22]. The predefined positions for the FixedBF are set in a circular form around the array with $10°$ distance, a radius of 1.5 m and 0.4 m height relative to the array, resulting in 36 positions. During the AU phase, an acoustic source localization is performed using the Steered Response Power - Normalized Arithmetic Mean (SRP-NAM) algorithm, as described in [23], and the beamforming vector corresponding to the estimated position is selected for source extraction. This method will be referred to as FixedBF.

2.2   Mask Estimation

In this section we describe the mask estimation required for SCM updates given in Eq. (6). It is a modified version of the SB source extraction network introduced in [11].

The neural network for mask estimation can be split in three parts: a recurrent neural network (RNN) layer, followed by an adaptation layer and a classification layer, consisting of two feed forward layers (FFs). In the adaptation layer one larger feed forward layer is split into several sub-layers. The outputs of these sub-layers are combined prior to the application of the non-linearity σ, using weights α:

$$h_k^{(\ell)} = \sigma\left(\sum_{j=1}^{N^{(\ell-1)}} h_j^{(\ell-1)} \sum_{m=1}^{M} \alpha_m W_{mjk}\right), \qquad k = 1, \ldots, N^{(\ell)} \quad (9)$$

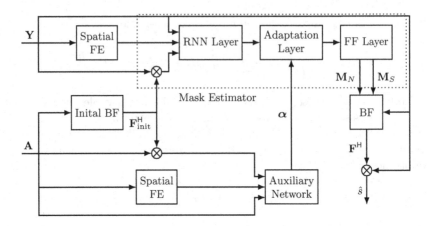

Fig. 1. System overview of the presented spatial speaker extractor.

where $h_j^{(\ell-1)}$ is the output of the j-th node in the preceding, $(\ell-1)$-st, layer, and $h_k^{(\ell)}$ the k-th node output in the ℓ-th layer. $N^{(\ell)}$ is the number of nodes in layer ℓ, W_{mjk} the learn-able weight matrix coefficients, where m indicates the sub-layer, and M the number of adaptation weights. Here, $\boldsymbol{\alpha} = [\alpha_1, ..., \alpha_M]^T$ is provided by an Auxiliary Network (AUX), to which the AU is used as input. This enables the mask estimator (ME) to focus on the speaker which was present during the AU.

The SB approach shows a degradation in performance when applied in a scenario with overlapping speakers with similar spectral characteristics as is observed in speakers of the same gender. To alleviate this problem spatial information is employed, assuming that the target speaker spoke the AU and his contribution to the speech mixture $\mathbf{Y}$ from the same position in the room. First, both the AU and the distorted signal $\mathbf{Y}$ are enhanced using a beamformer estimated from the SCM calculated on the AU as described above. Additionally, spatial features as described in [16] are extracted from both the AU and $\mathbf{Y}$:

$$\mathrm{cosIPD}(t,f,p,q) = \cos\left(\angle y_{t,f,p} - \angle y_{t,f,q}\right), \tag{10}$$
$$\mathrm{sinIPD}(t,f,p,q) = \sin\left(\angle y_{t,f,p} - \angle y_{t,f,q}\right), \tag{11}$$

where p, q are channel indices and $\angle$ is the phase operator. In the case of more than two channels all combinations of channel pairs are employed. However, at the output of the auxiliary network a mean pooling over the channel pairs is carried out to allow a more robust estimation in case of defective channels.

Furthermore, a beamformer is estimated on the AU. This beamformer, called "initial beamformer" in the following, is used to enhance the AU and the mixed speech to compute enhanced features.

To summarize, three sets of features are input to the AUX and mask estimation network: first, log-spectral features computed from the observed microphone signals, second, enhanced log-spectral features obtained after applying the initial beamformer to the microphone signals, and third, the aforementioned spatial features.

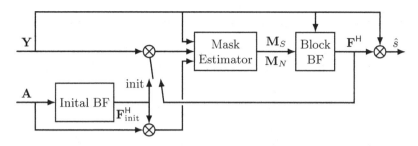

Fig. 2. System overview of the spatial speaker extractor reusing the estimated beamforming vector as initial beamformer for the next block of frames.

A block diagram of the presented system is depicted in Fig. 1.

Both the features computed from the initial beamformer and the spatial features computed on the AU are informative only under the assumption that both the speech of the target speaker in the speech mixture and the AU originate from the same position in the room. Therefore, a system dependent on these features will probably fail in a moving speaker scenario. However, the spatial information computed from the speech mixture can still be beneficial to extract the target speech, in particular if the competing speaker has similar spectral characteristics.

We propose to use a block-online recursive mask estimation system as depicted in Fig. 2. The initial beamformer estimated on the AU is used to enhance the first block of input frames which in turn are used to update the SCMs and estimate a new beamforming vector. This new beamforming vector then replaces the initial beamformer coefficients to compute the above mentioned set of enhanced features on the next block of frames. By this recursive update the enhanced feature set remains able to capture valid information in the presence of speaker movement or changes in the noise statistics.

3   Experiments

The presented systems are compared using four evaluation metrics: signal to distortion ratio (SDR) following the implementation presented in [24], an "invasive" SDR (InvSDR) [25], whereby the speech and the distortion are separately processed by the beamformer, and the SDR is computed as the power ratio of the resulting two outputs, the intelligibility measure STOI [26] and the perceptual speech quality metric PESQ [27]. All systems will be evaluated in terms of their gain compared to the signal at a reference microphone prior to the enhancement. Additionally, the systems are evaluated in terms of Word Error Rate (WER) of a subsequent Automatic Speech Recognition (ASR) system.

All signals are recorded or resampled with 8kHz. For the STFT computation, a 512-point FFT is used with a Hann window and an 75% overlap, resulting in 257 frequency bins for each time frame. The ME consists of an LSTM layer of 1024 units, two feedforward layers with 1024 units each and one output layer.

The first feedforward layer is split into 30 sub-layers for the SB approach. The auxiliary network has two feed-forward layers of 50 units each and an output layer of 30 units, as in [11]. Finally, for the block-online estimation we use a block size of $N = 5$ frames, corresponding to 80 ms.

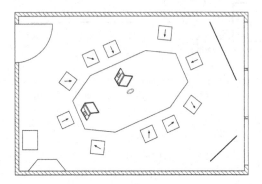

Fig. 3. Sketch of one of the meeting rooms the impulse responses and noises were recorded in. Room size approx. 4m × 6m. Drawn true to scale.

3.1 Database Description

We evaluate the proposed source extraction system on two databases. The first is the one described in [28], which consists of 30000 training, 500 development and 1500 evaluation examples. Each example is created by randomly choosing two utterances from the Wall Street Journal (WSJ) database and convolving the signals with six-channel room impulse responses (RIRs) with reverberation times $T_{60} \in [20\,\text{ms}, 500\,\text{ms}]$ simulated by the Image Methode [29]. The shorter of the generated multi-channel signals is padded with zeros to arbitrarily fall in the duration of the longer signal. The observation utterance then consists of the sum over both utterances, to which white Gaussian noise with an Signal to Noise Ratio (SNR) of 15 to 25 dB is added. The speaker sets of training, development and evaluation sets are mutually exclusive. Therefore, we characterize the database as open. For the AU we convolve a second utterance spoken by the target speaker with the same RIR and add white Gaussian noise. This database will be referred to as RirSim and is used for all parameter tuning and network training.

The second database is created similarly to the one described above, however the RIRs and the noise are replaced by real signals recorded in a conference scenario. The real RIRs and noises were recorded using a flat 8-channel Microelectromechanical systems (MEMS) microphone array, 7 cm × 10 cm in size and of elliptic shape. The recordings took place in two different meeting rooms with reverberation times of $T_{60} \approx 1$s at the premises of voice INTER connect GmbH in Dresden. Figure 3 shows the floor plan of one of these rooms. The microphone array was flush-mounted at the center of the meeting room table in both cases. The table height is 0.73 m. Impulse responses for ten different

Table 1. Gains of the beamformer output compared to the signal at a reference microphone w.r.t. different performance measures, and word error rate for different feature sets of the speaker extraction system on RirSim.

Method	Add. features		ΔSDR	ΔInvSDR	ΔSTOI	ΔPESQ	WER
	Enhanced	Spatial	dB	dB			%
Offline	–	–	6.48	6.49	0.10	0.26	32.66
	–	✓	9.54	9.36	0.14	0.40	29.43
	✓	–	10.16	10.22	0.16	0.46	27.32
	✓	✓	11.09	11.07	0.16	0.51	23.50
Online	✓	✓	7.57	9.00	0.15	0.41	30.61

lateral speaker positions per room were recorded using a coaxial loudspeaker at an assumed human speaker's mouth height of 1.15 m. The speaker positions for the depicted room, together with their directions of view, are shown as squares with arrows in Fig. 3. Four different types of typical meeting room noise sources (air-conditioning, paper shuffling, projector, typing noises) were recorded using the microphone array. The database thus created will be called RirReal.

3.2 ASR Backend

The Automatic Speech Recognition (ASR) backend used the wide residual network structure proposed in [30] with logarithmic mel filterbank input features and two Long-Short-Term-Memory (LSTM) layers. This acoustic model is combined with a trigram language model from the WSJ baseline script provided by the KALDI toolkit [31]. All hyper-parameters were taken from [30]. The same neural acoustic model, trained on the artificially reverberated WSJ utterances of RirSim, is used for both databases. The network is trained on alignments extracted with a HMM model trained in KALDI. The decoding is performed without language model rescoring.

3.3 Source Extraction in Static Speaker Scenario

In Table 1 the performance of different feature sets for the extraction systems described above are compared on the RirSim database. All systems use the log-spectral magnitude of the observation. As additional features we compare the log-spectral magnitude of the observation enhanced using an initial beamforming vector estimated on the AU, spatial features according to Eqs. (10) and (11), or both the spatial features and the enhanced signals. If the method is offline, both the beamforming vector and mask estimation are carried out in batch mode on the whole utterance.

All described features achieve better results than the original SpeakerBeam system, whose performance is given in the first results row of Table 1. Even the online system achieves better results using the additional features compared to the original offline SpeakerBeam system. Therefore, we conclude that using

Table 2. Gains of the beamformer output compared to the signal at a reference microphone w.r.t. different performance measures, and word error rate for non-stationary speaker on RirReal. Here Position (Pos.) 0 symbolizes the first speaker position which is equal to the position during the AU whereas Position 1 indicates a change in the position. "only ME" indicates that the additional spatial features are used as input to the mask estimation network only.

| Method | Add. features | | Pos. | ΔInvSDR | ΔSTOI | ΔPESQ | WER |
	Enhanced	Spatial		dB			%
FixedBF	–	–	0	−1.72	0.03	−0.04	63.27
			1	−6.71	−0.08	−0.05	94.51
Offline	✓	✓	0	2.76	0.05	0.11	36.26
			1	0.12	−0.02	0.03	88.82
Online	✓	✓	0	3.93	0.07	0.13	34.79
			1	1.41	0.01	0.05	63.94
		only ME	0	3.38	0.06	0.13	35.34
			1	1.71	0.02	0.06	62.18
	$\mathbf{F}(\ell - 1)$	only ME	0	3.43	0.05	0.11	35.12
			1	2.29	0.03	0.08	50.44

spatial information is beneficial for our source extraction system in case of static speakers. In [17] we present an in-depth evaluation of the described features in case of static speakers.

3.4 Source Extraction in the Presence of a Speaker Position Change

To simulate a change in speaker position, we divided the WSJ database in pairs of two utterances, where the first is convolved with the same set of RIRs as the AU and second is convolved with a different set of RIRs than the first, while keeping the competing speaker in the speech mixture and his/her position in the room fixed in both utterances.

The change of the target speaker position calls for adaptive beamforming. We thus expect the online beamformer to outperform the offline beamformer.

While the target speaker position in the first of the two utterances coincides with the one present in the AU, this no longer holds for the second. This renders the spatial information gained from the AUX incorrect. Table 2 displays the extraction results achieved with different features for online and offline systems. Note that neither the Acoustic Model (AM) nor the ME is retrained on the new RIR and noise.

Using spatial features during mask estimation but not in the AUX improves the extraction in case of changes in the target speaker position as can be seen in the entry with "only ME" in the column "spatial". Similarly, can be concluded that it is beneficial to update the initial beamforming vector for each block of frames, see the entry with $\mathbf{F}(\ell - 1)$ under the column "enhanced".

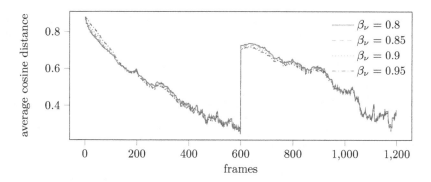

Fig. 4. Cosine distance between the block-online beamforming vector and an oracle offline beamforming vector calculated on the speech and noise image averaged over 500 utterances. Speaker positions changed at frame #600.

Additionally, the results confirm that the extraction achieved by a recursively updated beamforming vector is only slightly impeded by the change in speaker position, whereas a fixed beamformer estimated once for the concatenated utterances suffers significantly from changes in the speaker position. This is especially true for the fixed beamforming vector estimated on the AU since no information about the concurrent speaker is included in the noise SCM estimation.

To emphasize the benefits of recursive beamformer adaptation the cosine distance between the recursively estimated beamforming vector and an oracle offline beamformer is depicted in Fig. 4. Here, the coefficients of the offline beamformer have been obtained separately on the first and second utterance using the oracle speech and noise images at the microphones. The displayed tracking curves are averaged over multiple utterances.

The figure showcases the ability of the online beamforming vector to adapt to a change in speaker position. Furthermore, the recursive update displays an invariance concerning the forgetting factor β_ν

4   Conclusion

This paper offers a thorough investigation of speaker extraction systems guided by an AU in case of changes in the speaker position. We showcased the benefits of recursively updating a beamforming vector and investigated the usefulness of spatial features in case of target speaker position changes. While the spatial characteristics of the target speaker extracted from the adaptation utterance becomes outdated, the use of spatial features for mask estimation to extract a target speaker from a speech mixture remains beneficial. This can be attributed to the fact that they allow to separate speakers based on their spatial diversity, thus not relying solely on different spectro-temporal properties of the speakers.

Acknowledgements. The work was in part supported by DFG under contract number Ha3455/14-1. Computational resources were provided by the Paderborn Center for Parallel Computing.

References

1. Vincent, E., Watanabe, S., Nugraha, A.A., Barker, J., Marxer, R.: An analysis of environment, microphone and data simulation mismatches in robust speech recognition. Comput. Speech Lang. **46**, 535–557 (2017)
2. Sainath, T.N., Weiss, R.J., Wilson, K.W., Narayanan, A., Bacchiani, M.: Factored spatial and spectral multichannel raw waveform CLDNNs. In: IEEE International Conference on Acoustics, Speech, and Signal Processing (ICASSP) (March 2016)
3. Wang, Y., Fan, X., Chen, I.F., Liu, Y., Chen, T., Hoffmeister, B.: End-to-end anchored speech recognition. CoRR abs/1902.02383 (2019)
4. Heymann, J., Drude, L., Chinaev, A., Haeb-Umbach, R.: BLSTM supported GEV beamformer front-end for the 3rd CHiME challenge. In: Proceedings Workshop Automatic Speech Recognition, Understanding, pp. 444–451 (2015)
5. Higuchi, T., Ito, N., Yoshioka, T., Nakatani, T.: Robust MVDR beamforming using time-frequency masks for online/offline ASR in noise. In: 2016 IEEE International Conference on Acoustics, Speech and Signal Processing (ICASSP) (March 2016)
6. Yoshioka, T., Nakatani, T.: Generalization of multi-channel linear prediction methods for blind mimo impulse response shortening. IEEE Trans. Audio, Speech, Lang. Process. **20**(10), 2707–2720 (2012)
7. Yoshioka, T., Erdogan, H., Chen, Z., Xiao, X., Alleva, F.: Recognizing overlapped speech in meetings: a multichannel separation approach using neural networks. In: Interspeech (2018)
8. Luo, Y., Mesgarani, N.: Tasnet: surpassing ideal time-frequency masking for speech separation. CoRR abs/1809.07454 (2018)
9. Yu, D., Kolbaek, M., Tan, Z., Jensen, J.: Permutation invariant training of deep models for speaker-independent multi-talker speech separation. In: 2017 IEEE International Conference on Acoustics, Speech and Signal Processing (ICASSP), pp. 241–245 (March 2017)
10. Chen, Z., Luo, Y., Mesgarani, N.: Deep attractor network for single-microphone speaker separation. In: 2017 IEEE International Conference on Acoustics, Speech and Signal Processing (ICASSP), pp. 246–250 (March 2017)
11. Zmolíková, K., Delcroix, M., Kinoshita, K., Higuchi, T., Ogawa, A., Nakatani, T.: Speaker-aware neural network based beamformer for speaker extraction in speech mixtures. Proc. Interspeech **2017**, 2655–2659 (2017)
12. Kida, Y., Tran, D., Omachi, M., Taniguchi, T., Fujita, Y.: Speaker selective beamformer with keyword mask estimation. In: 2018 IEEE Spoken Language Technology Workshop (SLT), pp. 528–534 (Dec 2018)
13. Wang, Q., et al.: Voicefilter: targeted voice separation by speaker-conditioned spectrogram masking. arXiv e-prints arXiv:1810.04826 (2018)
14. Chen, Z., Xiao, X., Yoshioka, T., Erdogan, H., Li, J., Gong, Y.: Multi-channel overlapped speech recognition with location guided speech extraction network. In: 2018 IEEE Spoken Language Technology Workshop (SLT), pp. 558–565 (Dec 2018)
15. Liu, Y., Ganguly, A., Kamath, K., Kristjansson, T.: Neural network based time-frequency masking and steering vector estimation for two-channel MVDR beamforming. In: 2018 IEEE International Conference on Acoustics, Speech and Signal Processing (ICASSP) (April 2018)

16. Wang, Z., Le Roux, J., Hershey, J.R.: Multi-channel deep clustering: discrimina-tive spectral and spatial embeddings for speaker-independent speech separation. In: IEEE International Conference on Acoustics, Speech and Signal Processing (ICASSP), pp. 1–5 (April 2018)
17. Martín-Doñas, J.M., Heitkaemper, J., Haeb-Umbach, R., Gomez, A.M., Peinad, A.M.: Multi-channel block-online source extraction based on utterance adaptation. In: 20th Annual Conference of the International Speech Communication Associa-tion. Graz, Austria (September 2019)
18. Gannot, S., Vincent, E., Markovich-Golan, S., Ozerov, A.: A consolidated perspec-tive on multi-microphone speech enhancement and source separation. IEEE/ACM Trans. Audio, Speech, Lang. Process. **PP(99)**, 1 (2017)
19. Souden, M., Benesty, J., Affes, S.: On optimal frequency-domain multichannel linear filtering for noise reduction. IEEE Trans. Audio, Speech, Lang. Process. **18**(2), 260–276 (2007)
20. Wang, Z., Vincent, E., Serizel, R., Yan, Y.: Rank-1 constrained multichannel wiener filter for speech recognition in noisy environments. Comput. Speech Lang. **49**, 37–51 (2018)
21. Heitkaemper, J., Heymann, J., Haeb-Umbach, R.: Smoothing along frequency in online neural network supported acoustic beamforming. In: ITG 2018, Oldenburg, Germany (October 2018)
22. Fehér, T., Freitag, M., Gruber, C.: Real-time audio signal enhancement for hands-free speech applications. In: 16th Annual Conference of the International Speech Communication Association, pp. 1246–1250. Dresden, Germany (September 2015)
23. Salvati, D., Drioli, C., Foresti, G.L.: Incoherent frequency fusion for broadband steered response power algorithms in noisy environments. IEEE Signal Process. Lett. **21**(5), 581–585 (2014)
24. Raffel, C., et al.: mir_eval: a transparent implementation of common MIR metrics. In: Proceedings of the 15th International Society for Music Information Retrieval Conference, ISMIR (2014)
25. Tran Vu, D.H., Haeb-Umbach, R.: Blind speech separation employing directional statistics in an expectation maximization framework. In: 2010 IEEE International Conference on Acoustics, Speech and Signal Processing, pp. 241–244 (March 2010)
26. Taal, C.H., Hendriks, R.C., Heusdens, R., Jensen, J.: An algorithm for intelligibility prediction of time-frequency weighted noisy speech. IEEE Trans. Audio, Speech, Lang. Process. **19**(7), 2125–2136 (2011)
27. Rix, A.W., Beerends, J.G., Hollier, M.P., Hekstra, A.P.: Perceptual evaluation of speech quality (pesq)-a new method for speech quality assessment of telephone net-works and codecs. In: 2001 Proceedings IEEE International Conference on Acous-tics, Speech, and Signal Processing. (Cat. No.01CH37221). vol. 2, pp. 749–752 (2001)
28. Drude, L., Haeb-Umbach, R.: Integration of neural networks and probabilistic spa-tial models for acoustic blind source separation. In: IEEE Journal of Selected Topics in Signal Processing (2018)
29. Allen, J.B., Berkley, D.A.: Image method for efficiently simulating small-room acoustics. J. Acoust. Soc. Am. **65**(4), 943–950 (1979)
30. Heymann, J., Drude, L., Haeb-Umbach, R.: Wide residual BLSTM network with discriminative speaker adaptation for robust speech recognition. In: CHiME4 Workshop (2016)
31. Povey, D., et al.: The kaldi speech recognition toolkit. In: IEEE 2011 Workshop on Automatic Speech Recognition and Understanding. No. Idiap-RR-04-2012, IEEE Signal Processing Society, Rue Marconi 19, Martigny (Dec 2011)

Improving Speech Recognition with Drop-in Replacements for f-Bank Features

Sean Robertson[1,2](✉) (iD), Gerald Penn[1] (iD), and Yingxue Wang[1] (iD)

[1] Department of Computer Science, University of Toronto,
40 St. George St., Toronto, ON, Canada
{sdrobert,gpenn,yingxue}@cs.toronto.edu
[2] Vector Institute, 661 University Ave., Toronto, ON, Canada

Abstract. While a number of learned feature representations have been proposed for speech recognition, employing f-bank features often leads to the best results. In this paper, we focus on two alternative methods of improving this existing representation. First, triangular filters can be replaced with Gabor filters, a compactly supported filter that better localizes events in time, or with psychoacoustically-motivated Gammatone filters. Second, by rearranging the order of operations in computing filter bank features, the resulting coefficients will have better time-frequency resolution. By merely swapping f-banks with other types of filters in modern phone recognizers, we achieved significant reductions in error rates across repeated trials.

Keywords: Speech recognition · Phone recognition · Time-domain filter banks · Short-integration

1 Introduction

Time-frequency decomposition of speech has long been the dominant feature representation for speech recognition. It is often achieved through a Short-Time Fourier Transform (STFT). The STFT and its derivatives break up a speech signal into overlapping windows in time. Then, for a given window, a fixed number of coefficients corresponding to frequency are computed. Filtering the power spectrum, such as in the popular *f-bank* representation, can reduce the dimensionality of the resulting representation.

Though a number of different feature representations have been proposed, such as 3-D time-frequency-rotation filter banks [3,18,26], learned filter representations [17,23,25,34], and scattering transforms [2,21], state-of-the-art results still primarily employ f-banks [24,31]. Researchers are justifiably hesitant to change a formula that works so well.

Instead of learning a representation that is potentially drastically different, we propose making alterations to standard f-bank features. The alterations are intended to be fully compatible with existing speech recognizers, insofar as they encode similar time-frequency information as f-bank features, but lead

© Springer Nature Switzerland AG 2019
C. Martín-Vide et al. (Eds.): SLSP 2019, LNAI 11816, pp. 210–222, 2019.
https://doi.org/10.1007/978-3-030-31372-2_18

to improved recognition. First, we swap out triangular filters of f-banks with filters that have more desirable theoretical properties: the Gabor and Gammatone filters. Second, by swapping the order of operations in computation, we can generate a representation that bypasses some of the time-frequency resolution constraints of windowing. These adjustments are no more computationally expensive than f-banks, nor are they necessarily of a different dimension.

We explore the efficacy of these adjustments as drop-in replacements for f-banks in a phone recognition task on TIMIT. Evaluation spans three near state-of-the-art models from the literature: two hybrid recurrent models [23] and one end-to-end neural model [36]. Critically, we do not alter the architectures or any hyperparameters, we only make the above changes. With minimal effort on behalf of the developer – all code is available open source through the authors' repositories[1] or COVAREP [6] – we achieve significant improvements of similar size to novel neural architectures and learned feature representations.

2   Mel-Scaled Log Filter Banks

In the continuous domain, a filter bank coefficient k for a given frame of length T centered at sample c is popularly calculated for signal f as [22,33]:

$$\Psi_{T,f,w,h}[k,c] = \log \int_{-\infty}^{\infty} \widehat{h_k}^2(\omega) \left| \widehat{f w_{c\Delta,T}} \right|^2 (\omega) d\omega, c \in \mathbb{N} \qquad (1)$$

Where $\widehat{\cdot}$ indicates a signal's frequency spectrum, h_k is the k-th real filter in the filter bank $\{h_k\}_k$, $w_{c\Delta,T}$ is a windowing function of temporal support T centred at $c\Delta$, and Δ is the frame shift.

To calculate f-bank coefficients from Eq. (1), $\widehat{h_k^2}$ must be triangular in the Mel-scale power spectrum, and the pointwise square-root of a triangular filter in the Mel-scale magnitude spectrum $\widehat{h_k}$. The Mel scale [30], inspired by psychoacoustic experimentation, is roughly linear with respect to frequencies below 1kHz and logarithmic above. The frequencies captured by a coefficient of the f-bank are limited to the nonzero region of $\widehat{h_k}$. Like the human auditory system, f-banks have difficulty distinguishing between nearby frequencies, a problem which becomes exacerbated the further a filter's centre (apex) frequency is from 0 Hz.

The triangular filters are easy to calculate and have compact support in frequency. Nonetheless, other filters may be better suited to speech recognition. F-bank filters do not exhibit the asymmetry in frequency response (wherein high-frequency filters exhibit large "tails" towards zero) that can be found in human hearing [9]. We also find discontinuities at the triangle vertices, which may generate an unnecessarily severe preference towards certain frequencies.

---

[1] Features: https://github.com/sdrobert/pydrobert-speech   CNN-CTC: https://github.com/sdrobert/more-or-let   RNN-HMM: https://github.com/sdrobert/pytorch-kaldi.

Multiplying the signal with a window in time blurs the result's frequency responses. By the convolution theorem, $\widehat{fw_{c\Delta,T}} = \hat{f} * \widehat{w_{c\Delta,T}}$. For large T, $\hat{w}(\omega) \to \delta(\omega)$, where δ is the Dirac delta function ($\int_{0^-}^{0^+} \delta(\omega)d\omega = 1$ and $\delta(\omega) = 0$ for $\omega \neq 0$). Nonetheless, $\hat{w}$ will have a nonzero support. This behaves similarly to a rolling average over $\hat{f}$ and effectively increases the bandwidth of all $\hat{h}_k^2$. While not particularly detrimental to wideband filters, the bandwidth of the window can be similar to or greater than that of the narrowband filters, which can have drastically less discriminative power at low frequencies. In effect, all filters have a bandwidth above that of the window, limiting their overall resolution in frequency.

For $\hat{h}_k$ that are real-valued and positive, $\hat{h}_k^2|\widehat{fw}|^2 \equiv |\hat{h}_k\widehat{fw}|^2$. By the convolution theorem and Parseval's theorem, Eq. (1) can be rearranged into an integration over a windowed signal that has been filtered:

$$\Psi_{T,f,w,h}[k,c] = C + \log \int_{-\infty}^{\infty} |h_k * (fw_{c\Delta,T})|^2 (t)dt \tag{2}$$

where C is some constant. Equation (2) shows the insensitivity of filter bank coefficients to the duration of h_k, since it always integrates over the entire length of the analysis window. Compact h_k will not localize events in the signal any better than wide h_k. This is convenient for f-bank filters, as they have an extremely wide temporal support. Were we to widen the window so that $\hat{w}(\omega) \to \delta(\omega)$, we would also integrate over a greater duration in time, wiping out additional temporal dynamics.

We have established that the multiplication of the signal with an analysis window limits the resolution of f-bank filters in both time and frequency. In Sect. 5, we show how to modify Eq. (1) so that the time-frequency trade-off is closer to optimal and better reflects the frequency response of filters in the bank. Before that, we discuss two filters compatible with Eq. (1) that have more desirable properties than square-root triangular filters.

3   Gabor Filters

Gabor filters have been explored in a variety of contexts. 2-D Gabor filters are often applied to spectrograms (or log spectrograms) to produce a collection of 3-D spectral-temporal-rotational features [3,18,26]. The filters' 2-D construction allows the bank to capture meaningful geometric structures, such as formant movement. Likewise, 2-D Gabor bases have been learned as convolutional layers [4]. The present paper focuses on the design and evaluation of spectrogram-like features, rather than a set of features derived from a spectrogram. Dimitriadis et al. [8] designed a Mel-scaled Gabor filter bank much like the one presented here, but it was employed in an HMM-based architecture, not a neural network. Recently, Zeghidour et al. [34] trained end-to-end CNNs for phone recognition with weights initialized to a Gabor filter bank.

Gabor filters are simply Gaussian windows with a complex carrier. They are defined in time as

$$h_k(t) = Ce^{-\frac{t^2}{2s_k^2} + i\xi_k t} \tag{3}$$

To design the filter bank, centre frequencies ξ are sampled along the Mel-scale. Neighbouring filters' frequency responses intersect at their -3dB bandwidths.

Gabor filters have a provably optimal time-frequency trade-off [19]. Their regions of effective support in both time and frequency are bounded above by a Gaussian window. Equation (2) suggests that improvements to time resolution will be under-appreciated by standard filter bank representations due to the effects of windowing, but, with the modifications to filter banks described in Sect. 5, the effects of windowing will be minimized.

The Gabor filter has a symmetric frequency response. Log-linear scales of speech perception, such as the Mel scale, are decidedly asymmetric. The Gammatone filter helps mitigate this trade-off.

4   Gammatone Filters

Gammatone filters were derived by Flanagan [9], and formalized by Aertsen et al. [1]. Their skewed frequency responses lend themselves nicely to existing models of speech perception. Gammatone filters have been employed in ASR directly [27,29] and have been used as a starting point in learned feature representations [25,35].

The complex Gammatone filter is defined in time as

$$h_k(t) = Ct^{n-1}e^{-\alpha t + i\xi_k t}u(t) \tag{4}$$

Where n is the *order* of the Gammatone, usually set to 4, which controls the asymmetry of the envelope of h.

The Gammatone does not have an optimal time-frequency trade-off like the Gabor filter. It is still much more compact in duration than square-root triangular filters, but tapers very slowly to zero outside of the filter's bandwidth. The Gammatone replaces compact localization with biological plausibility.

5   Short Integration

As is shown in Sect. 2, windowing widens the bandwidth of the narrowband filters in the f-bank, a form of spectral leakage. Increasing the width of the window will decrease the magnitude of the spectral leakage. However, a wider window will capture more of the signal in time, decreasing its temporal resolution. Deriving inspiration from scattering transforms [2,20], we can modify Eq. (1) to mitigate the effects of windowing.

The following relationship was described by Mallat [20]. Assuming $h_k(t) = \phi_k(t)e^{i\xi_k t} \leftrightarrow \widehat{h_k}(\omega) = \widehat{\phi_k}(\omega - \xi)$, where ϕ_k is a low-pass filter, for an arbitrary signal, f, we find:

$$|(f * h_k)(t)| = \left|e^{i\xi_k t}(f_k * \phi_k)(t)\right| = |(f_k * \phi_k)(t)| \tag{5}$$

where $f_k(t) = f(t)e^{-i\xi_k t}$ shifts the frequency response of f such that $\widehat{f}(\xi_k) = \widehat{f_k}(0)$. Equation (5) shows that, for band-pass h_k, the point-wise modulus on filtered signals produces a low-frequency signal. A subsequent low-pass window can be used to capture much of that energy.

Existing work that uses scattering in speech has focused on wide windows and recursive filter bank computations [2,21]. Wide windows guarantee greater translation invariance in the representation [20], and recursive filtering quickly captures the energy that those wide windows fail to capture. Though later filtering does capture the time dynamics of a signal, it captures interferences which no longer resemble a spectrogram.

As translation invariance is an oxymoronic property of an accurate time-frequency feature representation, we forego wide windows and cascades of filtering operations. The following is the *short integration* method of feature computation:

$$\Psi_{f,w,h,p,T}[k,c] = \log\left(|f * h_k|^p * w_T\right)[c\Delta] \tag{6}$$

Where $p \in \mathbb{N}$, and $\{h_k\}$ is some set of filters. The low-pass quality of the modulus means that Eq. (6) will tend to vary more gradually than a direct convolution $f * h_k$. The primary role of windowing is to avoid any aliasing induced by sub-sampling vis-à-vis the frame shift Δ. For convenience, we choose $T = 2\Delta$, but T can be chosen so that filtering with w ensures the critical sampling rate is $\leq \Delta$.

w still limits the temporal resolution of the representation, but, for short w and wide h_k, the greater impact to resolution will come from the filter. For those low-frequency filters, Eq. (1) will have better temporal resolution than Eq. (6). This trade-off can be optimized by employing filters with compact temporal support, such as the Gabor filter.

Finally, short-integration coefficients have the same computational complexity as STFT-based coefficients. With the overlap-save method of convolution, short integration requires wider FFTs and must involve an inverse FFT, but, unlike STFT-based coefficients, the resulting modulated signal can be used to calculate the coefficients of additional frames.

6   Experiments

In order to explore the efficacy of filter types and methods of computation in speech recognition, we test them as drop-in replacements for f-banks in three modern recognizers and measured their effects on Phone Error Rates (PERs). The first, based on the work of Zhang et al. [36], is an end-to-end fully Convolutional Neural Network (CNN) with a Connectionist Temporal Classification (CTC) [14] loss function, hereinafter referred to as the CNN-CTC model.

The remainder are two hybrid Recurrent Neural Network (RNN) and Hidden Markov Model (HMM) hybrids proposed by Ravanelli et al. [24]. Hybrid models were originally proposed by Deng et al. [7]. The CNN-CTC model was implemented from scratch using Keras [5]. The RNN-HMMs are from an existing repository[2].

The TIMIT phone recognition task allows for fast experimental comparison and reduces the impact of language modelling on experimental results. Much of the data preparation was performed with Kaldi's *timit/s5* recipe [22]. The RNN-HMMs bootstrap the HMM topology of the speaker-dependent HMMs trained in the recipe.

A critical aspect of the experimental design is that no architectural or optimization decisions are based on the feature modifications. Both model types – CNN-CTC and RNN-HMM – were designed to work with f-banks. Debugging the architectures and any additional optimization required was performed solely on f-banks, though experiments had to be re-run when debugging feature implementations.

6.1 Models

The CNN-CTC model consists mostly of convolutional layers with maxout activations [13]. Maxout activations take the per-unit maximum of the output of at least two weight matrices that have received the same input. This (at least) doubles the number of trainable weights in memory. After discussion with Zhang et al. [36], we halved the weights listed in the paper so that the total number of parameters matched 4.3 million. The network has 10 convolutional layers, followed by 3 time-distributed fully-connected layers.

Two types of RNN cell are explored for RNN-HMM acoustic modelling: Long-Short Term Memory cells (LSTMs) [15], a mainstay in RNN architectures; and the modified Rectified Linear Unit-Gated Recurrent Unit, also called liGRU, proposed by Ravanelli et al. [24]. We use the repository's default architecture settings, which are slightly different from those reported in that paper, but purportedly achieve lower PERs. The LSTM model has 4 bidirectional layers for a total of 8 hidden layers, followed by a time-distributed fully-connected layer. With fewer parameters per cell, 5 bidirectional layers are employed in the liGRU model.

6.2 Training and Decoding

The training and decoding processes are similar to that of their source papers.

For the CNN-CTC model, training is broken up into two phases: the former with an aggressive learning rate and no weight regularization; the latter with a small learning rate and weight regularization. Each stage ends when a model's validation loss has not improved over 50 epochs. As large beam widths almost always lead to better error rates, a fixed width of 100 is employed.

---

[2] https://bitbucket.org/mravanelli/pytorch-kaldi-v0.0/src/master/.

The only difference between our experiments and those of the RNN-HMMs' source repository is the decoding process. The authors use a prior version of the Kaldi TIMIT scoring script, which removes silence tokens from both the reference and hypothesis transcripts. Removing silences reduces the PER of tested models by about 1–2% absolute. Second, a Kaldi standard is to tune decoding hyperparameters, specifically the language model rate, according to whichever setting is best on the test set. Instead, we use the best setting from the development set on the test set.

6.3 Features

The models are trained on four feature sets whose time-frequency matrices are of identical shape. *F-bank* is our implementation of the standard log Mel-scaled triangular filter bank. *G-bank* combines the STFT-based coefficients from Eq. (1) with Gabor filters from Eq. (3). *Tone-bank* is likewise for Gammatone filters from Eq. (4). and *sif-bank*, *sig-bank*, and *sitone-bank* are the short-integration analogues (Eq. (6)) of f-bank, g-bank, and tone-bank, respectively.

For the STFT-based computations, 40 log filters plus one energy coefficient are calculated every 10 ms over a frame of 25 ms. The short-integration filters' window size was chosen to be 20 ms. Filters are spaced uniformly on the Mel-scale between 20 Hz and 8000 Hz. For the CNN-CTC model, deltas and double deltas are included, totalling 123 dimensions. Only the original 40 filters plus energy are provided to the RNN-HMMs. Pre-emphasis, dithering, and compression were enabled at their standard Kaldi values. An additional baseline, *kaldifb*, was included to test Kaldi's built-in f-bank implementation as a sanity check. Ravanelli et al. reported much improved PERs by using *fMLLR* features, so we also test them for RNN-HMMs. Here, *fMLLR* is actually a composite of three linear transforms applied to the features: first, linear discriminant analysis on concatenated sequential frames; second, global semi-tied covariance [12]; and third, speaker-adaptive constrained maximum likelihood regression [11]. The latter two are iteratively trained via expectation maximization using tri-state diagonal-covariance GMMs.

6.4 Evaluation

To evaluate the performance of the proposed filters and computations, we perform a hybrid non-parametrical statistical analysis. To limit multiple comparisons, we treat changing the filter type and changing the computation as separate improvements. To compare filter types, we fix computations to the original STFT-based method. To compare computations (STFT versus short integration), we fix the square-root triangular filters. We correct for the 6 comparisons using the Holm-Bonferroni method [16].

10 independent training and testing cycles – trials – are repeated for each type of acoustic model: CNN-CTC, LSTM, and liGRU. Trials differ only in the choice of seed, which controls weight initialization and, in the case of CNN-CTC, utterance ordering. Within each architecture, the seed for a specific trial

number is fixed across feature types. Each trial PER is considered a sample for that filter/computation combination.

We use ranked non-parametrical statistics to evaluate significance. A rejected null hypothesis indicates that one or more variables are ranked consistently higher or lower than the rest, though not to what degree. We performed Wilcoxon tests [32] to compare computations and Friedman tests [10] to compare filters.

7 Results and Discussion

For the CNN-CTC architecture, we found that one seed – the same seed for each combination of filters and computations – failed to converge, with PERs around 70%. Those trials were removed from analysis, leaving 9 trials in the CNN-CTC condition.

Table 1 lists the six aforementioned experimental conditions, ordered by ascending p value. With $\alpha = 0.05$ and Holm-Bonferroni correction, only the highest-ranked comparison, namely comparing filter types in the liGRU STFT condition, is significant.

Since we found a significant difference between filter types in the liGRU condition, we performed pairwise comparisons of filter types in a post-hoc evaluation to tease out which filters differed significantly from the others. Table 2 lists the results of the comparisons. fMLLR-based features and tone-bank features differ significantly in rank from all other feature types. Combining these results from the descriptive statistics listed in Table 3, it is clear that fMLLR

Table 1. Experimental statistics, architectures, and comparison type, ordered by ascending p value. *filt* is a comparison of filter types, *comp* of computations

Arch	Comparison	Statistic
liGRU	filt	$Q = 22.243, p < 0.001$
LSTM	comp	$W = 6.000, p = 0.028$
liGRU	comp	$W = 8.000, p = 0.047$
LSTM	filt	$Q = 9.505, p = 0.050$
CNN-CTC	filt	$Q = 5.933, p = 0.115$
CNN-CTC	comp	$W = 13.000, p = 0.260$

Table 2. Post-hoc comparisons of PER rank. $**\,p < 0.01$; $*p < 0.05$, $\sim p \geq 0.05$

	fMLLR	kaldifb	f-bank	g-bank
kaldifb	**			
f-bank	**	$\sim$		
g-bank	**	$\sim$	$\sim$	
tone-bank	**	**	*	*

features performed much worse than the other features, whereas tone-bank features performed much better.

We may observe that, irrespective of Holm-Bonferroni correction, short integration performs much worse than STFT-based computations when combined with f-bank filters and RNN-HMMs. This result does not appear to apply to all short-integration filter banks: additional post-hoc comparisons reveal the significant differences are between *tone-bank* and *sitone-bank* with tone-bank preferred ($W = 5.000, p = 0.038$) and between *g-bank* and *sig-bank* with sig-bank preferred ($W = 4.000, p = 0.028$). Regardless, short integration does not appear to be a suitable drop-in replacement for STFT-based methods without additional considerations for both filter type and model architecture.

Table 1 tells us that the relative ranks of conditions did not vary as wildly in the CNN-CTC condition versus RNN-HMMs. The considerable variance in results is large enough to discount the differences in PERs in Table 3.

Table 3. Descriptive statistics of PER (%) over experimental conditions

Arch	Cond	N	Mean	Med	Std	Min	Max
CNN-CTC	f-bank	9	**18.60**	18.61	0.22	18.25	**18.96**
	sig-bank	9	18.68	**18.57**	0.36	18.31	19.33
	tone-bank	9	18.71	18.64	0.27	18.28	19.14
	sif-bank	9	18.74	18.75	0.26	**18.21**	19.11
	sitone-bank	9	18.75	18.81	0.30	18.30	19.17
	g-bank	9	18.77	18.82	0.26	18.48	19.17
	kaldifb	9	18.82	18.86	0.14	18.61	19.03
LSTM	g-bank	10	**16.13**	**16.15**	0.29	**15.60**	16.60
	kaldifb	10	16.20	16.20	0.22	15.90	16.60
	tone-bank	10	16.23	16.20	0.31	15.80	16.70
	f-bank	10	16.31	16.30	0.18	16.10	16.70
	sig-bank	10	16.32	16.40	0.26	15.80	16.60
	sitone-bank	10	16.33	16.35	0.16	16.10	**16.50**
	fMLLR	10	16.59	16.65	0.26	16.10	17.00
	sif-bank	10	16.70	16.75	0.34	16.20	17.20
liGRU	tone-bank	10	**15.75**	**15.70**	0.24	**15.50**	16.20
	sig-bank	10	15.89	15.90	0.10	15.70	**16.00**
	sitone-bank	10	16.01	15.95	0.19	15.80	16.40
	kaldifb	10	16.07	16.05	0.16	15.80	16.30
	f-bank	10	16.08	16.10	0.25	15.70	16.40
	g-bank	10	16.09	16.10	0.17	15.70	16.30
	sif-bank	10	16.51	16.50	0.34	16.00	16.90
	fMLLR	10	16.53	16.50	0.21	16.30	16.80

We hypothesize that end-to-end models spend much of their time learning to predict sequences of phones, rather than look for more evidence in the features. If so, end-to-end models will, in general, benefit less from feature engineering.

In contrast, the benefits of swapping features in the RNN-HMM cases are more obvious. Table 3 tells us that, by fixing the f-bank features and swapping from LSTM to liGRU architectures, we achieve an absolute error rate improvement of, on average, 0.23% ($W = 7.5, p = 0.041$). Switching from f-bank to tone-bank features for the liGRU architecture buys an average 0.33% ($W = 6.0, p = 0.028$) improvement. Admittedly, the improvement from switching architectures generalizes better across features than switching between two fixed features across architectures. Since the features are drop-in replacements for f-banks and require no additional computation cost, it is relatively easy to compare multiple features to f-banks and acquire similar significant gains (f-bank vs. g-bank in LSTMs leads to 0.18%, $W = 28.000, p = 0.007$). Hence, the feature engineering that we have undertaken in fact leads to significant gains of the same order as improved neural architectures without much effort.

Finally, we turn to how these results compare with recent literature on learned filter representations. By swapping f-bank features with two trainable layers that have been initialized to resemble a sig-bank, Zeghidour et al. [34] improved their PER from 18.1% to 18.0%. Schneider et al. [28] reported a 2.9% PER improvement over a baseline of 17.6% (silence phones removed), but required pre-training an addotopma; neural network on corpora which are tens and hundreds of times the size of TIMIT. Ravanelli and Bengio [23] reported an impressive average 1.1% PER improvement over an 18.1% baseline by learning the bandwidths of filters, though learning directly from the raw signal led to only marginal improvements over f-banks (0.1%). When applied to other corpora [17,25,35], we find that single-layer learned representations offer improvements to error rates which are similar to those we observed when swapping fixed filters. In order to obtain those improvements, moreover, learned representations often must initialize their weight matrices to resemble a fixed filter bank. We must therefore conclude that the low-hanging fruit borne merely of swapping filters using a better understanding of signal processing, while perhaps not as flamboyant as its savoury "deep-learning" cousins, has not yet entirely fallen. This remains an effective, relevant, and low-cost method of improving error rates.

8   Conclusion

In this work, we have presented alterations to the standard time-frequency representation for speech-recognition: the f-bank. First, we replace the square-root triangular filters of f-banks with Gabor filters or Gammatone filters. Second, we modify the traditional, STFT-based filter coefficients so that time and frequency information may better reflect the duration and bandwidths of filters in the bank. Merely swapping filter types lead to significant improvements in error rates. With minimal effort on the part of the developer, we achieve improvements of a similar order of magnitude to improving the model architecture or learning a new feature representation. We found no single best filter for all recognizers; it is critical to explore different features across architectures and data. To this end, all of our code is available open source and online.

The second improvement, short integration, did not appear beneficial as a drop-in replacement for STFT-based features. It should be emphasized, however, that this experiment focused on replicating existing recognition setups, including model architectures and hyperparameters like filter placement, frame shift, and learning rate. Tuning hyperparameters, permuting existing architectures, and developing new architectures that can exploit the high-resolution information that short integration provides remains an important topic for further investigation.

Acknowledgements. This research was funded by a Canada Graduate Scholarship and a Strategic Project Grant from the Natural Sciences and Engineering Research Council of Canada.

References

1. Aertsen, A.M.H.J., Olders, J.H.J., Johannesma, P.I.M.: Spectro-temporal receptive fields of auditory neurons in the grassfrog. Biol. Cybern. **39**(3), 195–209 (1981)
2. Andén, J., Mallat, S.: Deep scattering spectrum. IEEE Trans. Signal Process. **62**(16), 4114–4128 (2014)
3. Chang, S.-Y., Meyer, B.T., Morgan, N.: Spectro-temporal features for noise-robust speech recognition using power-law nonlinearity and power-bias subtraction. In: IEEE International Conference on Acoustics, Speech and Signal Processing. pp. 7063–7067 (2013)
4. Chang, S.Y., Morgan, N.: Robust CNN-based speech recognition with Gabor filter kernels. In: Proceedings Interspeech (2014)
5. Chollet, F., et al.: Keras. https://keras.io (2015)
6. Degottex, G., Kane, J., Drugman, T., Raitio, T., Scherer, S.: COVAREP — A collaborative voice analysis repository for speech technologies. In: IEEE International Conference on Acoustics, Speech and Signal Processing. pp. 960–964 (2014)
7. Deng, L., Acero, A., Dahl, G., Yu, D.: Context-dependent pre-trained deep neural networks for large vocabulary speech recognition. IEEE Trans. Audio, Speech, Lang. Process. **20**, 30–42 (2012)
8. Dimitriadis, D., Maragos, P., Potamianos, A.: On the effects of filterbank design and energy computation on robust speech recognition. IEEE Trans. Audio, Speech, Lang. Process. **19**(6), 1504–1516 (2011)
9. Flanagan, J.L.: Models for approximating basilar membrane displacement. Bell Syst. Tech. J. **39**(5), 1163–1191 (1960)
10. Friedman, M.: The use of ranks to avoid the assumption of normality implicit in the analysis of variance. J. Am. Stat. Assoc. **32**(200), 675–701 (1937)
11. Gales, M.J.F.: Maximum likelihood linear transformations for HMM-based speech recognition. Comput. Speech Lang. **12**(2), 75–98 (1998)
12. Gales, M.J.F.: Semi-tied covariance matrices for hidden Markov models. IEEE Trans. Speech Audio Process. **7**(3), 272–281 (1999)
13. Goodfellow, I.J., Warde-Farley, D., Mirza, M., Courville, A., Bengio, Y.: Maxout networks. In: Proceedings of the 30th International Conference on International Conference on Machine Learning. ICML 2013, vol. 28, pp. III-1319-III-1327 (2013)

14. Graves, A., Fernández, S., Gomez, F., Schmidhuber, J.: Connectionist temporal classification: labelling unsegmented sequence data with recurrent neural networks. In: Proceedings of the 23rd International Conference on Machine Learning. pp. 369–376. ICML 2006, ACM, New York, NY, USA (2006)
15. Hochreiter, S., Schmidhuber, J.: Long short-term memory. Neural Comput. 9(8), 1735–1780 (1997)
16. Holm, S.: A simple sequentially rejective multiple test procedure. Scand. J. Stat. 6(2), 65–70 (1979)
17. Hoshen, Weiss, R.J., Wilson, K.W.: Speech acoustic modeling from raw multichannel waveforms. In: IEEE International Conference on Acoustics, Speech and Signal Processing. pp. 4624–4628 (2015)
18. Kovács, G., Tóth, L., Gosztolya, G.: Multi-band processing with gabor filters and time delay neural nets for noise robust speech recognition. In: IEEE Spoken Language Technology Workshop. pp. 242–249 (2018)
19. Mallat, S.: A Wavelet Tour of Signal Processing: The Sparse Way. Elsevier Science, Amsterdam (2008)
20. Mallat, S.: Group invariant scattering. Commun. Pure Appl. Math. 65(10), 1331–1398 (2012)
21. Peddinti, V., Sainath, T., Maymon, S., Ramabhadran, B., Nahamoo, D., Goel, V.: Deep scattering spectrum with deep neural networks. In: IEEE International Conference on Acoustics, Speech and Signal Processing. pp. 210–214 (May 2014)
22. Povey, D., et al.: The Kaldi speech recognition toolkit. In: IEEE Workshop on Automatic Speech Recognition and Understanding. IEEE Signal Processing Society, Hilton Waikoloa Village, Big Island, Hawaii, US (Dec 2011)
23. Ravanelli, M., Bengio, Y.: Speech and Speaker Recognition from Raw Waveform with SincNet. CoRR abs/1812.05920 (2018)
24. Ravanelli, M., Brakel, P., Omologo, M., Bengio, Y.: Improving speech recognition by revising gated recurrent units. In: Proceedings Interspeech. pp. 1308–1312 (2017)
25. Sainath, T.N., Weiss, R.J., Senior, A., Wilson, K.W., Vinyals, O.: Learning the speech front-end with raw waveform CLDNNs. In: Proceedings Interspeech (2015)
26. Schädler, M.R., Kollmeier, B.: Separable spectro-temporal Gabor filter bank features: reducing the complexity of robust features for automatic speech recognition. J. Acoust. Soc. Am. 137(4), 2047–2059 (2015)
27. Schluter, R., Bezrukov, I., Wagner, H., Ney, H.: Gammatone features and feature combination for large vocabulary speech recognition. In: IEEE International Conference on Acoustics, Speech and Signal Processing. vol. 4, pp. IV-649 (Apr 2007)
28. Schneider, S., Baevski, A., Collobert, R., Auli, M.: Wav2vec: Unsupervised Pretraining for Speech Recognition. CoRR abs/1904.05862 (2019)
29. Shao, Y., Jin, Z., Wang, D., Srinivasan, S.: An auditory-based feature for robust speech recognition. In: IEEE International Conference on Acoustics, Speech and Signal Processing. pp. 4625–4628 (Apr 2009)
30. Stevens, S.S., Volkmann, J., Newman, E.B.: A scale for the measurement of the psychological magnitude pitch. J. Acoust. Soc. Am. 8(3), 185–190 (1937)
31. Tóth, L.: Combining time- and frequency-domain convolution in convolutional neural network-based phone recognition. In: IEEE International Conference on Acoustics, Speech and Signal Processing. pp. 190–194 (May 2014)
32. Wilcoxon, F.: Individual comparisons by ranking methods. Biometrics Bullet. 1(6), 80–83 (1945)
33. Young, S., et al.: The HTK book (for HTK version 3.4). Cambridge University Engineering Department 2(2), 2–3 (2006)

34. Zeghidour, N., Usunier, N., Kokkinos, I., Schaiz, T., Synnaeve, G., Dupoux, E.: Learning filterbanks from raw speech for phone recognition. In: IEEE International Conference on Acoustics, Speech and Signal Processing. pp. 5509–5513 (Apr 2018)

35. Zeghidour, N., Usunier, N., Synnaeve, G., Collobert, R., Dupoux, E.: End-to-end speech recognition from the raw waveform. In: Proceedings Interspeech. pp. 781–785 (2018)

36. Zhang, Y., et al.: Towards end-to-end speech recognition with deep convolutional neural networks. In: Proceedings Interspeech. pp. 410–414 (2016)

Investigation on N-Gram Approximated RNNLMs for Recognition of Morphologically Rich Speech

Balázs Tarján[1,2]([⊠]), György Szaszák[1], Tibor Fegyó[1,2], and Péter Mihajlik[1,3]

[1] Department of Telecommunications and Media Informatics,
Budapest University of Technology and Economics, Budapest, Hungary
{tarjanb,szaszak}@tmit.bme.hu
[2] SpeechTex Ltd., Budapest, Hungary
fegyo@speechtex.com
[3] THINKTech Research Center, Budapest, Hungary
mihajlik@thinktech.hu

Abstract. Recognition of Hungarian conversational telephone speech is challenging due to the informal style and morphological richness of the language. Recurrent Neural Network Language Model (RNNLM) can provide remedy for the high perplexity of the task; however, two-pass decoding introduces a considerable processing delay. In order to eliminate this delay we investigate approaches aiming at the complexity reduction of RNNLM, while preserving its accuracy. We compare the performance of conventional back-off n-gram language models (BNLM), BNLM approximation of RNNLMs (RNN-BNLM) and RNN n-grams in terms of perplexity and word error rate (WER). Morphological richness is often addressed by using statistically derived subwords - morphs - in the language models, hence our investigations are extended to morph-based models, as well. We found that using RNN-BNLMs 40% of the RNNLM perplexity reduction can be recovered, which is roughly equal to the performance of a RNN 4-gram model. Combining morph-based modeling and approximation of RNNLM, we were able to achieve 8% relative WER reduction and preserve real-time operation of our conversational telephone speech recognition system.

Keywords: Speech recognition · Neural language model · RNNLM · LSTM · Conversational telephone speech · Morphologically rich language

1 Introduction

Recognition of conversational telephone speech poses great challenge due to the low acoustic quality (limited bandwidth, speaker noises, lossy compression etc.) on the one hand and high perplexity of spontaneous speaking style on the other hand. The less constrained grammar and word order of informal speech make the language model estimation less accurate due to the increased variability of both the individual words and their possible sequential combinations. Data sparsity

© Springer Nature Switzerland AG 2019
C. Martín-Vide et al. (Eds.): SLSP 2019, LNAI 11816, pp. 223–234, 2019.
https://doi.org/10.1007/978-3-030-31372-2_19

issues caused by morphological richness of the language or lack of sufficient training data make the problem even harder.

In the last few years neural networks have been successfully applied in the field of language modeling [2,11]. Recurrent networks have proved particularly efficient for the task [11] especially if they exploit Long Short-Term Memory (LSTM) units [8,17]. However, the RNNLMs have a vast amount of inner states that makes their usage in the first-pass of an Automatic Speech Recognition (ASR) system computationally infeasible. RNNLMs hence are usually utilized in a second decoding pass for rescoring the hypotheses obtained with a less heavy LM. The two-pass decoding, however, introduces a considerable processing delay [7,17].

Various techniques have been proposed to address direct applicability of RNNLMs in the single-pass decoding scheme. A possible solution is to approximate the probability distributions of RNNLMs with conventional back-off n-gram language models [1,2,6]. Although the converted model (RNN-BNLM) loses its ability to model long contexts and distributed input features, it can be directly applied for first-pass decoding that makes these techniques attractive. Recently another approach called RNN n-gram has also been introduced [3]. RNN n-gram language models are special recurrent networks trained on n-grams sampled from the training data. As a consequence, the size of the modeled context here is also limited, but RNN n-gram models are able to learn word embeddings just like standard RNNLMs.

Our ambition in this paper is to compare conventional BNLMs, RNNLMs and n-gram approximated RNNLMs in a morphologically very rich language, Hungarian. The rich morphology of Hungarian allows for a weakly constrained word order, and per se, results in an extreme large vocabulary. We, moreover, go for spontaneous speech. All of these three effects – varying word order, large vocabulary and spontaneity – hamper statistic models' ability to yield consistent estimates by high confidence. Since data sparsity issues can be often handled by estimating language models on statically derived subword units (morphs) [5,9, 10] in morphologically rich languages, we extended our investigation to morph-based language models, as well.

Besides the related work already cited, another paper, written by Tüske et al. [20] is also closely related to our work. In this comprehensive study a RNNLM, RNN n-gram models and BNLMs are compared on various English and German ASR tasks. RNN n-gram models were found to be superior to BNLMs both in terms of word perplexity and WER, whereas high order RNN n-grams were close to the performance of an unrestricted RNNLM. However, in [20] ASR results were obtained with two-pass decoding, and German (and obviously English) morphology is less complex than Hungarian.

Although subword language modeling has been used in morphologically rich Finnish ASR systems for more than a decade now [5,9], it was not found beneficial for spontaneous conversational speech until recently. In [7], subword RNNLMs were trained on Finnish and Estonian conversations and used for rescoring lattices generated with conventional back-off models. Subword language models have already been applied successfully for recognition of Hungarian conversational speech [10, 19], but neural language models have not been used before to the best of our knowledge. We found only one mention of application of morph-based approximated RNNLMs in the first pass of an ASR system [13]. This is, however, a preprint paper from which the most relevant subword results are missing.

Overall, we consider the main contributions of our work are (1) presenting the first ASR results with using morph-based RNN-BNLMs in single-pass decoding; (2) comparing the performance of BNLMs and n-gram approximations of RNNLM (RNN n-gram models, RNN-BNLMs); (3) carrying out for the first time an evaluation of neural language models on very rich morphology Hungarian for speech recognition tasks on spontaneous speech; and (4) doing this preserving real-time operation capabilities by low delay.

In next section the experimental database is introduced along with the applied preprocessing techniques. In Sect. 3 we describe the techniques we used for training our different types of language models. Next Sect. 4 presents the experimental results, while in the conclusions we highlight the most impactful outcome of our work.

2 Database

2.1 Training Data

Original Data. Our experiments were performed on anonymised manual transcripts of telephone customer service calls which were collected from the Hungarian Call Center Speech Database (HCCSD). We selected 290 h of recordings from HCCSD for training purposes. The corresponding transcripts that were used for building the language models consisted of 3.4 million word tokens and contained 100,000 unique word forms. In order to accelerate the training of recurrent networks only the most frequent 50,000 word forms were retained in the final vocabulary. Out-Of-Vocabulary (OOV) words and sentence endings were replaced with ⟨unk⟩ and ⟨eos⟩ symbols respectively.

Morph Segmented Data. Language modeling of morphologically rich languages poses a great challenge, since the large number of word forms cause data sparseness and high OOV rate. A common remedy is to segment words into smaller parts and train language models on these subword sequences [9, 10]. One of the most popular statistical word segmentation algorithm is Morfessor [5], which was specifically designed for processing morphologically rich languages. We applied the Python implementation of the original algorithm called Morfessor 2.0 [21]. Hyperparameters of the segmentation were optimized on the validation test set (see Sect. 2.2).

Morph segmentation increased the number of tokens in the training text with around 12% (from 3.4 million to 3.8 million). However, number of types decreased to around 32,000 from 100,000. In order to provide sufficient amount of training samples to ⟨unk⟩, the vocabulary size of morph-based models was limited in 30,000 morphs. Sentence endings were replaced with ⟨eos⟩ just like in the case of word-based text data. Non-initial morphs of every word were tagged to provide information to the ASR decoder for the reconstruction of word boundaries (see left-marked style in [14]).

2.2 Test Data

Almost 20 h of conversations were selected from HCCSD for testing purposes. The test dataset was split into two disjoint parts (see Table 1). The validation set (~7.5 h) and the corresponding text transcripts were used for optimization of the hyperparameters (e.g. learning rate control, early stopping), whereas evaluation set (~12 h) was used to test the models and report experimental results. Morph-based segmentation of evaluation dataset was performed with Morfessor 2.0 toolkit using the segmentation model we optimized on the validation set.

Table 1. Test database statistics

	Validation	Evaluation
Duration [h:m]	7:31	12:12
# of word tokens	45773	66312
# of morph tokens	57849	84385
word OOV rate [%]	2.7	2.5
morph OOV rate [%]	0.07	0.08

3 Language Modeling

3.1 Back-Off N-Gram Models

The conventional, count-based, back-off language models (BNLMs) were trained using the SRI language modeling toolkit [15]. In order to maximize their performance, the baseline BNLMs applied neither count-based n-gram cut-offs nor entropy-based pruning [16]. All BNLMs were estimated on cross-sentence n-grams and smoothed with Chen and Goodman's modified Kneser-Ney discounting [4].

3.2 Recurrent Language Model

The 2-layered LSTM RNNLM structure we used in our experiments is illustrated in Fig. 1. This type of network has already been successfully applied for other language modeling tasks [3,22]. Our implementation[1] is based on the TensorFlow sample code of the Penn Tree Bank language model presented in [22].

---

[1] https://github.com/btarjan/stateful-LSTM-LM

The hyperparameters of the neural network were optimized on the validation set. One batch consists of 32 sequences containing 35 tokens each (words or morphs). LSTM states are preserved between the batches, so *stateful* recurrent networks are trained according to TensorFlow terminology. The 650 dimension word/morph embedding vectors are trained on the input data, since we did not find any benefit of Hungarian pretrained embeddings. In order to match the dimensionality of embeddings the output dimension of LSTM neurons is also set to 650. After testing several optimization algorithms, we decided on the momentum accelerated, Stochastic Gradient Descent (SGD). The initial learning rate was set to 1, which is halved after every epoch where the cross entropy loss increases. For regularization purposes, dropout layers with keep probability of 0.5 and early stopping with patience of 3 epochs are applied.

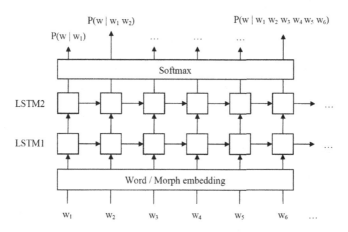

Fig. 1. The recurrent LSTM language model structure used in our experiments

3.3   RNN N-Grams

Although RNNLM can model word sequences with outstanding accuracy [11,17], the need for large context prevents its practical use in many cases. The modeled context can be reduced if we organize training data into n-grams [3]. It was shown that this limitation of history length does not necessarily have a drastic impact on perplexity [20].

Two examples for the many-to-one structure of our RNN n-gram implementations are illustrated in Fig. 2. The hyperparameters and optimization used in RNN n-gram training were the same as those we applied for the RNNLM – except for the sequence length and batch size. Sequence length of RNN n-grams depends on the actual n-gram order (n-1), whereas – thanks to the shorter sequences – RNN n-grams can use a larger batch size (512). An additional important difference compared to RNNLM is that RNN n-gram models do not apply dropout between the two LSTM layers.

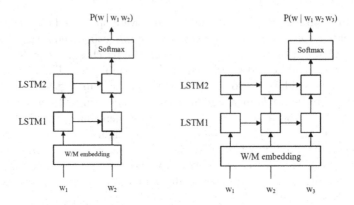

Fig. 2. Two examples for the applied RNN n-gram structures (3-gram and 4-gram)

3.4 Approximation of RNNLM with BNLM

There are various approaches for the approximation of an RNNLM with a back-off ngram language model [1,6]. In [1] three methods are compared and a text generation based approximation is suggested. The main idea of this approach is that the BNLM is estimated from a large text which was generated with the RNNLM. For training the RNN-BNLM models we generated 100 million words/morphs with the corresponding RNNLM (RNN-BNLM 100M). In order to assess the importance of corpus size, we generated a text with 1 billion morphs (RNN-BNLM 1B), as well. The generation of 1 billion morphs took around one week with four NVIDIA GTX 1080 Ti GPUs. Note, that perplexity results in Sect. 4.1 were measured with unpruned RNN-BNLM models, whereas RNN-BNLMs used in ASR decoding are pruned to limit runtime memory usage.

4   Experimental Results

In the first part of this section, we present perplexities measured on the evaluation text set of our conversational speech database. We compare the performance of the language modeling techniques that were described in Sect. 3. Our aim is to measure the perplexity reduction that can be achieved with RNNLM compared to BNLMs and how much of this reduction can be preserved with the n-gram approximated models. In the second part, we utilize these language models in an ASR system to show whether the application of subwords and approximated RNNLMs can turn to reduction in WER.

4.1   Perplexity Results

All perplexity results were measured with cross-sentence language models as it was discussed in Sect. 3. Note that BNLMs and RNN-BNLMs were estimated only up to 6-grams as larger model order did not result significant reduction in perplexity.

Word-Based Models. Perplexity results of word-based models are shown in Table 2. What can be clearly seen at the first glance is the superiority of RNN-based language models over BNLMs. Perplexity of RNN n-gram models improves step by step as we increase the modeling context, while results of BNLMs saturates at around 5-gram. This can be explained by the fact that recurrent models can provide a more accurate probability estimate for unobserved n-grams with the help of distributed modeling of input tokens.

Table 2. Perplexities of word-based backoff n-gram, RNN n-gram and backoff approximated RNN language models as function of n-gram order

Order	BNLM	RNN-BNLM 100M	BNLM + RNN-BNLM 100M	RNN n-gram
2	130.8	136.3	125.0	124.4
3	92.4	94.5	82.5	77.8
4	85.7	86.8	74.4	64.2
5	84.4	85.5	72.8	58.3
6	84.1	85.4	72.5	54.9
8				52.4
10				49.5
14				47.1
18				46.4
inf				44.6

In the last row of Table 2, where the order of context is indicated with infinite (inf.), we can find the perplexity of the LSTM RNN language model (see Sect. 3.2). This implies that this model takes (theoretically) all previous words into account to estimate probability. RNNLM can halve the perplexity of conventional BNLM, however as RNN n-gram results suggest it is only partly due to the modeling of large context, but also due to the previously mentioned generalization abilities of RNNs [20].

The perplexity of the BNLM approximation of the RNNLM (RNN-BNLM 100M) is slightly worse, but very close to the perplexity of the original BNLM. The interpolated model (BNLM + RNN-BNLM 100M), however, improves perplexity with around 10–15% which suggests that there are different n-gram probability distributions behind the similar perplexities. If we would like to capture the effectiveness of RNNLM approximation, we could say that the performance of a pure BNLM model is bit worse than a RNN 3-gram, while the interpolated language model is slightly better than the RNN 3-gram.

We can get an even better insight to the benefit of n-gram approximated RNNLMs, if we estimate the perplexity reduction associated with each approach. Assuming that we utilize 4-gram language models which usually represent a good trade-off between precision and memory consumption, the total amount of perplexity improvement between the baseline 4-gram BNLM (85.7) and the

LSTM RNNLM (44.6) is 41.1. After the interpolation of the BNLM and the RNN-BNLM models perplexity decreases with 11.3. This means that around 27% of potential perplexity reduction can be recovered during the conversion of RNNLM to BNLM. If we were able to utilize RNN 4-gram (64.2) in the downstream task, this recovery rate could go up to around 52%.

Morph-Based Models. Just like in the case of word-based models, morph-based RNN language models significantly outperform BNLMs for every context size (see Table 3) as BNLM perplexities saturate at around 5 or 6-grams. Although morph-based perplexities are lower than word-based ones, note that the two perplexities can not be directly compared, since the vocabulary size of the two model types differs (50k vs. 30k).

Table 3. Perplexities of morph-based backoff n-gram, RNN n-gram and backoff approximated RNN language models as function of n-gram order

Order	BNLM	RNN-BNLM 100M	BNLM + RNN-BNLM 100M	RNN-BNLM 1B	BNLM + RNN-BNLM 1B	RNN n-gram
2	120.7	127.6	114.9	122.6	113.4	112.2
3	83.0	87.7	74.1	80.2	71.1	69.7
4	76.2	80.9	66.6	71.8	62.7	57.5
5	74.7	79.6	64.9	70.1	60.8	52.1
6	74.4	79.4	64.5	69.8	60.3	48.7
8						45.7
10						43.4
14						43.2
18						40.7
inf						40.2

The morph-based results related to RNN-BNLM 100M model are also very similar to the word-based ones. The approximated model itself is a bit worse than the original BNLM; however, the interpolated model reduces perplexity with around 10–15%. The question naturally arises: what if a much larger corpus is generated with the morph-based RNNLM. In order to answer this question we generated a ten times bigger corpus containing 1 billion morphs. As it can be seen in Table 3 RNN-BNLM 1B significantly outperforms not just the RNN-BNLM 100M model but also the original BNLM. Moreover, the interpolated model (BNLM + RNN-BNLM 1B) further decreases perplexity, which suggests that in the future it may be useful to generate even larger corpora.

We calculated the perplexity recovery rate for the morph-based language models, as well. Interpolation of the morph-based BNLM and RNN-BNLM 100M models recover almost the same proportion of the potential perplexity reduction as the word-based models (~29%). The 1 billion-morph-corpus, however, increases this rate to 40%, which means almost half of the RNNLM-based perplexity improvement can be utilized in the ASR system. This way, morph-based BNLM approximations of RNNLMs got much closer to RNN 4-grams than in the case of word-based models.

4.2   Speech Recognition Experiments

Perplexity is a useful measure to compare language models with a shared vocabulary. However, to assess the impact of different language modeling approaches on the ASR task, the best is to directly compare the automatic transcripts.

Experimental Setup. Classical hybrid Hidden Markov-Model (HMM) approach with Deep (feed-forward) Neural Network (DNN) probability distributions were used with three hidden layers consisting of 2500 neurons and output layer with 4907 neurons (senones). The acoustic model was trained on the 290 h of the HCCSD 8 kHz sampled training data using the KALDI toolkit [12]. As for acoustic features 13 dimensional MFCC (Mel-Frequency Cepstral Coefficients) were applied followed by LDA+MLLT [12]. Shared-state context-dependent phone models were used, three states per phones. Acoustic and language model resources were compiled into weighted finite-state transducers and decoded with VoXerver [18] ASR decoder.

Speech Recognition Results. We performed single-pass decoding with the BNLM and RNN-BNLM models and calculated WER of each output (see Table 4). In order to ensure the fair comparison among the modeling approaches, we pruned each RNN-BNLM so that they had similar runtime memory footprint as the baseline BNLM models (~1GB). The interpolated language models (BNLM + RNN-BNLM) are also evaluated in a setup, where larger memory consumption is allowed.

ASR results of word-based language models show similar trends as perplexity results. The BNLM approximation of RNNLM (RNN-BNLM 100M) has a slightly higher WER than the baseline BNLM; however, the interpolated model (BNLM + RNN-BNLM 100M) outperforms both. The relative WER improvement of interpolated model compared to baseline BNLM is only around 2%. Memory limit does not seem to have significant impact on the results.

Replacing words with subwords in the baseline BNLM yields 2% relative WER reduction, which is in accordance with our former findings [19]. The morph-based BNLM trained on the 100-million-morph corpus (RNN-BNLM 100M) has larger WER than the original BNLM, just like in the case of word-based models. Using a ten times larger corpus to train the approximated model, however, seems to change the trend. Morph-based RNN-BNLM 1B model is the first approximated RNN model that outperforms a baseline BNLM by itself without interpolation. This observation underlines the importance of the size of the generated text. The difference between 100M and 1B models are also reflected in their interpolated counterparts. BNLM + RNN-BNLM 1B model can reduce

Table 4. Word Error Rate of the ASR system using the proposed language models

Token type	Model	# of n-grams [million]	Memory usage [GB]	WER [%]
Word	BNLM	5.0	1.3	29.2
	RNN-BNLM 100M	4.8	0.9	30.2
	BNLM+RNN-BNLM 100M	7.0	1.5	28.5
		29.7	6.1	28.4
Morph	BNLM	5.1	1.0	28.7
	RNN-BNLM 100M	8.5	1.1	28.9
	RNN-BNLM 1B	7.2	0.9	28.6
	BNLM+RNN-BNLM 100M	7.9	1.1	27.7
		31.8	4.2	27.5
	BNLM+RNN-BNLM 1B	7.2	1.1	27.3
		46.6	5.9	27.0

WER of morph-based BNLM by 5% or even 6% if runtime memory consumption is not a restricting factor.

All in all, the performance of morph-based BNLM approximations of RNN language models have exceeded our expectations. We managed to reduce the word error rate of our speech transcription system by 8% relative by preserving real-time operation.

5   Conclusions

In this paper our aim was to improve our Hungarian conversational telephone speech recognition system by handling morphological richness of the language and transferring information from a recurrent neural language model to the back-off n-gram model used in the single-pass decoding. We compared various types of word-based and subword-based n-gram approximated RNNLMs and found that by generating a text with 1 billion morphs around 40% of the perplexity improvement associated with the RNNLM can be transferred to the BNLM model. With the combination of subword modeling and RNNLM approximation, we were able to achieve 8% relative WER reduction and preserve real-time operation of our conversational telephone speech recognition system. The perplexity we achieved with BNLM approximation of RNNLMs is roughly equal to the performance of an RNN 4-gram. The fact that RNN-BNLM was able to keep up with RNN n-gram until 4-gram is a quite promising result, but it also suggests that there is room for further improvement in utilizing higher order RNN n-grams in ASR decoding.

We consider the main contributions of our work are (1) presenting the first ASR results with using morph-based RNN-BNLMs in single-pass decoding; (2)

comparing the performance of BNLMs and n-gram approximations of RNNLM (RNN n-gram models, RNN-BNLMs); (3) carrying out for the first time an evaluation of neural language models on very rich morphology Hungarian for speech recognition tasks on spontaneous speech; and (4) doing this preserving real-time operation capabilities by low delay.

In the future, we plan to place more emphasis on the study of OOV words. We would like to measure the recognition rate of OOV words and compare it among the word and morph-based language modeling approaches proposed in this paper. Moreover, we would like to evaluate models that extract features with character-based convolutional neural networks. Extending our work to other ASR tasks and share knowledge among them utilizing transfer learning methods is also a very promising direction of further research.

Acknowledgments. The research was partly supported by the DANSPLAT (EUREKA_15_1_2016-0019) project.

References

1. Adel, H., Kirchhoff, K., Vu, N.T., Telaar, D., Schultz, T.: Comparing approaches to convert recurrent neural networks into backoff language models for efficient decoding. Interspeech **2014**, 651–655 (2014)
2. Arisoy, E., Chen, S.F., Ramabhadran, B., Sethy, A.: Converting neural network language models into back-off language models for efficient decoding in automatic speech recognition. IEEE Trans. Audio, Speech Lang. Process. **22**(1), 184–192 (2014)
3. Chelba, C., Norouzi, M., Bengio, S.: N-gram Language Modeling using Recurrent Neural Network Estimation. CoRR 1703.10724 (Mar 2017)
4. Chen, S.F., Goodman, J.: An empirical study of smoothing techniques for language modeling. Comput. Speech Lang. **13**(4), 359–393 (1999)
5. Creutz, M., Lagus, K.: Unsupervised discovery of morphemes. In: Proceedings of the ACL-02 Workshop on Morphological and Phonological Learning. vol. 6, pp. 21–30. Association for Computational Linguistics, Morristown, NJ, USA (2002)
6. Deoras, A., Mikolov, T., Kombrink, S., Karafiat, M., Khudanpur, S.: Variational approximation of long-span language models for lVCSR. In: 2011 IEEE International Conference on Acoustics, Speech and Signal Processing (ICASSP). pp. 5532–5535. IEEE (may 2011)
7. Enarvi, S., Smit, P., Virpioja, S., Kurimo, M.: Automatic speech recognition with very large conversational finnish and estonian vocabularies. IEEE/ACM Trans. Audio Speech Lang. Process. **25**(11), 2085–2097 (2017)
8. Hochreiter, S., Schmidhuber, J.: Long short-term memory. Neural Comput. **9**(8), 1735–1780 (1997)
9. Kurimo, M., et al.: Unlimited vocabulary speech recognition for agglutinative languages. In: Proceedings of the main conference on Human Language Technology Conference of the North American Chapter of the Association of Computational Linguistics, pp. 487–494. Association for Computational Linguistics, Morristown, NJ, USA (2007)

10. Mihajlik, P., Tüske, Z., Tarján, B., Németh, B., Fegyó, T.: Improved recognition of spontaneous hungarian speech-morphological and acoustic modeling techniques for a less resourced task. IEEE Trans. Audio, Speech Lang. Process. **18**(6), 1588–1600 (2010)
11. Mikolov, T., Karafiat, M., Burget, L., Cernocky, J., Khudanpur, S.: Recurrent neural network based language model. Interspeech **2010**, 1045–1048 (2010)
12. Povey, D., et al.: The Kaldi speech recognition toolkit. In: IEEE 2011 Workshop on Automatic Speech Recognition and Understanding. IEEE Signal Processing Society (2011)
13. Singh, M., Smit, P., Virpioja, S., Kurimo, M.: First-pass decoding with n-gram approximation of RNNLM : The problem of rare words (2018), working paper
14. Smit, P., Virpioja, S., Kurimo, M.: Improved subword modeling for WFST-based speech recognition. In: Interspeech 2017. pp. 2551–2555. ISCA, ISCA (Aug 2017)
15. Stolcke, A.: SRILM - an extensible language modeling toolkit. In: Proceedings International Conference on Spoken Language Processing, pp. 901–904. Denver, US (2002)
16. Stolcke, A.: Entropy-based pruning of backoff language models. In: Proceedings of the DARPA Broadcast News Transcription and Understanding Workshop. pp. 270–274 (2000)
17. Sundermeyer, M., Schlueter, R., Ney, H.: LSTM neural networks for language modeling. Interspeech **2012**, 194–197 (2012)
18. Tarján, B., Mihajlik, P., Balog, A., Fegyó, T.: Evaluation of lexical models for Hungarian Broadcast speech transcription and spoken term detection. In: 2nd International Conference on Cognitive Infocommunications (CogInfoCom), pp. 1–5. Budapest, Hungary (2011)
19. Tarján, B., Sarosi, G., Fegyo, T., Mihajlik, P.: Improved recognition of Hungarian call center conversations. In: 2013 7th Conference on Speech Technology and Human - Computer Dialogue. SpeD 2013, pp. 1–6. IEEE, Cluj-Napoca, Romania (Oct 2013)
20. Tüske, Z., Schlüter, R., Ney, H.: Investigation on LSTM recurrent N-gram language models for speech recognition. In: Interspeech 2018, pp. 3358–3362. ISCA, ISCA (Sep 2018)
21. Virpioja, S., Smit, P., Grönroos, S.A., Kurimo, M.: Morfessor 2.0: Python Implementation and Extensions for Morfessor Baseline. Technical Report September, Aalto University (2013)
22. Zaremba, W., Sutskever, I., Vinyals, O.: Recurrent Neural Network Regularization. CoRR **1409**, 2329 (2014)

Tuning of Acoustic Modeling and Adaptation Technique for a Real Speech Recognition Task

Jan Vaněk📁, Josef Michálek(✉)📁, and Josef Psutka📁

University of West Bohemia, Univerzitní 8, 301 00 Pilsen, Czech Republic
{vanekyj,orcus,psutka}@kky.zcu.cz

Abstract. At the beginning, we had started to develop a Czech telephone acoustic model by evaluating various Kaldi recipes. We had a 500-h Czech telephone Switchboard-like corpus. We had selected the Time-Delay Neural Network (TDNN) model variant "d" with the i-vector adaptation as the best performing model on the held-out set from the corpus. The TDNN architecture with an asymmetric time-delay window also fulfilled our real-time application constrain. However, we were wondering why the model totally failed on a real call center task. The main problem was in the i-vector estimation procedure. The training data are split into short utterances. In the recipe, 2-utterance pseudospeakers are made and i-vectors are evaluated for them. However, the real call center utterances are much longer, in order of several minutes or even more. The TDNN model was trained from i-vectors that did not match the test ones. We propose two ways how to normalize statistics used for the i-vector estimation. The test data i-vectors with the normalization are better compatible with the training data i-vectors. In the paper, we also discuss various additional ways of improving the model accuracy on the out-of-domain real task including using LSTM based models.

Keywords: Neural networks · Acoustic model · Automatic speech recognition · Adaptation · I-vectors

1 Introduction

Deep neural networks (DNNs) have been successfully applied to acoustic modelling for automatic speech recognition (ASR). ASR systems are now capable of real-world applications, especially if we have plenty of data from the target domain. However, there can be a performance degradation due to the mismatch between training and testing conditions, such as speaker, recording channel, speaking style, and acoustic environment [3,12]. Many approaches have been proposed in recent years to achieve a robust ASR or to improve the DNN adaptability. Generally, there are two ways that can also be combined [1,15]. First, a boost of the training data variability [6,7]. Second, an adaptation of the acoustic model [5,11,13]. Variability of the training data can be improved by an addition

© Springer Nature Switzerland AG 2019
C. Martín-Vide et al. (Eds.): SLSP 2019, LNAI 11816, pp. 235–245, 2019.
https://doi.org/10.1007/978-3-030-31372-2_20

of more sources of real speech or by an artificial modification of the existing data itself – usually called data augmentation. In this paper, we have evaluated these approaches on a real call center speech recognition task. We have identified that a proper use of the i-vector adaptation is crucial. Especially an i-vector statistics normalization. Therefore, we focus on these techniques in detail below.

2   I-Vector Calculation

We only outline the main points of the i-vector calculation here. More detail can be found in [4,11]. The acoustic feature vectors $\mathbf{x}_t \in \mathcal{R}^D$ are seen as samples generated from a universal background model (UBM) represented as a GMM with K diagonal covariance Gaussians

$$\mathbf{x}_t \sim \sum_{k=1}^{K} c_k \mathcal{N}(\cdot; \mu_k(0), \boldsymbol{\Sigma}_k) \tag{1}$$

with mixture coefficients c_k, means $\mu_k(0)$ and diagonal covariance matrices $\boldsymbol{\Sigma}_k$. Moreover, data $\mathbf{x}_t(s)$ belonging to speaker s are drawn from the distribution

$$\mathbf{x}_t(s) \sim \sum_{k=1}^{K} c_k \mathcal{N}(\cdot; \mu_k(s), \boldsymbol{\Sigma}_k) \tag{2}$$

where $\mu_k(s)$ are the means of the GMM adapted to speaker s. The basis of the i-vector algorithm is to assume a linear dependence between the speaker-adapted means $\mu_k(s)$ and the speaker-independent means $\mu_k(0)$ of the form

$$\mu_k(s) = \mu_k(0) + \mathbf{T}_k \mathbf{w}(s), \qquad k = 1 \dots K \tag{3}$$

$\mathbf{T}_k$, of size $D \times M$, is called the factor loading submatrix corresponding to component k and $\mathbf{w}(s)$ is the speaker identity vector (i-vector) corresponding to s. Each $\mathbf{T}_k$ contains M bases which span the subspace with important variability in the component mean vector space.

For the i-vector estimation, we assume a fixed soft alignment of frames to mixture components. We estimate the posterior distribution of $\mathbf{w}$ given speaker data as

$$p(\mathbf{w}|\mathbf{x}_t(s)) = \mathcal{N}(\mathbf{w}; \mathbf{L}^{-1}(s) \sum_{k=1}^{K} \mathbf{T}_k^T \boldsymbol{\Sigma}_k^{-1} \boldsymbol{\Theta}_k(s), \mathbf{L}^{-1}(s)) \tag{4}$$

with precision matrix $\mathbf{L}(s)$ of size $M \times M$ expressed as

$$\mathbf{L}(s) = \mathbf{I} + \sum_{k=1}^{K} \gamma_k(s) \mathbf{T}_k^T \boldsymbol{\Sigma}_k^{-1} \mathbf{T}_k \tag{5}$$

The quantities that appear in (4) and (5) are the zero-order and centered first-order statistics and are defined as

$$\gamma_k(s) = \sum_t \gamma_{kt}(s), \tag{6}$$

$$\Theta_k(s) = \sum_t \gamma_{kt}(s)(\mathbf{x}_t(s) - \mu_k(0)) \tag{7}$$

with $\gamma_{kt}(s)$ being the posterior probability of mixture component k given $\mathbf{x}_t(s)$. The i-vector that we are looking for is simply the MAP point-estimate of the variable $\mathbf{w}$ which is the mean of the posterior distribution from (4), i.e.

$$\mathbf{w}(s) = \mathbf{L}^{-1}(s) \sum_{k=1}^{K} \mathbf{T}_k^T \mathbf{\Sigma}_k^{-1} \Theta_k(s). \tag{8}$$

The inversion of matrix $\mathbf{L}^{-1}(s)$ could be numerically unstable. More robust variant to i-vector estimation is based on a linear solver. E.g. Kaldi implementation uses a linear conjugate gradient solver.

Because of nature of the MAP point-estimate, low amount of accumulated data leads to an i-vector close to the central zero point. A DNN trained from short utterances works with i-vectors that are not far from the central zero point. Long utterances in the test set produce precise i-vectors far from the central zero point that the DNN never saw. In that case, the long test utterance i-vectors cause failure of the recognizer. Also, we have investigated the online scenario, where first several words of the utterance are recognized well, but when the utterance length starts to be significantly longer than the typical training length, recognition errors ramp up. The solution is an i-vector statistics normalization comparable with the training utterances length distribution.

3   I-Vector Statistics Normalization Methods

The most efficient method of i-vectors normalization is to scale statistics (Eqs. (5), (6), and (7)) before the i-vector calculation because of the additive nature of statistics and real meaning of them (amount of accumulated data in seconds).

3.1   Length Normalization

The simplest way is to scale down statistics to some predefined length of data in seconds. The proper length may be derived from training utterance (pseudospeaker) lengths. Disadvantage of this approach is a convergence to a constant i-vector for very long utterances. This i-vector calculated from long-term statistics is not precisely compatible with the short-term i-vectors in training data.

3.2   Exponential Forgetting

To focus more on the local short-term information, a local time-window may be used. Similar, but simple to implement, is the exponential forgetting of the statistics. Every time-step, the statistics accumulators are multiplied by a constant less than one. The constant value α_e is set compatible to a time-window length T_w, thus easy to understand

$$\alpha_e = 1 - \frac{1}{T_w}, \tag{9}$$

where T_w is the time-window length in time-step units, e.g. frames.

4 Experiments

4.1 Training Data

As a main source of our training data, we have a 500-h Czech telephone Switchboard-like corpus. It is called *Bezplatne Hovory (BH)* (eng. Toll-free Calls). The corpus consists of unrestricted spontaneous speech in variable conditions and speaking styles. The training data part was filtered a bit, we omitted very short utterances and utterances with majority of non-speech events. The total length of data for training was 406.6 h. The total number of calls was 5,535. It does not match perfectly with the number of speakers because some speakers may call more than once.

To add more variability and robustness into the training data, we added three additional corpora:

- *Siemens* – a read speech corpus recorded through a telephone channel. It is a small corpus of 10.7 h of speech in total, but with a large number of speakers: 1,121.
- Czech part of *SpeechDat-E (SD-E)* – a read speech telephone corpus [8]. The Czech part used for training consists of 739 speakers and 20.8 h of speech in total.
- *Telephone Quality Speech Corpus (TQSC)* – a read speech telephone corpus recorded at our department. Each of 1,929 speakers uttered 40 sentences. These sentences were uttered by native Czech male and female speakers, and they contain a large number of silent parts and low–level noises. The used training data has 26.2 h in total.

All added corpora are read speech ones, because we were interested in testing whether the addition of a read speech into the training data improves the spontaneous speech recognition results.

4.2 Test Data and Recognizer Setup

We have prepared two tests. First, an in-domain test called *BH Test*. The data for this test were taken from a held-out part of the main training corpus (BH). We have selected 221 utterances of 26 speakers. The test data has 17.3 min in total. The in-domain test showed how an acoustic model performed in the matching conditions.

Second, an out-of-domain test from a real call center task. We have split an Operator and Customer part of the test for the evaluation. Operators are skilled speakers with more formal speech in contrast with more spontaneous customers. We selected 20 operator and 20 customer calls/utterances. The calls were about several minutes long and both groups were balanced to have 1.2 h of speech in total.

We have used our proprietary recognizer optimized for online speech recognition [10]. We have used two trigram language models. A general telephone conversation model with 550k words for the *BH Test* and a task-specific 33k-words model for the Call center test. OOV words were added explicitly to the vocabulary to avoid disturbing the recognition results.

4.3   Acoustic Model Training

Two kinds of acoustic models were trained.

TDNN Models. were trained in Kaldi [9,14]. The procedure followed a Kaldi Switchboard example recipe. In our prior work, we tested various recipes and setups. Note that not every model architecture is usable for a real-time recognition. Therefore, we are restricted with absence of any offline technique and limited by the total latency. From models that are real-time recognition compatible, we have selected the "s5c" TDNN Switchboard recipe "tdnn_d".

The model is trained in three stages. First, a base GMM-HMM is trained. The triphone clustered states (senones) and alignments produced by the GMM-HMM were then used for further TDNN training. Low resolution MFCC features were used for the entire GMM-HMM training. The GMM-HMM was trained with LDA, MLLT, SAT techniques. Second and third stages used high resolution MFCC features.

The second stage was estimation of an i-vector model and extraction of i-vectors for the entire training set. The i-vector model was based on pseudospeakers that were made by setting two utterances from a speaker as a pseudospeaker data. It boosted the speaker variability and enriched the i-vectors space. For even more speaker-space variability, a speed perturbation with 0.9 and 1.1 speed ratios was used [6]. Thus, the total amount of training data was tripled to almost 1400 h.

The third stage was the TDNN training itself. The TDNN has 6 hidden layers with 1024 ReLU neurons. The time-delay coefficients were as follows: $(-2, -1, 0, 1, 2)$, $(-1, 2)$, $(-3, 3)$, $(-3, 3)$, and $(-7, 2)$. The total delay of the network is 12 frames. The final softmax layer has 7,149 outputs – senones – triphone states. The net was trained by Kaldi GPU parallel implementation with momentum SGD.

LSTM Models. follow the first two stages of the TDNN training [2]. They share the same features, triphone states, alignments, and i-vectors. Only a NN architecture and training procedure were different. We used Chainer 5.0 as a NN training framework for all LSTM models.

During our preliminary tests, we had found out that a residual neural network (ResNet) with LSTM layers worked best for our data. The neural network architecture used in this paper was 6-layer LSTM network with 1024 units in each layer. The skip connections were between 4-th and the last (linear) layer (see Fig. 1). First, we trained each network using Adam optimizer and then we performed 3 training stages with the momentum SGD with momentum equal to 0.9 and learning rate equal to $1e-3$, $1e-4$, and $1e-5$, respectively. In all training stages, we used batch size 128 and dropout $p = 0.2$. We used early stopping, where each training stage was stopped when a validation data criterion increased in comparison to the last epoch.

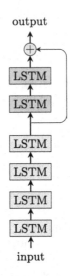

Fig. 1. Our LSTM ResNet architecture

4.4 Results

First, we evaluated the impact of i-vectors on the TDNN network performance. We trained several TDNN models. The models were trained on the BH only dataset and on a combined datasets BH, Siemens, SD-E, and TQSC (called "All" in the figures). We also trained each TDNN model with and without i-vectors. The models with i-vectors used exponential forgetting with α_e equivalent to 5s time window length. All TDNN models were trained on speed perturbed (SP) data with speed ratios 0.9 and 1.1.

From the Fig. 2 it can be seen, that models with i-vectors have worse WER than models without i-vectors on the in-domain test. However, on the out-of-domain tests, models with i-vectors perform better. The biggest impact of adding i-vectors on the in-domain test was for model trained on BH only, the resulting WER increased by 1.39%. The biggest improvement on out-of-domain tests was for model trained on All datasets, where call center customer WER improved by 3.37%. The use of All datasets compared to only BH had negligible impact on models without i-vectors, but on models with i-vectors it noticeably improved WER for all tests. The biggest improvement was 1.03% lower WER on call center customer test.

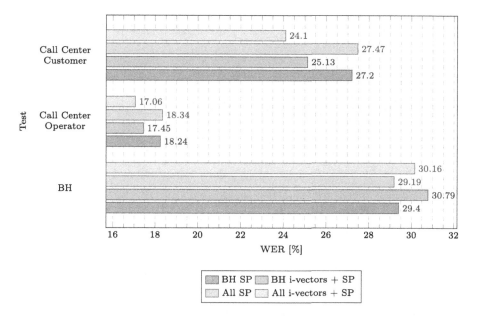

Fig. 2. Word Error Rate [%] of TDNN Models on In-domain and Out-of-domain Recognition Task

Then, we trained LSTM models as described in Sect. 4.3. As with TDNN, all LSTM models were trained on BH only dataset and also on combined datasets called "All". We also trained all models with and without i-vectors. And we trained all models with and without speed perturbation (SP).

The results for LSTM models are in Fig. 3. From the figure, it is obvious that for all tests, in-domain and out-of-domain, all models with i-vectors perform worse than models without i-vectors. The biggest difference in WER with i-vectors compared to without is for a model trained on All datasets without SP, where the call center customer test WER got worse by even 17.96%. Speed perturbation improved results for all test cases except for models trained on BH with i-vectors. In-domain test in this case has lower WER, but out-of-domain tests have higher WER, with call center customer WER increasing by 2.5%. The best performing model for out-of-domain call center operator test was a model trained on All datasets with SP without i-vectors, having 17.36% WER. The best models for other out-of-domain test (customer) and in-domain test were both trained on BH only with SP without i-vectors, but their WER were very close in comparison to the best model for call center operator test mentioned above.

Next, we evaluated the performance of two i-vector normalization techniques. When we started working with TDNN models with i-vectors, we found out that the models performed well on in-domain test, but failed on out-of-domain tests. We found out, that the reason was an utterance length. Our out-of-domain test (real call center data) contained utterances much longer than our training data and some form of i-vector normalization had to be employed.

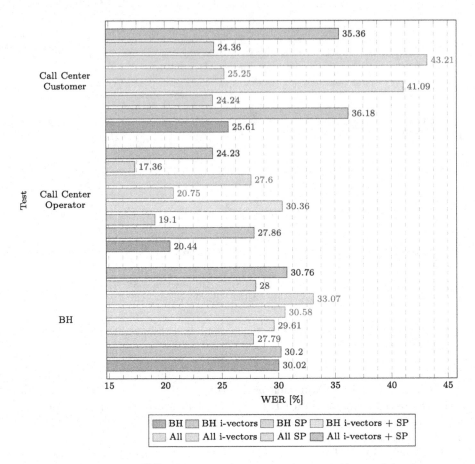

Fig. 3. Word error rate [%] of LSTM models on in-domain and out-of-domain recognition task

The image Fig. 4 shows a histogram of a pseudospeaker utterance length for our training and out-of-domain test data. It can be seen that training pseudospeaker data are generally up to 10 s (typically 3–5 s) long while test utterances are mostly longer than one minute.

We have evaluated two i-vector normalization techniques, the exponential forgetting and the length normalization as described in Sect. 3. We have trained a model on various normalized data lengths and evaluated WER of each normalization technique. Used model was TDNN trained on all datasets with SP. The results can be seen in Fig. 5. From the figure it can be said that the best results for all the normalization techniques are generally obtained using data length from 3 to 6 s. Length normalization gives best WER on both tests for data length of 3 s and exponential forgetting gives best WER on operator test for data length of 4 s and on customer test for data length of 6 s, although WER for all tests in this range are very similar. Exponential forgetting gives better

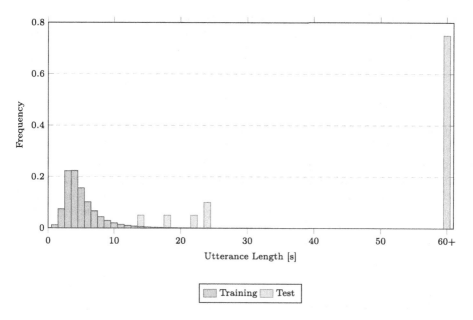

Fig. 4. Histogram of pseudospeaker utterance length in training and test data

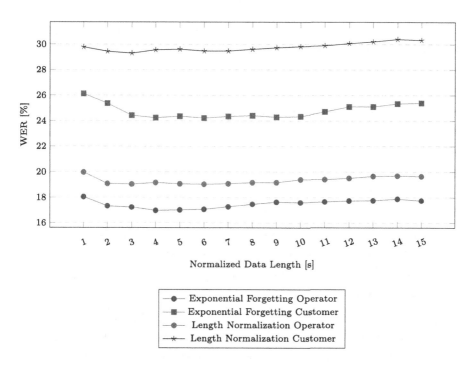

Fig. 5. Impact of data length on the word error rate [%] of TDNN models with different i-vector normalization techniques

WER than length normalization on all our tests for all normalized data lengths. The WER obtained using exponential forgetting compared to the length normalization is improved in average by 7.3% for operator test and by 10.4% for customer test.

5   Conclusion

In this paper, we described our experiences with the development of the Czech telephone acoustic model. We have tested models based on two architectures, TDNN and LSTM, suited for the real-time recognition task. At first, the models totally failed on a real call center task. We have identified the main problem in the i-vector estimation procedure and propose and evaluate two i-vector statistics normalization methods. The use of the exponential forgetting compared to the length normalization was far better. The forgetting constant α_e was robust to set and a value matching a typical training pseudospeaker data length worked well.

We also tested various additional techniques: data augmentation, addition of real data, and i-vector adaptation. Generally, we may recommend using the speed perturbed data augmentation. Other techniques behaviour was model architecture dependent. TDNN models worked well with i-vectors on the call center test and addition of the out-of-domain real data did not help. In contrast, with LSTM models, the i-vector adaptation failed, speed perturbation helped in all cases, and adding the read speech data helped only on the call center operator test.

Summarized, LSTM based model worked better by more than 1% absolutely than TDNN on the in-domain test. In contrast, the TDNN model with the proper i-vector normalization was more robust and worked slightly better on the out-of-domain test. It seems, that the LSTM model has an ability to outperform the TDNN one. However, some other adaptation technique needs to be developed to improve its robustness on the out-of-domain data.

Acknowledgement. This research was supported by the LINDAT/CLARIN, project of the Ministry of Education of the Czech Republic No. CZ.02.1.01/0.0/0.0/16_013/0001781.

References

1. Chen, X., et al.: Recurrent neural network language model adaptation for multi-genre broadcast speech recognition. In: Sixteenth Annual Conference of the International Speech Communication Association (2015)
2. Hochreiter, S., Schmidhuber, J.: Long short-term memory. Neural Comput. **9**(8), 1735–1780 (1997)
3. Hsu, W.N., Zhang, Y., Glass, J.: Unsupervised domain adaptation for robust speech recognition via variational autoencoder-based data augmentation. In: 2017 IEEE Automatic Speech Recognition and Understanding Workshop (ASRU), pp. 16–23. IEEE (2017)
4. Karafiát, M., Burget, L., Matějka, P., Glembek, O., Černocký, J.: iVector-based discriminative adaptation for automatic speech recognition. In: 2011 IEEE Workshop on Automatic Speech Recognition & Understanding, pp. 152–157. IEEE (2011)

5. Karanasou, P., Wang, Y., Gales, M.J., Woodland, P.C.: Adaptation of deep neural network acoustic models using factorised I-Vectors. In: Fifteenth Annual Conference of the International Speech Communication Association (2014)
6. Ko, T., Peddinti, V., Povey, D., Khudanpur, S.: Audio augmentation for speech recognition. In: Sixteenth Annual Conference of the International Speech Communication Association (2015)
7. Park, D.S., et al.: Specaugment: a simple data augmentation method for automatic speech recognition. arXiv preprint arXiv:1904.08779 (2019)
8. Pollak, P., et al.: SpeechDat(E) - Eastern European telephone speech databases. In: Proceedings LREC 2000 Satelite workshop XLDB - Very Large Telephone Speech Databases, pp. 20–25. European Language Resources Association, Athens (2000)
9. Povey, D., et al.: The kaldi speech recognition toolkit. In: IEEE 2011 Workshop on Automatic Speech Recognition and Understanding. IEEE Signal Processing Society (Dec 2011), iEEE Catalog No.: CFP11SRW-USB
10. Pražák, A., Müller, L., Šmídl, L.: Real-time decoder for LVCSR system. In: The 8th World Multi-Conference on Systemics, Cybernetics and Informatics: vol. VI : Image, Acoustic, Signal Processing and Optical Systems, technologies and applications, pp. 450–454. International Institute of Informatics and Systemics, Orlando, Florida (2004). http://www.kky.zcu.cz/en/publications/PrazakA_2004_Real-timedecoderfor
11. Saon, G., Soltau, H.: Speaker adaptation of neural network acoustic models using I-Vectors. In: IEEE Workshop on Automatic Speech Recognition and Understanding pp. 55–59 (2013). https://doi.org/10.1109/ASRU.2013.6707705,http://ieeexplore.ieee.org/xpls/abs_all.jsp?arnumber=6707705
12. Sim, K.C., et al.: Domain adaptation using factorized hidden layer for robust automatic speech recognition. In: Proceedings of the INTERSPEECH (2018)
13. Sim, K.C., Qian, Y., Mantena, G., Samarakoon, L., Kundu, S., Tan, T.: Adaptation of deep neural network acoustic models for robust automatic speech recognition. In: Watanabe, S., Delcroix, M., Metze, F., Hershey, J.R. (eds.) New Era for Robust Speech Recognition, pp. 219–243. Springer, Cham (2017). https://doi.org/10.1007/978-3-319-64680-0_9
14. Waibel, A., Hanazawa, T., Hinton, G., Shikano, K., Lang, K.J.: Phoneme recognition using time-delay neural networks. Backpropagation: Theory, Architectures and Applications, pp. 35–61 (1995)
15. Yu, D., Li, J.: Recent progresses in deep learning based acoustic models. IEEE/CAA J. Automatica Sin. 4(3), 396–409 (2017)

Text Analysis and Classification

Automatic Identification of Economic Activities in Complaints

Luís Barbosa[1], João Filgueiras[1,2], Gil Rocha[1,2],
Henrique Lopes Cardoso[1,2(✉)], Luís Paulo Reis[1,2], João Pedro Machado[3],
Ana Cristina Caldeira[3], and Ana Maria Oliveira[3]

[1] Departamento de Engenharia Informática, Faculdade de Engenharia da
Universidade do Porto, Porto, Portugal
{up201405729,gil.rocha,filgueiras,hlc,lpreis}@fe.up.pt
[2] Laboratório de Inteligência Artificial e Ciência de Computadores (LIACC),
Rua Dr. Roberto Frias, s/n, 4200-465 Porto, Portugal
[3] Autoridade de Segurança Alimentar e Económica (ASAE),
Rua Rodrigo da Fonseca, 73, 1269-274 Lisboa, Portugal
{jpmachado,accaldeira,amoliveira}@asae.pt

Abstract. In recent years, public institutions have undergone a progressive modernization process, bringing several administrative services to be provided electronically. Some institutions are responsible for analyzing citizen complaints, which come in huge numbers and are mainly provided in free-form text, demanding for some automatic way to process them, at least to some extent. In this work, we focus on the task of automatically identifying economic activities in complaints submitted to the Portuguese Economic and Food Safety Authority (ASAE), employing natural language processing (NLP) and machine learning (ML) techniques for Portuguese, which is a language with few resources. We formulate the task as several multi-class classification problems, taking into account the economic activity taxonomy used by ASAE. We employ features at the lexical, syntactic and semantic level using different ML algorithms. We report the results obtained to address this task and present a detailed analysis of the features that impact the performance of the system. Our best setting obtains an accuracy of 0.8164 using SVM. When looking at the three most probable classes according to the classifier's prediction, we report an accuracy of 0.9474.

Keywords: Text categorization · Natural language processing ·
User-generated text · Complaint analysis

1 Introduction

Several countries have public administration institutions that provide public services electronically. Moreover, such institutions are responsible for processing citizen requests, also performed by electronic means, often materialized through

© Springer Nature Switzerland AG 2019
C. Martín-Vide et al. (Eds.): SLSP 2019, LNAI 11816, pp. 249–260, 2019.
https://doi.org/10.1007/978-3-030-31372-2_21

email contacts or by filling-in contact forms in so-called virtual counters. In specific types of public institutions, such as those in charge of enforcing compliance of citizens or economic agents, a significant number of such requests are in fact complaints that need to be appropriately dealt with.

The amount of complaints received can reach the thousands in a short period of time, depending on the size of the country/administrative region. The Portuguese Economic and Food Safety Authority (ASAE), for instance, receives more than 20 thousand complaints annually, more than 30% of which are usually found not to be in the jurisdiction of ASAE; the rest are sent to the ASAE Operational Units. Given the high amount of complaints, the use of human labor to analyze and properly handle them quickly becomes a bottleneck, bringing the need to automate this process to the extent possible. One of the obstacles to do it effectively is the fact that contact forms typically include free-form text fields, bringing high variability to the quality of the content written by citizens.

This work focuses on automatically identifying economic activities in complaints written in Portuguese, through the use of natural language processing (NLP) and machine learning (ML) techniques. Portuguese is a low-resourced language in terms of NLP. We employ different features and analyze which ones give the best results using different ML algorithms. We start by discussing related work in Sect. 2. Section 3 describes the dataset used in this work. We detail the employed preprocessing and feature extraction techniques in Sect. 4. Using different ML models, Sect. 5 describes several experiments, including those related with feature selection and data balancing techniques. In Sect. 6, we provide an error analysis and make pertinent observations on the difficulty of the task. Finally, Sect. 7 concludes and presents some lines of future work.

2   Related Work

Although several works exist on analyzing user-generated content, they mostly study social media data [1], focusing on tasks such as sentiment analysis and opinion mining [15], or predicting the usefulness of product reviews [4]. Forte and Brazdil [6] focus on sentiment polarity of Portuguese comments, and use a lexicon-based approach enriched with domain specific terms, formulating specific rules for negation and amplifiers. Literature on (non-social media) complaint analysis is considerably more scarce, mainly due to the fact that such data is typically not publicly available. Nevertheless, the problem has received significant attention from the NLP community, as a recent task on consumer feedback analysis shows [11]. Given the different kinds of analysis one may want to undertake, however, the task concentrates on a single goal: to distinguish between comment, request, bug, complaint, and meaningless. In our work, we want to further analyze the contents of complaints, with a finer granularity.

Ordenes et al. [14] propose a framework for analyzing customer experience feedback, going beyond sentiment analysis and using a linguistics-based text mining model. The approach explores the identification of activities, resources and context, so as to automatically distinguish compliments from complaints, regarding different aspects of the customer feedback. This is made possible through a manual annotation process. The work focuses on a single activity domain, and in the end aims at obtaining a refined sentiment analysis model. In our case, we aim at distinguishing amongst a number of economic activities, without entering into a labor-intensive annotation process of domain-specific data.

Traditional approaches to text categorization employ feature-based sparse models, using bags-of-words and TF-IDF metrics. In the context of insurance complaint handling, Dong and Wang [17] make use of synonyms and Chi-square statistics to reduce dimensionality.

Dealing with complaints as a multi-label classification problem can be effective, even when the original problem is not, due to the noisy nature of user-generated content. Ranking algorithms [10,12] are a promising approach in this regard, providing a set of predictions sorted by confidence. These techniques have been applied in complaint analysis [5], although with modest results.

Kalyoncu et al. [9] approach customer complaint analysis from a topic modeling perspective, using techniques such as Latent Dirichlet Allocation (LDA) [2]. This work is not so much focused on automatically processing complaints, but instead on providing a visualization tool for mobile network operators.

3 Data

The dataset under study has been provided by ASAE. It contains a total of 48,850 complaints received by this governmental entity between 2014 and 2018, submitted by citizens, economic operators, public organizations or other organizations either by email or through a contact form in an official website. Each complaint contains its textual content and is classified with a single economic activity. This is the focus of this work, i.e., to train a classifier that is able to predict this activity (or a generalization thereof).

The economic activity taxonomy used by ASAE is hierarchical in nature. The first level contains 11 classes, and its imbalanced distribution is shown in Table 1. Generally, each class is composed of a number of sub-classes, which have a further decomposition level. Given the large number of second and third-level classes, we decided to train our classifiers to predict first-level classes only.

Since our goal is to aid ASAE staff in handling complaints, we have decided to base our classifications on their textual contents alone. The average complaint is 1,664 characters long after removing HTML tags and other artifacts, containing information on its subject matter, the targeted economic agent and contact information of the claimant.

4 Data Preprocessing and Feature Extraction

We have gone through a typical preprocessing pipeline, including tokenization and lemmatization. Based on [13], we have chosen to use NLTK[1], StanfordNLP[2]

---

[1] https://www.nltk.org/.
[2] https://stanfordnlp.github.io/stanfordnlp/.

252 L. Barbosa et al.

Table 1. Distribution per classes

Class	# examples	# 2nd level subclasses
I - Primary Production	134	7
II - Industry	2031	26
III - Restoration and beverages	20899	4
IV - Wholesalers	299	4
V - Retail	5951	23
VI - Direct selling establishments	1	1
VII - Distance selling (by Catalog and Internet)	2856	1
VIII - Production and Trade	3335	69
IX - Service Providers	9933	85
X - Safety and Environment	696	62
Z - No activity identified	2715	N/A

and spaCy[3]. Given the lack of conclusive data on their performance for Portuguese, the non-exhaustive experiments shown in Table 2 were performed to analyze which were better to identify the economic activity of a complaint.

StanfordNLP was chosen for most experiments, given its competitive contribution to the task and because it is able to identify punctuation marks. Additionally, StanfordNLP provides specific and complete support for Portuguese and presents the data using the CoNLL-U format [16], which increases interoperability with other tools. After obtaining the lemmas, we remove punctuation marks and stop words (using NLTK's stop word list for Portuguese) before performing TF-IDF counts. Given that we have a single example for class VI, as per Table 1, we decided to leave it out of our classification problem.

To perform feature extraction, different data representation techniques were used: count, hashing and TF-IDF, as provided by scikit-learn [3]. The count technique transforms a collection of texts into a matrix of token counts. Hashing obtains a matrix of either token counts or binary occurrences, depending if we want counts or one-hot encoding. We used it to obtain token counts and compare the difference with the count technique because it has a few advantages, like low memory scalability. TF-IDF obtains features representing the importance of each token in the collection of all documents. For these three techniques, we present results obtained by using bags-of-words of 1-grams, 2-grams, 3-grams and intervals of 1 to 2-grams, 1 to 3-grams and 2 to 3-grams.

5 Predicting Economic Activity

The classification task addressed in this paper concerns predicting the economic activity targeted in a complaint. We focus on the first level of the hierarchy, as explained in Sect. 3. In order to find out which classifiers would allow us to obtain

---

the best results, we decided to use Random Forests, Bernoulli NB, Multinomial NB, Complement NB, k-Nearest Neighbors, SVM, Decision Tree, Extra Tree and Random (stratified). The latter will be used as baseline. All of them were implemented using scikit-learn[4] and the default parameters are used (for version 0.22), except those explicitly stated.

To split the original dataset into training and test set, we use 30% of the data for testing, while keeping the distribution of classes of the original dataset in both training and test sets. Following this procedure, we ensure that the trained classifier learns the real distribution of the data, and that the distribution is kept in the test set. Cross-validation was considered but given the considerable amount of training data it was deemed unnecessary to ensure consistency. This is important not only to ensure proper training but also to ensure that, when applying over/under sampling, no over/underfitting occurs in a class.

Our main performance metric was the accuracy score instead of the average macro-F1 score. We aim to provide a list of classes sorted by confidence and it is not critical to correctly classify minority classes. As a baseline we used a stratified random classifier that yielded an accuracy of 0.2504.

Table 2. Economic Activity Multiclass Classification accuracy scores using different tokenizers/lemmatizers

Classifier	StanfordNLP (baseline)	NLTK	spaCy - pt_core_news_sm	spaCy - xx_ent_wiki_sm
Random Forests	0.6787	0.6924	0.6818	0.6911
Bernoulli NB	0.5115	0.5363	0.5110	0.5185
Multinomial NB	0.4603	0.4719	0.4613	0.4648
Complement NB	0.5914	0.6263	0.5965	0.6066
K-Neighbors	0.6283	0.3146	0.6328	0.6180
SVM (linear)	**0.8075**	**0.8164**	**0.8093**	**0.8135**
Decision Tree	0.6659	0.6698	0.6669	0.6703
Extra Tree	0.5056	0.5228	0.5185	0.5166

In Table 2 we present the accuracy scores obtained using different tokenizers and lemmatizers to preprocess the text of the examples in the dataset. For this experiment, we used 1-gram TF-IDF to represent the features extracted. NLTK obtains the best scores overall, followed by spaCy and, finally, StanfordNLP. Nevertheless, we chose to continue using StanfordNLP because the performance loss is negligible and it provides PoS information, including punctuation marks. This proved useful to remove punctuation on all experiments and also experiment with removing adjectives. Furthermore, it has the advantage of having specific support for several languages, several more than the ones supported by NLTK and spaCy (although for now we are focusing on Portuguese).

---

[4] https://scikit-learn.org/stable/.

Table 3. Economic Activity Multiclass Classification accuracy scores using different feature extraction techniques

Classifier	Count	Hashing	TF-IDF
Random Forests	0.6958	0.6561	0.6787
Bernoulli NB	0.5115	0.4415	0.5115
Multinomial NB	0.6329	Error[a]	0.4603
Complement NB	0.6790	Error[a]	0.5914
K-Neighbors	0.5359	0.5750	0.6283
SVM (linear)	**0.7784**[b]	**0.7953**	**0.8075**
Decision Tree	0.6786	0.6671	0.6659
Extra Tree	0.4968	0.4865	0.5056

[a] Hashing may generate negative feature values, not supported by some classifiers.
[b] Failed to converge after 1,000 iterations.

Table 3 presents accuracy scores obtained using the different feature representation techniques discussed in Sect. 4. We used StanfordNLP for preprocessing and represent only 1-grams. Accuracy scores vary considerably depending on the classifier used, the best being obtained using SVM and TF-IDF. For that reason, subsequent experiments make use of TF-IDF.

Table 4. Economic Activity Multiclass Classification accuracy scores using different n-grams

Classifier	1-gram	1 to 2-grams	2-grams	1 to 3-grams	2 to 3-grams	3-grams
Random Forests	0.6737	0.6503	0.6323	0.6230	0.6127	0.5663
Bernoulli NB	0.5115	0.4763	0.4703	0.4622	0.4561	0.4495
Multinomial NB	0.4603	0.4568	0.4700	0.4568	0.4683	0.4733
Complement NB	0.5914	0.5432	0.5978	0.5381	0.5922	0.6320
K-Neighbors	0.6283	0.6152	0.5821	0.5950	0.5631	0.5413
SVM (linear)	**0.8075**	**0.8098**	**0.7640**	**0.8004**	**0.7396**	**0.6532**
Decision Tree	0.6659	0.6729	0.6121	0.6717	0.6120	0.5413
Extra Tree	0.5056	0.5338	0.5462	0.5541	0.5398	0.5298

In Table 4 we present the accuracy scores obtained using different n-grams when performing feature extraction with TF-IDF. It is not possible to conclude which is the best interval of n-grams because it depends on the classifier, but, for SVM, 1 to 2-grams is the best choice, followed by 1-gram. Because the difference between 1-grams and 1 to 2-grams in small for SVM, but higher for Random Forests, the following experiments use only 1-grams.

Taking into account the potential usage of the classifier, which is meant to help humans on analyzing complaints by providing likely classification labels

Table 5. Economic Activity Multiclass Classification accuracy scores for top-k predictions. Acc@k: accuracy scores considering the top-k (Acc@k) predicted classes, according to the confidence of the classifier predictions

Classifier	Acc@1	Acc@2	Acc@3
Random Forests	0.6787	0.8322	0.8885
Bernoulli NB	0.5115	0.7533	0.7913
Multinomial NB	0.4603	0.6790	0.7936
Complement NB	0.5914	0.8214	0.8873
K-Neighbors	0.6283	0.7699	0.8447
SVM (linear)	**0.8075**	**0.9031**	**0.9474**
Decision Tree	0.6659	0.7086	0.7226
Extra Tree	0.5056	0.5627	0.5703

(as opposed to imposing a definitive one), we looked at the performance of the classifier considering the ranking provided. In Table 5 we present accuracy scores obtained by accepting the 1st, 2nd and 3rd best probabilities. The second column shows the accuracy scores accepting as correct only the option with the highest probability. The third/fourth column shows the accuracy scores when accepting as correct one of the two/three options with the highest probabilities. For most classifiers, the accuracy of the top-2 is considerably higher than the accuracy considering the top-1. The 0.9474 score with SVM and top-3 demonstrates that presenting a set of classes sorted by confidence will be an effective help.

5.1 Feature Selection

We noticed that TF-IDF using 1-gram extracted 252,000 features, while only 101,159 are of interest when analyzing feature importance with Random Forests. As such, although a lot of features are extracted, a considerable part will be of no use to a classifier. For that reason, we explored feature selection via Latent Dirichlet Allocation (LDA), with the aim of bringing the number of features down while improving classification and training speed by clustering the features that are more important for the classification problem. However, as shown in Table 6, the use of LDA largely reduces the effectiveness of the classifiers. Moreover, although Random Forests presents an increase of 6% when raising the number of LDA components, most other classifiers maintain or even decrease accuracy scores. For this experiment, we used StanfordNLP and TF-IDF, extracting only 1-grams and analyzing top-1 predictions.

Based on these results, we concluded that performing LDA is not effective for this classification task. Applying Principal Component Analysis (PCA) [18] has led to a similar result.

Finally, we performed experiments using the Recursive Feature Elimination and Cross-Validated selection (RFECV) approach [7]. This technique consists in training a classifier multiple times with different features and yielding the

Table 6. Economic Activity Multiclass Classification accuracy scores using LDA

Classifier	No LDA (baseline)	10 components	100 components
Random Forests	0.6787	0.4053	**0.4612**
Bernoulli NB	0.5115	0.4278	0.4278
Multinomial NB	0.4603	-	0.4306
Complement NB	0.5914	0.4157	0.3561
K-Neighbors	0.6283	0.3995	0.4146
SVM (linear)	**0.8075**	**0.4484**	0.4424
Decision Tree	0.6659	0.3320	0.3525
Extra Tree	0.5056	0.3300	0.3419

feature matrix that generated the best classifier according to a chosen metric. RFECV was tested with Complement NB because it is fast to train, resulting in a classifier with significantly better accuracy. On the other hand, testing with SVM has shown that this classifier does not benefit from further optimization.

5.2   Over and Under Sampling

As shown in Table 1, the class distribution for our problem is very imbalanced. To improve the overall classification performance and, more specifically, the performance on minority classes, we explore two widely used techniques to deal with imbalanced datasets [8]: *random under sampling* and *random over sampling*.

We have chosen to use the "imblearn" Python package[5]. There were three alternatives to perform the over sampling: RandomOverSampler (ROS), SMOTE and ADASYN. ROS duplicates some of the examples of the classes, increasing the number of examples of all classes to the number of examples of the class with the highest number of examples, as indicated in the documentation of "imblearn". SMOTE generates new samples by interpolation, not distinguishing between easy and hard examples. ADASYN generates new samples by interpolation, focusing on generating samples based on the original samples which are incorrectly classified using a k-Nearest Neighbors classifier. Because we were testing several different classifiers, including a k-Nearest Neighbors classifier, we decided to use the RandomOverSampler to reduce bias in the results. For random under sampling, RandomUnderSampler (RUS) was chosen to be comparable to the RandomOverSampler. RandomUnderSampler randomly selects a subset of data for the targeted classes, reducing the number of examples of each class to the number of examples of the class with the smallest number of examples.

Table 7 presents the accuracy and average macro-F1 scores obtained by performing random over sampling and random under sampling on the dataset. For these experiments, we used StanfordNLP for preprocessing and TF-IDF to represent the features extracted. Only 1-grams were extracted and only the top-1

---

[5] https://imbalanced-learn.readthedocs.io/en/stable/.

was analyzed. As shown in Table 7, when performing random over sampling the accuracy scores related to Naive Bayes increased significantly and the accuracy of Random Forests also increased, but for all others it decreased. A similar situation can be observed regarding the corresponding average macro-F1 score. This is demonstrative that repeating the same data in the classes with a lower number of examples does not help distinguishing the different classes (except for Naive Bayes) and indicates that the classifiers are not predicting mostly the more frequent classes due to their amount of examples.

Table 7. Accuracy scores and average macro-F1 score using over or under sampling

Classifier	Accuracy (baseline)	Accuracy ROS	Accuracy RUS	Avg macro-F1 (baseline)	Avg macro-F1 ROS	Avg macro-F1 RUS
Random Forests	0.6787	0.7137	0.4402	0.42	0.49	0.30
Bernoulli NB	0.5115	0.6477	0.4703	0.18	0.48	0.28
Multinomial NB	0.4603	0.7299	0.5223	0.09	0.56	0.37
Complement NB	0.5914	0.7130	0.5258	0.28	0.52	0.37
K-Neighbors	0.6283	0.5456	0.3959	0.46	0.46	0.29
SVM (linear)	**0.8075**	**0.7985**	**0.5555**	**0.63**	**0.62**	**0.43**
Decision Tree	0.6659	0.6294	0.3678	0.45	0.44	0.26
Extra Tree	0.5056	0.4942	0.2074	0.33	0.32	0.15

On the other hand, when performing random under sampling, only the accuracy scores related to Bernoulli NB and Multinomial NB increased, while for all the other classifiers it has decreased significantly. All average macro-F1 score are relatively low, but 6 of them decreased and 3 of them increased. This is demonstrative that reducing the amount of examples for the classes with a higher number of examples reduces the ability of distinguishing the different classes.

5.3   Additional Experiments

An experiment performed to analyze the impact of the removal of adjectives identified by StanfordNLP was performed to identify if they were important for the classification task. This experiment was interesting because strong adjectives are apparently important for the classification task, but other weaker adjectives should not be. Depending on the amount and type of adjectives present in the dataset, their removal could reduce the amount of features that are irrelevant for the problem. Comparing the accuracy scores of all classifiers with the accuracy scores obtained by not removing the adjectives (baseline), as is the case in Table 5, the percentage was always the same, differing only on the permillage. These results are indicative that adjectives are partially important for the classifiers, although most of them have a low or even null importance/coefficient.

Experiments performed to increase the accuracy of SVM (with linear kernel) generating different class weights and balanced class weights (hyperparameterization) [8] obtained accuracy and average macro-F1 scores close to the ones obtained using the default parameters: a maximum accuracy of 0.8096 with a macro-F1 score of 0.63. Also, the different kernels available for SVM (linear, poly, rbf, sigmoid, precomputed) were tested and it was found that the linear kernel is the best in terms of accuracy, immediately followed by the sigmoid kernel, and that the sigmoid kernel is the best in terms of average macro-F1 score, immediately followed by the linear kernel. Finally, experiments performed to test the use of ensembles based on decision trees, which usually have interesting performances, provided accuracy scores higher than the ones obtained using Random Forests, but considerably lower than the ones provided by SVM.

6   Error Analysis

Based on the different accuracy and average macro-F1 scores obtained, we decided to focus on SVM for the sake of error analysis. We show the obtained confusion matrix in Table 8, when considering top-1 classification only. The influence of the majority class III is visible, but also of the second majority class IX. Class Z, where there is no identified economic activity, seems to be the most ambiguous for the classifier.

Table 8. Confusion matrix of the baseline SVM (Top-1)

						Predicted				
	I	II	III	IV	V	VII	VIII	IX	X	Z
I	14	5	10	1	5	0	1	0	0	4
II	1	324	155	2	61	2	8	26	0	30
III	0	37	5935	1	72	7	30	160	2	24
IV	0	9	16	22	22	0	3	7	0	11
V	1	26	184	3	1454	16	32	42	1	26
VII	0	0	16	0	7	722	26	62	1	23
VIII	1	18	126	1	61	31	596	114	6	46
IX	0	5	314	0	26	30	83	2479	10	33
X	0	0	17	1	6	8	31	52	81	12
Z	2	35	181	3	72	55	93	163	6	204

(Actual)

To better understand in which situations the classifier was making erroneous predictions, we randomly sampled 50 examples from the dataset where the classifier was not capable of correctly predicting (from the top-3 predictions) the gold-standard class. Based on a manual analysis of such cases, we were able to draw the following observations:

- The dataset includes some short text complaints, not providing enough information to classify their target economic activity. Furthermore, a small number of complaints are not written in Portuguese. Some complaint texts are followed by non complaint-related content, sometimes in English.[6]

---

[6] Complaints received by e-mail often include "think twice before printing" appeals.

- Some classes exhibit semantic overlap (to a certain degree), thus confusing the classifier. For example, class VIII apparently overlaps with classes II and V. Moreover, while being labeled with a given class, some complaints contain words that are highly related with a different class.
- A non-negligible number of examples refer to previously submitted complaints, either to provide more data or to request information on their status. These cases do not contain the complaint itself, the same happening when a short text simply includes meta-data or points to an attached file.
- Finally, we were able to identify some complaints that have been misclassified by the human operator.

7   Conclusions and Future Work

For the imbalanced complaints dataset of ASAE, SVM with a linear kernel proved to be the best option among the experimented models. It is reasonably fast, allows to get probability scores and gives the best accuracy scores and average macro-F1. It is particularly valuable if we need a ranked output, given its high accuracy when aggregating the top-3 predicted classes. It is interesting to note that removing punctuation and stop words after lemmatization, using TF-IDF and training the SVM generates better accuracy scores than using additional techniques like feature selection and different quantities of n-grams.

After analyzing misclassified examples, several improvements have been planned. Non-Portuguese complaints need to be ignored, as the number of examples is too low to warrant a multilingual classifier. Furthermore, we aim to further assess how to discard texts that are simply not informative enough to consider as valid complaints (besides empty complaints, which the system correctly classifies). We also aim to tackle additional classification problems exploring this rich dataset. The ideas presented in this work will be the baseline for these future classifiers. We intend to explore recent advances on word embeddings approaches and deep learning techniques, and compare the results obtained with the models presented in this paper. The end goal is to create a system that will greatly assist ASAE personnel when handling these complaints.

Acknowledgements. This work is supported by project IA.SAE, funded by Fundação para a Ciência e a Tecnologia (FCT) through program INCoDe.2030. Gil Rocha is supported by a PhD studentship (with reference SFRH/BD/140125/2018) from FCT.

References

1. Batrinca, B., Treleaven, P.C.: Social media analytics: a survey of techniques, tools and platforms. AI Soc. **30**(1), 89–116 (2015)
2. Blei, D.M., Ng, A.Y., Jordan, M.I.: Latent dirichlet allocation. J. Mach. Learn. Res. **3**, 993–1022 (2003)

3. Buitinck, L., et al.: API design for machine learning software: experiences from the scikit-learn project. In: ECML PKDD Workshop: Languages for Data Mining and Machine Learning, pp. 108–122 (2013)
4. Diaz, G.O., Ng, V.: Modeling and prediction of online product review helpfulness: a survey. In: Proceedings of the 56th Annual Meeting of the Association for Computational Linguistics, pp. 698–708 (2018)
5. Fauzan, A., Khodra, M.L.: Automatic multilabel categorization using learning to rank framework for complaint text on bandung government. In: 2014 International Conference of Advanced Informatics: Concept, Theory and Application (ICAICTA), pp. 28–33. Institut Teknologi Bandung, IEEE (2014)
6. Forte, A.C., Brazdil, P.B.: Determining the level of client's dissatisfaction from their commentaries. In: Silva, J., Ribeiro, R., Quaresma, P., Adami, A., Branco, A. (eds.) PROPOR 2016. LNCS (LNAI), vol. 9727, pp. 74–85. Springer, Cham (2016). https://doi.org/10.1007/978-3-319-41552-9_7
7. Guyon, I., Weston, J., Barnhill, S., Vapnik, V.: Gene selection for cancer classification using support vector machines. Mach. Learn. **46**(1), 389–422 (2002)
8. He, H., Garcia, E.A.: Learning from imbalanced data. IEEE Trans. Knowl. Data Eng. **21**(9), 1263–1284 (2009). https://doi.org/10.1109/TKDE.2008.239
9. Kalyoncu, F., Zeydan, E., Yigit, I.O., Yildirim, A.: A customer complaint analysis tool for mobile network operators. In: 2018 IEEE/ACM International Conference on Advances in Social Networks Analysis and Mining (ASONAM), pp. 609–612. IEEE (2018)
10. Li, H.: Learning to rank for information retrieval and natural language processing. In: Synthesis Lectures on Human Language Technologies, 2nd edn. Morgan & Claypool Publ., San Rafael (2014)
11. Liu, C.H., Moriya, Y., Poncelas, A., Groves, D.: IJCNLP-2017 task 4: customer feedback analysis. In: Proceedings of the IJCNLP 2017, Shared Tasks. Asian Federation of Natural Language Processing, Taipei, Taiwan, pp. 26–33, December 2017
12. Momeni, E., Cardie, C., Diakopoulos, N.: A survey on assessment and ranking methodologies for user-generated content on the web. ACM Comput. Surv. **48**(3), 41:1–41:49 (2015)
13. Omran, F.N.A.A., Treude, C.: Choosing an NLP library for analyzing software documentation: a systematic literature review and a series of experiments. In: Proceedings of the 14th International Conference on Mining Software Repositories, pp. 187–197. MSR 2017. IEEE Press, Piscataway (2017)
14. Ordenes, F.V., Theodoulidis, B., Burton, J., Gruber, T., Zaki, M.: Analyzing customer experience feedback using text mining: a linguistics-based approach. J. Serv. Res. **17**(3), 278–295 (2014)
15. Petz, G., Karpowicz, M., Fürschuß, H., Auinger, A., Stříteský, V., Holzinger, A.: Opinion mining on the web 2.0 – characteristics of user generated content and their impacts. In: Holzinger, A., Pasi, G. (eds.) HCI-KDD 2013. LNCS, vol. 7947, pp. 35–46. Springer, Heidelberg (2013). https://doi.org/10.1007/978-3-642-39146-0_4
16. Qi, P., Dozat, T., Zhang, Y., Manning, C.D.: Universal dependency parsing from scratch. In: Proceedings of the CoNLL 2018 Shared Task: Multilingual Parsing from Raw Text to Universal Dependencies, pp. 160–170. Association for Computational Linguistics, Brussels, Belgium, October 2018
17. Dong, S., Wang, Z.: Evaluating service quality in insurance customer complaint handling throught text categorization. In: 2015 International Conference on Logistics, Informatics and Service Sciences (LISS), pp. 1–5. IEEE (2015)
18. Tipping, M.E., Bishop, C.M.: Probabilistic principal component analysis. J. Roy. Stat. Soc. B **61**(3), 611–622 (1999)

Automatic Judgement of Neural Network-Generated Image Captions

Rajarshi Biswas[1]([✉]), Aditya Mogadala[3], Michael Barz[1,2], Daniel Sonntag[1], and Dietrich Klakow[3]

[1] German Research Center for Artificial Intelligence (DFKI),
Saarland Informatics Campus D3 2, 66123 Saarbrücken, Germany
{rajarshi.biswas,michael.barz,daniel.sonntag}@dfki.de
[2] Saarbrücken Graduate School of Computer Science,
Saarland Informatics Campus D3 2, 66123 Saarbrücken, Germany
[3] Spoken Language Systems (LSV), Saarland Informatics Campus D3 2, 66123
Saarbrücken, Germany
{amogadala,dietrich.klakow}@lsv.uni-saarland.de
http://iml.dfki.de

Abstract. Manual evaluation of individual results of natural language generation tasks is one of the bottlenecks. It is very time consuming and expensive if it is, for example, crowdsourced. In this work, we address this problem for the specific task of automatic image captioning. We automatically generate human-like judgements on grammatical correctness, image relevance and diversity of the captions obtained from a neural image caption generator. For this purpose, we use pool-based active learning with uncertainty sampling and represent the captions using fixed size vectors from Google's Universal Sentence Encoder. In addition, we test common metrics, such as BLEU, ROUGE, METEOR, Levenshtein distance, and n-gram counts and report $F1$ score for the classifiers used under the active learning scheme for this task. To the best of our knowledge, our work is the first in this direction and promises to reduce time, cost, and human effort.

Keywords: Active learning · NLP · NLG ·
Automated human judgement · Image captioning · Neural networks

1 Introduction

Recently, automatic image caption generation has received a lot of attention in scientific natural language processing (NLP) and applications of natural language generation (NLG) in particular. It has attracted a lot of attention from the machine learning (ML) community as well—because of far reaching NLG-ML-applications ranging from assisting the visually impaired to the development of socially interactive robots [10,16,20,31].

Although significant progress has been made in dealing with the caption generation problem [3,18,24,30], we still need to perform manual human evaluation for assessing the quality of the generated descriptions. This is both expensive

© Springer Nature Switzerland AG 2019
C. Martín-Vide et al. (Eds.): SLSP 2019, LNAI 11816, pp. 261–272, 2019.
https://doi.org/10.1007/978-3-030-31372-2_22

and time consuming. In this work, we have modified [3] to remedy this situation by automating human judgement on the quality of the generated descriptions (see examples in Fig. 1) through an active learning scheme (Fig. 2). Specifically, we infer human judgement on grammatical correctness, image relevance and diversity of the generated captions in an automatic manner.

For this purpose, we employ standard ML classifiers, SVM and logistic regression, under a pool based active learning scheme [29]. First, we generate diverse captions for images in the MSCOCO dataset [22] using the neural architecture in [32] along with beam search [23]. A small number of these captions are randomly selected and binary labels on their grammatical correctness, image relevance and diversity are crowdsourced to train the mentioned classifiers for each task. Using the learned classifiers we predict grammatical correctness, image relevance and diversity labels for the unlabeled captions. Subsequently, a batch of 200 instances which lie close to the decision boundary of the classifiers are selected and annotated using the same crowdsourcing platform. We incorporate them in the training set and re-train the classifiers on the new training set. We repeat this cycle 4 times and report the $F1$ scores for the classifiers on a separate human labeled test set.

To summarize, our primary contributions are: first, a new approach in the direction of automatic human evaluation of machine generated image captions; and second, a computational model that uses a fixed size vector representation for sentences, obtained from a pre-trained network and standard metrics which produce a good baseline for automating human evaluation. The paper is organized as follows: Sect. 2 describes related work in the field of image caption generation. Section 3 describes in details the method used in our work for automatically inferring human judgement on the three quality aspects discussed above. Section 4 provides experiments and results followed by a short discussion in Sect. 5. Section 6 provides the conclusion.

2   Related Work

Although there has been considerable interest in language grounding in perceptual data [9,25,28], in the recent past there has been an explosion of interest in the problem of image captioning. As a matter of fact, this is part of a broader effort to investigate the boundary between vision and language. The caption generation method in our work uses the neural framework proposed in [6] where, instead of translating text from one language to another, an image is translated into a caption or sentence that describes it. The neural architecture for image caption generation consists of a deep convolutional network [13] and a recurrent neural network [12]. The first approach in this direction is credited to Kiros et al. [18,19] who proposed to construct a joint multimodal embedding space and provide a natural way to perform both ranking and generation. Works [7,30] offer slight contrast as the authors adopt LSTM RNNs instead of stock RNNs. Karpathy et al. [15] proposes to learn a joint embedding space for both ranking and generation. In fact, their model learns to score sentence and image similarity as a function of convnet object detections with outputs of a bidirectional RNN.

The caption generation problem also is a structured learning problem since both the input and output of this problem have a rich structure. That is, the

1. a wooden table topped with different types of food
2. there are many plates of food on the table
3. a variety of different types of food on a table

1. a baseball player taking a swing at a ball
2. a batter catcher an umpire during a baseball game
3. an image of a professional baseball game being played

1. a view of a living room and dining room
2. the living room is clean and ready for us to use
3. a living room filled with lots of furniture and a flat screen tv

1. a man holding a smart phone in his hand
2. a man taking a picture with his cellphone
3. a close up of a person holding a phone

Fig. 1. Diverse image captions generated using beam search

image of a natural scene is made up of multiple random variables, such as, position of objects, their inter-relationship and all of them have a rich joint distribution. Moreover, there needs to be an alignment between the output words of a caption with the spatial regions of the input image. So, to properly address the structured nature of this problem, we make use of attention mechanism in our work. Hence, we have adopted the show, attend and tell architecture by Xu et al. [32] which uses attention to generate the captions for images.

In addition to being an important task in the area of computer vision, image caption generation is also a major problem in the area of Natural Language Generation (NLG) where proper evaluation of such a system is a core issue. The methods for evaluation can be divided into intrinsic and extrinsic methods. Human Judgement falls under the category of intrinsic evaluation methods and one of the most important requirement for new applications such as [2,26]. The common criteria here include readability or fluency, which refer to the linguistic quality of the text, and also accuracy or relevance relative to the input which shows the NLG system's ability to satisfactorily reproduce content. However, none of the image captioning or NLG methods described above have tried to automatically generate human judgement on the quality of their generations and instead relied on conducting time and cost intensive human evaluation through public surveys. It is worth mentioning here that standard metrics, such as, BLEU [27], ROUGE [21] aim to emulate human judgement but often fall short as they suffer from low correlation between them and human judgements, a fact which is

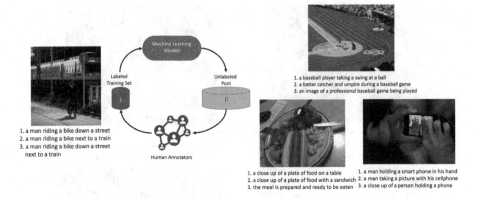

Fig. 2. Pool based active learning scheme

widely reported in the NLG community. In addition, these standard metrics are dependent on groundtruth information since they measure the overlap between a generation and its groundtruth for quality assessment. This in our view is a severe limitation and prevents true automatic evaluation of NLG tasks.

To the best of our knowledge, we believe our attempt which uses fixed size vectors from pretrained sentence encoders, is the first one in the direction of automated human judgement for quality assessment which does not require groundtruth information and thus reduces cost, boosts productivity.

3 Method

We aim at automating human judgements on neural network generated image captions using active learning. In the following, we describe the caption generator, the features that we consider for modeling human judgements and our active learning approach.

3.1 Image Caption Generation

For generating the image captions we use the Show, Attend and Tell [32] approach on the MSCOCO dataset [22] as depicted in Fig. 4. In this approach instead of using a single fixed dimensional vector to represent the image, a set of fixed dimensional vectors from a lower convolution layer of the CNN architecture is used. This helps to maintain a fine-grained correspondence between the different portions of a $2D$ image represented through the corresponding vectors. With this the decoder becomes more powerful as it can focus selectively on different parts of an image during the generation process by selecting a subset of the feature vectors.

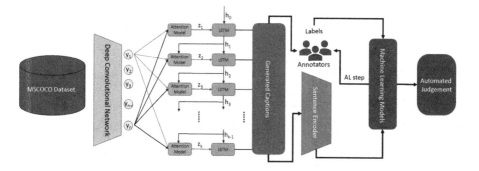

Fig. 3. Full schema for generating automated human judgement

The detailed operations of the LSTM based decoder, used in [32] for generating the captions, are described through the following equations,

$$
\begin{bmatrix} i_t \\ f_t \\ o_t \\ g_t \end{bmatrix} = \begin{bmatrix} \sigma \\ \sigma \\ \sigma \\ tanh \end{bmatrix} T_{D+m+n,n} \begin{bmatrix} E_{y(t-1)} \\ h_{t-1} \\ \hat{z}_t \end{bmatrix} \tag{1}
$$

$$
c_t = f_t \odot c_{t-1} + i_t \odot g_t \tag{2}
$$

$$
h_t = o_t \odot tanh(c_t) \tag{3}
$$

where, i_t, f_t, c_t, o_t, h_t denote input, forget, memory, output gates and the hidden state respectively. It is to be noted that T represents a mapping of the form $f_{s,t}$: $\mathbb{R}^s \rightarrow \mathbb{R}^t$. Thus, $T_{D+m+n,n}$ is a mapping from $\mathbb{R}^{(D+m+n)}$ to $\mathbb{R}^n$. $\hat{z} \in \mathbb{R}^D$ denotes the context vector responsible for capturing the visual information related to a specific location in the input image. E denotes the embedding matrix and has the dimension $m \times k$. The dimension of the embedding vector is given by m while the dimension of the LSTM hidden state is denoted by n. Furthermore, σ and $\odot$ represent the logistic sigmoid and element-wise multiplication respectively.

For handling the MSCOCO data, we adopt the data splits proposed in [14] in which the training set contains $113,287$ images with each having 5 corresponding captions while the validation and test sets contain $5,000$ images with each having 5 corresponding groundtruth captions. For our work, we build a vocabulary by dropping a word which has a frequency below 5 leading to a vocabulary size of $10,000$ words. We use image features obtained from the RESNET-101 architecture with 101 layers [11]. The dimensions for the LSTM hidden state, image, word and attention embeddings are set to 512 for our model. We train our model under the cross entropy objective, using beam search for the decoder and ADAM [17] as the preferred optimizer. We use beam search with a beam width of 200 from which we select the top three captions for each image. We use this setup to generate captions for all images in the MSCOCO test set and use them for evaluating our proposed approach for automatically inferring human judgement on them.

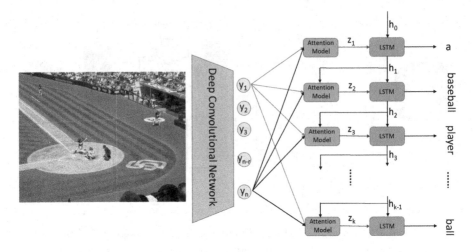

Fig. 4. Neural caption generation mechanism [32]

3.2 Features

We consider two different representations of the generated image captions for the purpose of training different classifiers for automating human judgement using active learning. First, we generate a dense vector representation of the captions using the pre-trained Universal Sentence Encoder [4]. It is a 512-dimensional vector, representing each caption, which promises to capture the context and semantic meaning of the sentence. We consider this representation to be useful for identifying syntactic or grammatical accuracy, image relevance and for identifying diverse captions, i.e., the ones which are more informative compared to the other describing the same image. The second representation for captions that we test is a 10 dimensional feature vector formed from different metrics which are popular in the caption generation community. These include overlap scores, such as, BLEU [27], ROUGE [21], METEOR [1] between the model generated captions and their corresponding groundtruths. Also Levenshtein distance, Levenshtein ratio and the ratio of number of unique unigrams, unique bigrams in the set of generated captions compared to the total number of words in the set of the generated captions.

3.3 Active Learning

We use pool-based active learning with uncertainty sampling for automating judgement on the quality of generated captions. We model the tasks of automatic human judgement on grammatical correctness, image relevance and diversity as binary classification problems. We initially select a random batch of generated captions and obtain human judgement labels for them using the crowdsourcing platform (Figure Eight https://www.figure-eight.com/). For each task, we train different classifiers with this initial labeled data and then apply them to the unlabeled pool of captions to predict their labels. For every active learning iteration we select 200 instances on which prediction probabilities for the binary labels, for each task, are between the threshold 0.45 to 0.55. These instances are annotated by crowdworkers and incorporated into the training set for re-training the respective classifiers. This cycle is repeated 4 times. For each iteration, we report the performances of the classifiers on a completely separate human labeled test set. Figure 3 provides a schematic diagram for the entire process.

We use a SVM classifier with three different oversampling techniques for handling data imbalance in grammatical correctness and relevance estimation: Random Oversampling (ROS), Synthetic Minority Oversampling [5] technique (SMOTE) and Adaptive Synthetic [8] oversampling (ADASYN). We use the SVM and logistic regression without any oversampling for inferring human judgments on diversity, because the labels are balanced. The data imbalance for grammatical correctness and image relevance stems from the fact that most of the model generated captions are grammatically correct and relevant to their corresponding images compared to the few which are incorrect. Whereas, for diversity the data is balanced as for each image there is only one caption which is diverse and another which is not diverse.

In brief, ROS employs oversampling randomly to handle the issue of class imbalance whereas SMOTE is an oversampling approach where the minority class is oversampled by creating synthetic samples instead of oversampling with replacement. Oversampling for the minority class is done by considering each observation in the minority class and then generating synthetic examples along the line segments joining any or all of the k minority class nearest neighbors. The k nearest neighbors are chosen randomly depending upon the amount of oversampling needed. ADASYN on the other hand, aims to reduce the learning bias introduced by the original imbalance in the data distribution and at the same time, it adaptively shifts the decision boundary to focus on those samples which are difficult to learn.

4   Experiments and Results

For automatically determining human judgement on the three quality aspects of the generated captions, we first conduct surveys on a crowdsourcing platform to obtain the labels for an initial batch of randomly selected captions. We train different classifiers using these labels under an active learning scheme and report their performances on a separate test set. The labels for the test set are obtained separately using the same crowdsourcing platform.

We show that the performance of the classifiers, under active learning, using the 512-dimensional feature vector representation obtained from the sentence encoder [4] is much better compared to the representation using standard metrics based vector for all the tasks. This also establishes a new baseline for generating automatic human judgements without groundtruth information.

4.1 Results of Active Learning for Grammatical Accuracy

It is important to note that the dataset for grammatical accuracy is highly imbalanced since most of the model generated captions are grammatically correct. So, we combine different oversampling techniques (ROS, SMOTE and ADASYN) with a SVM and report the $F1$ score on the test set for initial (*Base*) and subsequent active learning iterations (*Iter 1-4*) for which the classifier is retrained. Table 1 shows the scores for models trained with the vector representations from the Universal Sentence Encoder and Table 2 for models based on the 10-dimensional metric vector. $F1$ scores from the two tables establish that standard metrics perform poorly in comparison to the features obtained from the universal sentence encoder for automating judgement on grammatical accuracy.

Table 1. Grammatical accuracy: $F1$ score of SVM using vector representation from Universal Sentence Encoder.

Classifier	Base	Iter1	Iter2	Iter3	Iter4
ROS + SVM	0.6650	0.6925	0.6922	0.6911	0.6821
SMOTE + SVM	0.6440	0.6711	0.6711	0.6794	0.6828
ADASYN + SVM	0.6651	0.6446	0.6505	0.6757	0.6559

Table 2. Grammatical accuracy: $F1$ score of SVM with sentence representation using metric scores.

Classifier	Base	Iter1	Iter2	Iter3	Iter4
ROS + SVM	0.4473	0.4445	0.4373	0.3998	0.3233
SMOTE + SVM	0.4722	0.4401	0.4202	0.3880	0.4115
ADASYN + SVM	0.3746	0.3839	0.2886	0.4444	0.4444

4.2 Results of Active Learning for Image Relevance

The dataset for image relevance also suffers from data imbalance, which is why we use SVMs in combination with oversampling, as well. We report the $F1$ score obtained with each combination for initial and subsequent active learning iterations on the test set for caption representations using Google's Universal Sentence Encoder [4] (see Table 3) and the one using a vector of overlap metrics discussed above (see Table 4). For automatic human judgment on image relevance of the generated captions, we see that the features from the sentence encoder produce superior results compared to the standard metric based features.

Table 3. Image relevance: $F1$ score of SVM with sentence representation from Universal Sentence Encoder.

Classifier	Base	Iter1	Iter2	Iter3	Iter4
ROS + SVM	0.5863	0.5982	0.6028	0.5807	0.6005
SMOTE + SVM	0.5940	0.6098	0.5886	0.5757	0.6110
ADASYN + SVM	0.5901	0.6024	0.6214	0.6075	0.6254

Table 4. Image relevance: $F1$ score of SVM with sentence representation using metric scores.

Classifier	Base	Iter1	Iter2	Iter3	Iter4
ROS + SVM	0.5709	0.5389	0.5389	0.5399	0.5306
SMOTE + SVM	0.5706	0.5446	0.5315	0.5306	0.5122
ADASYN + SVM	0.4002	0.4138	0.5709	0.5306	0.5211

4.3 Results of Active Learning for Diversity

Finally, we report the $F1$ scores for predicting human judgement on the diversity of the generated captions using logistic regression and SVM models. Since the dataset for diversity is balanced, we do not use any of the oversampling techniques. Table 5 shows the scores for models using the Universal Sentence Encoder, Table 6 the scores for models using the metric vector.

From the tables below, we see that for automatically determining human judgement on diversity of the generated captions, we see that feature vectors obtained from the sentence encoder do not provide significant advantage over the metric based vectors.

Table 5. Diversity: $F1$ score of classifiers with sentence representation from Universal Sentence Encoder.

Classifier	Base	Iter1	Iter2	Iter3	Iter4
Log. Reg.	0.5294	0.5175	0.5411	0.5400	0.5288
SVM	0.5288	0.5116	0.4642	0.4630	0.4658

Table 6. Diversity: $F1$ score of classifiers with sentence representation using metric scores.

Classifier	Base	Iter1	Iter2	Iter3	Iter4
Log. Reg.	0.529	0.558	0.482	0.490	0.57
SVM	0.523	0.530	0.52	0.50	0.58

5 Discussion

The results from our experiments show that feature vectors obtained from the pretrained sentence encoder [4] produce much higher $F1$ scores compared to standard overlap metrics when employed for the task of automatically inferring human judgement on neural network generated image captions. We believe the reason behind this performance increase is that the vectors from the sentence encoder capture the semantic and syntactic information present in the captions more than the standard overlap metrics such as BLEU, ROUGE, METEOR etc. Moreover, representing the generated captions with fixed size feature vectors, obtained from the pretrained sentence encoder [4], do not require corresponding groundtruth information for the captions. In our opinion, this is a major advantage over standard metrics which are completely dependent on groundtruth information.

The results further indicate that we can automate human judgement on grammatical accuracy and image relevance more successfully compared to automatically determining human judgement on diversity. However, we believe our approach, which combines feature vectors and standard ML classifiers under the active learning scheme, can significantly reduce annotation cost. In addition, the requirement for groundtruth information for automating human judgement on different quality aspects of neural network generated captions and NLG evaluation in general is reduced.

6 Conclusion

We implemented a technical architecture and conducted experiments to demonstrate that active learning can be used for automatically generating human judgement on the quality of the captions generated by a neural image caption generator. For this purpose, we tested sentence representations obtained from Google's Universal Sentence Encoder and another one obtained using standard metrics computed between the generated captions and their corresponding groundtruths. Subsequently, we trained SVM and logistic regression classifiers under an active learning framework and reported the $F1$ scores for a separate test set.

The $F1$ scores of the used classifiers show that under active learning better results are obtained using the 512 dimensional vectors from Universal Sentence Encoder across all three tasks. Also, we found that under active learning better results are obtained for the task of automating judgement on grammatical correctness and image relevance compared to the performance of automating judgement on diversity. Note that automatic human judgement on quality assessment is novel and an important step towards automated quality assessments in the evaluation of image captions and natural language generation in general. Our approach will be tested in future experiments as we believe it can reduce manual evaluation costs thereby simplifying NLG evaluation significantly.

Acknowledgement. This research was funded in part by the German Federal Ministry of Education and Research (BMBF) under grant number 01IS17043 (project SciBot). Aditya Mogadala was supported by the German Research Foundation (DFG) as part of SFB1102.

References

1. Banerjee, S., Lavie, A.: METEOR: an automatic metric for MT evaluation with improved correlation with human judgments. In: Proceedings of the ACL Workshop on Intrinsic and Extrinsic Evaluation Measures for Machine Translation and/or Summarization, pp. 65–72 (2005)
2. Barz, M., Polzehl, T., Sonntag, D.: Towards hybrid human-machine translation services. EasyChair Preprint (2018)
3. Biswas, R.: Diverse Image Caption Generation And Automated Human Judgement through Active Learning. Master's thesis, Saarland University (2019)
4. Cer, D., et al.: Universal sentence encoder. arXiv:1803.11175 (2018)
5. Chawla, N., Bowyer, K., Hall, L., Kegelmeyer, W.: Smote: synthetic minority oversampling technique. J. Artif. Intell. Res. **16**, 321–357 (2002)
6. Cho, K., Merrienboer, B., Gulcehre, C., Bougares, F., Schwenk, H., Bengio, Y.: Learning phrase representations using RNN encoder-decoder for statistical machine translation. In: EMNLP (2014)
7. Donahue, J., et al.: Long-term recurrent convolutional networks for visual recognition and description. In: CVPR (2015)
8. Haibo, H., Bai, Y., Garcia, E., Li, S.: ADASYN: adaptive synthetic sampling approach for imbalanced learning. In: IEEE International Joint Conference on Neural Networks, pp. 1322–1328 (2008)
9. Harnad, S.: The symbol grounding problem. Physica **42**, 335–346 (1990)
10. Harzig, P., Brehm, S., Lienhart, R., Kaiser, C., Schallner, R.: Multimodal image captioning for marketing analysis, February 2018
11. He, K., Zhang, X., Ren, S., Sun, J.: Deep residual learning for image recognition. In: CVPR (2016)
12. Hochreiter, S., Schmidhuber, J.: Long short term memory. Neural Comput. **9**, 1735–1780 (1997)
13. Ioffe, S., Szegedy, C.: Batch normalization: Accelerating deep network training by reducing internal covariate shift. arXiv:1502.03167 (2015)
14. Karpathy, A., Fei-Fei, L.: Deep visual-semantic alignments for generating image descriptions. In: CVPR (2015)
15. Karpathy, A., Joulin, A., Fei-Fei, L.: Deep fragment embeddings for bidirectional image sentence mapping. In: NIPS (2014)
16. Kim, J., Rohrbach, A., Darrell, T., Canny, J., Akata, Z.: Textual explanations for self-driving vehicles. In: Ferrari, V., Hebert, M., Sminchisescu, C., Weiss, Y. (eds.) ECCV 2018. LNCS, vol. 11206, pp. 577–593. Springer, Cham (2018). https://doi.org/10.1007/978-3-030-01216-8_35
17. Kingma, D., Ba, J.: Adam: A method for stochastic optimization. In: ICLR (2015)
18. Kiros, R., Salahutdinov, R., Zemel, R.: Multimodal neural language models. In: ICLR, pp. 595–603 (2014)
19. Kiros, R., Salahutdinov, R., Zemel, R.: Unifying visual-semantic embeddings with multimodal neural language models. arXiv:1411.2539 (2014)
20. Kisilev, P., Sason, E., Barkan, E., Hashoul, S.Y.: Medical image captioning : learning to describe medical image findings using multitask-loss CNN (2016)

21. Lin, C.: Rouge: a package for automatic evaluation of summaries. In: Text Summarization Branches Out (2004)
22. Lin, T.-Y., et al.: Microsoft COCO: common objects in context. In: Fleet, D., Pajdla, T., Schiele, B., Tuytelaars, T. (eds.) ECCV 2014. LNCS, vol. 8693, pp. 740–755. Springer, Cham (2014). https://doi.org/10.1007/978-3-319-10602-1_48
23. Lowerre, B., Reddy, R.: The harpy speech understanding system. In: Readings in Speech Recognition, pp. 576–586 (1990)
24. Mao, J., Xu, W., Yang, Y., Wang, J., Yuille, A.: Deep captioning with multimodal recurrent neural networks (m-RNN). arXiv:1412.6632 (2014)
25. Oviatt, S., Schuller, B., Cohen, P., Sonntag, D., Potamianos, G.: The Handbook Of Multimodal-Multisensor Interfaces, Volume 1: Foundations, User Modeling, and Common Modality Combinations. ACM, New York (2017)
26. Oviatt, S., Schuller, B., Cohen, P., Sonntag, D., Potamianos, G., Kruger, A.: Introduction: scope, trends, and paradigm shift in the field of computer interfaces, pp. 1–15. ACM, New York (2017)
27. Papineni, K., Roukos, S., Ward, T., Zhu, W.: Bleu: a method for automatic evaluation of machine translation. In: Association for Computational Linguistics, pp. 311–318 (2002)
28. Roy, D., Reiter, E.: Connecting language to the world. Artif. Intell. **167**, 1–12 (2005)
29. Settles, B.: Active Learning Literature Survey, vol. 52, no. 55-66, p. 11. University of Wisconsin, Madison (2010)
30. Vinyals, O., Toshev, A., Bengio, S., Erhan, D.: Show and tell: a neural image caption generator. In: CVPR (2015)
31. Xu, A., Liu, Z., Guo, Y., Sinha, V., Akkiraju, R.: A new chatbot for customer service on social media. In: Proceedings of the 2017 CHI Conference on Human Factors in Computing Systems, pp. 3506–3510 (2017)
32. Xu, K., er al.: Show, attend and tell: neural image caption generation with visual attention. In: ICML (2015)

Imbalanced Stance Detection by Combining Neural and External Features

Fuad Mire Hassan[1,2]([⊠]) and Mark Lee[1]

[1] School of Computer Science, University of Birmingham, Birmingham, UK
f.mire@pgr.bham.ac.uk, m.g.lee@cs.bham.ac.uk
[2] Faculty of Computing, Simad University, Mogadishu, Somalia

Abstract. Stance detection is the task of determining the perspective "or stance" of pairs of text. Classifying the stance (e.g. *agree, disagree, discuss* or *unrelated*) expressed in news articles with respect to a certain claim is an important step in detecting fake news. Many neural and traditional models predict well on *unrelated* and *discuss* classes while they poorly perform on other minority represented classes in the Fake News Challenge-1 (FNC-1) dataset. We present a simple neural model that combines similarity and statistical features through a MLP network for news-stance detection. Aiding augmented training instances to overcome the data imbalance problem and adding batch-normalization and gaussian-noise layers enable the model to prevent overfitting and improve class-wise and overall accuracy. We also conduct additional experiments with a light-GBM and MLP network using the same features and text augmentation to show their effectiveness. In addition, we evaluate the proposed model on the Argument Reasoning Comprehension (ARC) dataset to assess the generalizability of the model. The experimental results of our models outperform the current state-of-the-art.

Keywords: Text categorization · Stance detection · Fake news

1 Introduction

Fake news detection has become one of the important research directions in Natural Language Processing (NLP). One of the key challenges faced by social media users or online news communities is that anyone can share fake news/false claims and easily propagate fake news through the Internet for financial or political gain [25]. Pomerleau and Rao [22] organized the first Fake News Challenge (FNC-1) and introduced a dataset for stance detection task which is an extension of the work of Ferreira & Vlachos [10]. Acknowledging the complexity of fake news detection task, the challenge organizers noted that tackling the fake-news stance detection problem could be the first step to help prevent the spread of misinformation. It could also assist human fact-checkers to identify incorrect claims by detecting the stance of relevant articles in knowledge-bases. The goal of this task is to determine the "perspective" stance of two pieces of text (e.g.

© Springer Nature Switzerland AG 2019
C. Martín-Vide et al. (Eds.): SLSP 2019, LNAI 11816, pp. 273–285, 2019.
https://doi.org/10.1007/978-3-030-31372-2_23

article headline and body). Specifically, the focus of this task is to detect the stance of a headline by predicting its class as *agree*, *disagree*, *discuss* or *unrelated* in relation to an article body.

Recent advances on news-stance detection have mostly employed an ensemble of feature engineering with MLPs [4,23,27], CNNs [2,29], RNNs [6,13] and Memory Networks [20]. These complex ensembles with deep neural layers [6,13], attention networks [6] and memory networks [20] have been shown to achieve state-of-the-art performance. But the increase in complexity of these neural models tends to overfitting on smaller datasets. In addition to that, we observe that previous classic and deep neural models [2,13,18,22] achieved better classification accuracy on *unrelated* and *discuss* classes with above 97% and 76% accuracy respectively but they also struggle at predicting the *disagree* class (e.g. 0–18%) and the predictions of *agree* class often fall short (e.g. 0–58%) as well because of the dataset's imbalanced class distribution.

Inspired by related work that applies machine learning models in combination with external features to news-stance detection, we explore other potential methods which can reduce textual noise and generate more training examples for minority classes in order to avoid overfitting and improve the models' robustness. By addressing a document-level stance detection problem over a FNC-1 dataset, this work explores two aspects in addition with deep learning: (a) an assessment of important external features and their predictive power with respect to also machine learning model; and (b) the performance of using regularization (gaussian and batch normalization) and text augmentation (text summarization and synonym replacement methods).

Our contributions are summarized as follows: (i) We combine a simple 1-layer Gated Recurrent Unit (GRU) model with various important features, fine-tuned using batch-normalization and gaussian-noise, to better make predictions for news-stance detection task. (ii) Experimental results show that this combined model with GloVe embeddings and text augmentation outperforms all previous models in both of the evaluation settings (e.g. the FNC-1 metric and the Macro-F1 metric) on the FNC-1 dataset. (iii) We also provide a cross-domain validation using ARC dataset and a comparative report about proposed feature-based models (e.g. light-GBM and MLP) as they achieve state-of-the-art performances.

The rest of the paper is structured as follows. In Sect. 2, we elaborate the related work for news-stance detection. Section 3 presents the details of the proposed models. Section 4 discusses our experimental procedure and results. Finally, Sect. 5 draws some conclusions of our work.

2   Related Work

The problem of news-stance detection has emerged recently at the FNC-1 lab challenge, which was organized by Pomerleau and Rao [22]. So far, researchers in the competition and the wider NLP community have built a range of models using traditional machine learning, deep learning (DL) or a combined model. Some of the works used hand-crafted features with classical methods [11,18,22] to detect the stance of an article-headline towards an article-body.

Different deep learning architectures, including MLPs, CNNs and RNNs together with hand-crafted features were also proposed to capture the semantic and the contextual similarity of the headline and body pairs. SOLAT team [2] presented the best performing model in the FNC-1 contest by combining a gradient-boosted decision trees classifier (with various classical features) and a deep CNN. The two other best performing teams [14,23] used different architectures of MLP classifiers along with different hand-engineered features to predict the stance. Thorne *et al.* [27] also used a stacked ensemble of five models including the baseline model of the contest in addition with another three MLP architectures.

Subsequent to the FNC-1 contest, Bhatt *et al.* [4] combined some statistical, external and neural features as they also employed an ensemble of MLP classifiers for the stance detection. A recent trend in deep learning towards Memory Neural Networks encouraged Mohtarami *et al.* [20] to deploy an end-to-end memory network (MemN2N) combined with Bag-of-Words (BoW) and its cosine similarity features to improve the performance of the stance detection task. A thorough feature ablation analysis of the FNC-1's top three systems was conducted by Hanselowski *et al.* [13] where they also proposed a stackLSTM architecture using the best features from the analysis. Xu *et al.* [29] conducted a study of transfer learning, called adversarial domain adaptation, from the FEVER domain to stance detection domain as they tried to improve on *agree* and *disagree* classes respectively. Recently, Borges *et al.* [6] proposed a deep neural network model for stance detection that is a combination of Bi-directional RNNs, an attention mechanism, max-pooling as well as external hand-crafted features. A recent work used conditional encoding and co-matching attention neural models [9] to classify the related-part (*agree, disagree* and *discuss*) of the stance detection pipeline.

Different from previous models, we use a simple 1-layer GRU model enhanced with statistical and similarity features, regularization methods as well as text augmentation techniques that was previously found useful to make models more robust to overfitting for sentence modeling tasks.

3 Methodology

In this section, we describe methods that we adopt for news-stance detection. Our methodology is a combined model of GRU [7] and feature-engineering heuristics fed through a fully connected MLP network to predict the stance of four-class classifications. We also made use of regularization and text augmentation techniques to improve the overall performance of the model. Moreover, we apply the same feature-engineering to train additional proposed models include a light-GBM (gradient boosting) and an MLP neural model.

The GRU network architecture [7] is a simplified version of a LSTM [15] and widely adopted in sequence modeling tasks [1,5]. The key difference is that the GRU network adaptively capture sequence dependencies over different time scales using an update gate and a reset gate as opposed to an LSTM model with three gates (i.e. forget, input and output gates) for the same purpose. The GRU network has fewer parameters to be trained compared to an LSTM and it can be able to train on smaller datasets efficiently while achieving comparable performances to an LSTM [8]. In addition, we use word embeddings to capture the semantic relations among similar words. The goal of an embedding layer is to transform the input text into a lower dimensional vectors. Following initial evaluation of five different word-embeddings, we choose 50-d GloVe and 300-d GloVe [21] for use in the later experiments described in this paper.

As a regularization techniques for our model, we also add Batch-Normalization (BN) layer and Gaussian-Noise (GN) layer with a standard deviation 0.1 to the output vectors before the fully connected layer. These methods can help prevent overfitting and optimize the neural network performance [28]. The GN layer adds noise to the input values while creating more samples but the output shape remains the same as the input. This will help for the neural network to learn and better able to generalize and improve the performance [28,30].

3.1   Features

In our proposed approach, we apply different preprocessing methods to create number of features composed in different groups from the dataset. To normalize words and remove noise in sentences, the text of news articles are tokenized, lowercased, and removed the stopwords, non-alphabets and punctuation in order to clean the dataset as we create external features and embedding vectors.

Based on our experimental analysis, the following external features in Table 1 were proven to convey additional relevant information that neural models cannot easily represent, so we combine them with the output of simple neural model. We include features taken from the FNC-1 official baseline and the top three performing models [2,14,23] in the competition such as Single Value Decomposition (SVD), Latent Dirichlet Allocation (LDA) and Non-negative Matrix Factorization (NMF). We also made use of a feature [13] that adds a "_NEG" tag to negate words that appear after a negative keyword in both of the 500 most frequent 2-grams BoW headline and body vectors. Besides the use of cosine distance to compare the similarity between headline and body 3000 most frequent BoW vectors, we also add Jaccard and Euclidean similarity distances that have been successfully applied in detecting Duplicate Questions [3]. Word mover's distance [17] feature of the word2vec [19] embeddings between the headline and the article also being generated. Common entities between the pairs are also being extracted as a feature by using the SpaCy toolkit. We also incorporate "agreeing word" count (a.g. confirm, support, valid, correct, etc.) and "hedging word"[1] count features as well as polarity and subjectivity features generated by using textblob library.

---

[1] https://en.wiktionary.org/wiki/Category:English_hedges.

Table 1. Set of features used in this study

Statistical Features
Overlapping (CW) character and word ngrams between headline and body
Overlapping (WT) word ngrams of top 25 TF-IDF-body-vectors in the headline
(RC) refuting, (AC) agreeing and (HC) hedging word counts of both text
(CE) common entities between headline and body text
(TP) polarity and (TS) subjectivity in the heads and top 25 body BoW
(ES) emotional scores [26] and (PS) polarity scores [16] in each of the text.
(BW) 3000 most common 3-gram BoW vectors from the headline and body text
(BW) negated-sign feature for 500 most frequent BoW of both text
Similarity Features
(BW) cosine, jaccard and euclidean similarities between head and body BoWs
(TC) cosine similarity feature between heads and body TF-IDF vectors
(SC) cosine similarity feature between 100 headline and body SVD components
(LC) cosine similarity feature between 100 headline and body LDA topics
(NC) cosine similarity feature between 50 headline and body NMF topics
(WM) WMD similarity between the headline and body word-embedding vectors

3.2 Data Augmentation

Data augmentation is a common way to expand the number of training instances in order to avoid overfitting and control generalization error for machine learning models. As the amounts of *disagree* and *agree* training instances in FNC-1 dataset are very small compared to *unrelated* and *discuss* pairs, we paid a special attention to text augmentation techniques to enlarge the minority classes through label-preserving transformations. We hypothesize that summarizing news article body will help reduce the noisy text as extractive summarization [24] can produce a summary while preserving the meaning of the original text. In addition, the summarized version of the news body can be used to create new training instances by replacing random words with their synonyms from the thesaurus [31].

Our approach involves using an extractive centroid-based text summarization technique [24] to summarize text pairs of minority classes. The centroid is the document vector which is computed as the average word embeddings of the most common words occurring in the document. To generate the summary sentences, this algorithm selects the closest vectors (sentences in the document) which have vectors similar to centroid embeddings. In contrast to other methods [31], we use GloVe embeddings to find synonyms for randomly-chosen words from the summary and then we replace 30% of the original words with their synonyms.

4   Experiments

4.1   Dataset

We evaluate the proposed approach on the FNC-1 dataset [22] composed of textual documents for the task of Stance Detection. There are 49972 instances of headlines and news documents in the training set whereas the test-set comprises of 25,413 pairs. The distribution of the classes of headline/article pairs is highly imbalanced where the *unrelated* (UNR) pairs are approximately 73% while the other three classes share only the remaining 27% of the whole dataset (e.g. *agree* (AGR): 7.4%, *disagree*(DSG): 2% and *discuss* (DSC): 17.7%). To test the generalizability of the proposed model, we also use an Argument Reasoning Comprehension (ARC) dataset [12]. This dataset, composed of 17,792 claims and multi-sentence user posts (e.g. *agree*: 8.9%, *disagree*: 10%, *discuss*: 6.1% and *unrelated*: 75%), is designed for stance detection by Hanselowski *et al.* [13]. The dataset is divided 80/20 for training and test sets.

4.2   Metrics

The FNC-1 organizers introduced a mechanism to evaluate the performance of the models in the competition. The evaluation metric weights the score of 25% for correctly classifying *related/unrelated* pairs and 75% for correctly classifying the related instances into further three-class classifications (e.g. *agree, disagree* and *discuss*). Also, following the previous work by Hanselowski *et al.* [13], they suggested using a Macro-F1 metric because the FNC-1 evaluation metric does not take into account the imbalanced distribution of related classes and undermines the fair evaluation of class-wise performance as explained in [4,13]. We use both evaluation metrics to show the performance of our models.

4.3   Experimental Procedure

We estimate the best hyper-parameters using a grid search and we finally set the hyper-parameters of light-GBM as: learning rate - 0.09, number of leaves - 50, number of boosting rounds - 1000 and early stopping rounds - 50. The MLP model consists of the external features with three layers of 600 neurons and rectified linear unit (ReLu) activation function followed by a softmax classifier. To improve the performance, we train a GRU model on top of pre-trained word embeddings to generate 100-vector (e.g. headlines and bodies) input sequences. We use 50-d GloVe embedding together with a single-layer GRU of 64 neurons for each of the headline and body vectors as we set the probability of dropout and recurrent_dropout to 0.2 and 0.1 respectively. This is followed by a concatenation layer with handcrafted features, a batch-normalization layer, a gaussian-noise layer (e.g. set to 0.1) and then, three fully connected layers with 600 neurons and ReLU optimizer. Finally, the outputs of the 4 classes are decoded by a softmax classifier. To evaluate the effectiveness of our text augmentation methods, we trained the same light-GBM, MLP and the combined GRU models as we add more training instances of 3678 and 840 for *agree* and *disagree* classes respectively. In addition to these models, we provide a weighted-sum average of the predictions from the experiments of each model with and without text augmentation.

The implementation of our models are based on Keras with Tenserflow as a backend and a light-GBM library. We use the Adam weight optimizer and we set epochs to 10 and batch size to 100 for the both of the combined-GRU and the MLP experiments. 20% of the training-set is being used as a development-set.

Table 2. Comparison with the state-of-the-art traditional models

Models	UNR-F1	AGR-F1	DSG-F1	DSC-F1	FNC-1	Macro-F1
GBT (FNC-1 baseline)	97.98	9.09	1.00	79.66	75.20	46.93
3-step LR/RF [18]	98.00	52.00	1.00	76.00	82.10	56.75
SVM [11]	–	–	–	–	–	58
Prposed Light-GBM						
Light-GBM	**99.13**	57.75	2.87	**80.00**	**83.40**	59.94
Light-GBM-Augment	98.88	**68.47**	4.16	71.42	82.27	60.73
Ensemble-Model	98.91	67.89	**4.30**	72.22	82.44	**60.83**

4.4 Results

In this section, we empirically evaluate our proposed models with the literature.

Light-GBM: Table 2 compares the results of our light-GBM model against the state-of-the-art traditional approaches with different feature-groups for this task including: Gradient Boost Decision Tree (GBT) baseline from the FNC-1 contest [22], 3-step Logistic Regression (LR) and Random Forest (RF) [18] as well as Support Vector Machine (SVM) [11]. The light-GBM model performs better than the FNC-1's baseline and other state-of-the-art classical models for this task, with our light GBM's performance increased by a significant margin in each of the evaluation metric as presented in Table 2. The Light-GBM model with text augmentation effectively improves the class *agree* and the overall Macro-F1 while the weighted ensemble model yields a slight increase on the Macro-F1 score.

Figure 1 reveals the ablation study of this model as we remove one feature-set (refer to Table 1) in each run of an experiment. We find that removing any of the feature-set from the model leads to a reduced performance in terms of FNC-1 and F1 scores. We also observe that leaving-out BW and TC features produces the worst FNC and F1 scores of 81.97% and 57.98% respectively.

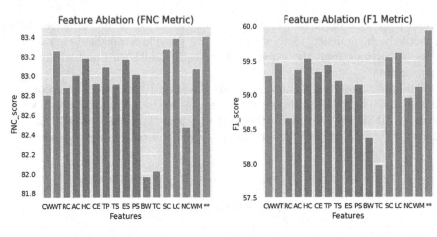

Fig. 1. Feature Ablation Study (** denotes All features)

Table 3. Comparison with the state-of-the-art MLP models

Models	UNR-F1	AGR-F1	DSG-F1	DSC-F1	FNC-1	Macro-F1
1-layer MLP [23]	97.90	44.04	6.60	81.38	81.72	57.48
7-layer MLP [14]	**99.25**	44.72	9.47	80.89	81.97	58.58
Ensemble-MLP [27]	–	–	–	–	78.04	–
Ensemble-MLP [4]	98.04	43.82	6.31	**85.68**	**83.08**	58.46
2-layer MLP [11]	–	–	–	–	–	59.60
Proposed 3-layer MLP						
MLP	98.95	53.13	12.20	79.61	**83.08**	60.97
MLP-Augment	96.71	**60.64**	**21.52**	72.13	81.53	62.75
Ensemble-Model	96.93	60.33	21.23	73.30	81.82	**62.95**

Multi-layer Perceptron (MLP): We also train a MLP neural network model on the same external features. Variations of MLP approaches have been employed on this task with different combination of features including: 1-layer MLP neural network [23], 7-layers MLP architecture [14], ensemble of three MLP networks (among five classifiers) [27], ensemble of three MLP architectures [4] and 2-layer MLP neural networks [11]. The proposed MLP achieved 83.08% of FNC-1 and 60.97% of Macro-F1 scores as shown in Table 3 outperforming all previous MLPs. Moreover, MLP with text augmentation model shows a substantial increase (more than 2% improvement compared to other methods) on Macro-F1 and the weighted ensemble model improves a bit on the Macro-F1 score.

Ensemble-GRU Model: Table 4 compares the evaluation results of the proposed model with the state-of-the-art neural models. GRU-MLP Baseline performs the worst because it does not use any of the external features, batch-normalization and gaussian-noise layers. GRU-MLP-External baseline performs comparably good in terms of FNC-1 (82.88%) and Macro-F1 (59.59%) metrics but it does not improve the results for *disagree* and *agree* classes. The proposed ensemble GRU-MLP-External-BN-GN model with 50-d GloVe shows higher Macro-F1 score for the stance detection setting.

Table 4. Comparison with the state-of-the-art neural models

Models	UNR-F1	AGR-F1	DSG-F1	DSC-F1	FNC-1	Macro-F1
Ensemble-CNN/GBT [2]	98.70	58.50	1.86	76.18	82.02	58.81
MemoryNN-BoW [20]	–	–	–	–	81.23	56.88
stackLSTM [13]	**99.50**	50.10	18.00	75.70	82.10	60.90
Transfer learning [29]	97.70	54.60	15.10	72.60	80.30	60.00
DeepNet (best encoder) [6]	96.74	51.34	10.33	**81.52**	82.23	59.98
Proposed model baselines						
GRU-MLP	86.94	26.54	0.86	49.96	60.37	41.07
GRU-MLP-External	98.98	56.28	4.30	78.79	**82.88**	59.59
Proposed Model: GRU-MLP-External-BN-GN						
GloVe-6B-50d	98.48	60.43	15.64	74.33	82.36	**62.22**
GloVe-6B-50d-Augment	96.62	**66.47**	23.39	65.32	80.26	**62.95**
Ensemble-Model	97.22	65.32	**24.53**	68.86	81.46	**63.98**

However, when text augmentation is added with the proposed model, it demonstrates the best results on *agree* and *disagree* scores of 66.47% (from 58.50% [2]) and 23.39% (from 18.0% [13]) as it performs comparably well to previous approaches in other two classes. Table 4 shows that a weighted ensemble model achieves higher overall accuracy, 63.98% on the Macro-F1 metric, compared to the previous models' highest Macro-F1 score 60.90% [13]. This shows that the simple ensemble GRU model optimized with text augmentation methods and regularization layers is robust in terms of class-wise accuracy compared to other complex ensemble models [2,6,13,20,29] and the overall Macro-F1 metric was improved by more than 3% from 60.90% to 63.98%.

ARC Dataset for Cross-domain Validation: we test the combined GRU model on the ARC dataset to determine its generalizability. Table 5 presents the performance of the cross-domain evaluations. For in-domain ARC-ARC training and test scenario, our model gives a slight improvement on both of the evaluation metrics over the stackLSTM model [13]. For the cross-domain test, we find that the proposed model trained on the ARC training-set and tested on the FNC-1 test-set outperforms by a large margin (11.82%) on Macro-F1 compared to the stackLSTM model in [13]. We also observe that the cross-domain FNC-ARC test performance on our proposed approach is lower than the stackLSTM in [13] but our approach still receives higher performance than the other model on the *unrelated* and *discuss* classes.

Table 5. Cross-domain evaluation using ARC dataset

Models	UNR-F1	AGR-F1	DSG-F1	DSC-F1	FNC-1	Macro-F1
ARC-ARC$_1$						
stackLSTM [13]	93.50	45.10	**51.80**	19.4	68.50	52.40
GloVe-6B-300d-BN-GN	**95.18**	**47.60**	46.51	**27.37**	**69.95**	**54.17**
ARC-FNC-1						
stackLSTM [13]	**95.00**	34.30	11.60	8.20	61.30	37.30
GloVe-6B-300d-BN-GN	94.57	**71.36**	**21.09**	**9.48**	**63.94**	**49.12**
FNC-ARC						
stackLSTM [13]	91.00	**32.10**	**19.10**	18.20	**59.10**	**40.10**
GloVe-6B-300d-BN-GN	**91.14**	26.35	5.38	**28.49**	56.05	37.84

4.5 Discussion

Our proposed light-GBM with all the external features achieves the highest FNC-1 weighted accuracy 83.40%. It is interesting to see that our proposed light-GBM variations with all the external features perform on par, i.e. 60.83%, with previous state-of-the-art (e.g. 60.90%) stackLSTM model [13] in terms of Macro-F1 as shown in Tables 2 and 4. Light-GBM with text augmentation model also performs the best in terms of *agree* class (68.47%). But, we observe that the light-GBM models lack the semantic understanding required in improving the detection of class *disagree* as the other traditional models have difficulties making better predictions due to the imbalance of training data. Furthermore, our experiments show the importance of the proposed feature-based MLP and the combined GRU models where they strengthened the semantic-understanding ability to predict and improve class-wise and overall Macro-F1 score by a significant margin. Our experiments also demonstrate that text augmentation and regularization methods such as batch-normalization and gaussian-noise are useful methods that can help prevent overfitting to overcome the class imbalance problem. We have seen that it is possible to outperform the state-of-the-art results with these simple models compared with complex deep learning ensembles [6,13,20]. Our simple 1-layer GRU model improves on the current state-of-the-art over 3% on Macro-F1 metric as illustrated in Table 4. The proposed GRU-MLP-External-BN-GN is also applied to ARC dataset for cross-domain validation and it has shown better performance compared to previous stackLSTM model [13].

5   Conclusion

In this paper, we presented a simple combined model of deep learning with hand-crafted features for automating the stance detection on news headline and body pairs which improves on the state-of-the-art accuracy for the FNC-1 dataset. We first generated different groups of classical features from headlines and bodies.

We then integrate the external features and a simple 1-layer GRU neural network with 50-d GloVe pre-trained embedding to boost the semantic understanding of the model. This combined model is optimized with text augmentation, batch-normalization and gaussian-noise regularization methods to provide significant improvement. We also show that light-GBM and MLP models with the same hand-engineered features provide state-of-the-art results on this task. As future work, we will investigate how different state-of-the-art neural sentence modeling architectures can be applied to understand the deeper semantics of the article sentences and their interactions with the headlines so as to improve the performance of the fake news detection. We will also consider multitask and transfer learning methodologies to exploit knowledge from other related domains.

References

1. Attardi, G., Carta, A., Errica, F., Madotto, A., Pannitto, L.: Fa3l at semeval-2017 task 3: a three embeddings recurrent neural network for question answering. In: Proceedings of the 11th International Workshop on Semantic Evaluation (SemEval-2017), pp. 299–304 (2017)
2. Baird, S., Sibley, D., Pan, Y.: Talos targets disinformation with fake news challenge (2017). https://blog.talosintelligence.com/2017/06/talos-fake-news-challenge.html
3. Baldwin, T., Liang, H., Salehi, B., Hoogeveen, D., Li, Y., Duong, L.: UniMelb at semeval-2016 task 3: identifying similar questions by combining a CNN with string similarity measures. In: Proceedings of the 10th International Workshop on Semantic Evaluation (SemEval-2016), pp. 851–856 (2016)
4. Bhatt, G., Sharma, A., Sharma, S., Nagpal, A., Raman, B., Mittal, A.: Combining neural, statistical and external features for fake news stance identification. In: Companion Proceedings of the Web Conference 2018, pp. 1353–1357. International World Wide Web Conferences Steering Committee (2018)
5. Bogdanova, D., Foster, J., Dzendzik, D., Liu, Q.: If you can't beat them join them: handcrafted features complement neural nets for non-factoid answer reranking. In: Proceedings of the 15th Conference of the European Chapter of the Association for Computational Linguistics: Volume 1, pp. 121–131. Long Papers (2017)
6. Borges, L., Martins, B., Calado, P.: Combining similarity features and deep representation learning for stance detection in the context of checking fake news. J. Data Inf. Qual. (JDIQ) **11**, 14 (2019). ACM
7. Cho, K., Van Merriënboer, B., Bahdanau, D., Bengio, Y.: On the properties of neural machine translation: Encoder-decoder approaches. arXiv preprint arXiv:1409.1259 (2014)
8. Chung, J., Gulcehre, C., Cho, K., Bengio, Y.: Empirical evaluation of gated recurrent neural networks on sequence modeling. arXiv preprint arXiv:1412.3555 (2014)
9. Conforti, C., Pilehvar, M.T., Collier, N.: Towards automatic fake news detection: cross-level stance detection in news articles. In: Proceedings of the First Workshop on Fact Extraction and VERification (FEVER), pp. 40–49 (2018)
10. Ferreira, W., Vlachos, A.: Emergent: a novel data-set for stance classification. In: Proceedings of the 2016 Conference of the North American Chapter of the Association for Computational Linguistics: Human Language Technologies, pp. 1163–1168 (2016)

11. Ghanem, B., Rosso, P., Rangel, F.: Stance detection in fake news a combined feature representation. In: Proceedings of the First Workshop on Fact Extraction and VERification (FEVER), pp. 66–71 (2018)

12. Habernal, I., Wachsmuth, H., Gurevych, I., Stein, B.: The argument reasoning comprehension task: Identification and reconstruction of implicit warrants. arXiv preprint arXiv:1708.01425 (2017)

13. Hanselowski, A., et al.: A retrospective analysis of the fake news challenge stance-detection task. In: Proceedings of the 27th International Conference on Computational Linguistics, pp. 1859–1874 (2018)

14. Hanselowski, A., PVS, A., Schiller, B., Caspelherr, F.: Description of the system developed by team Athene in the FNC-1 (2017). https://medium.com/@andre134679/team-athene-on-the-fake-news-challenge-28a5cf5e017b

15. Hochreiter, S., Schmidhuber, J.: Long short-term memory. Neural Comput. **9**(8), 1735–1780 (1997). MIT Press

16. Hutto, C.J., Gilbert, E.: Vader: a parsimonious rule-based model for sentiment analysis of social media text. In: Eighth International AAAI Conference on Weblogs and Social Media (2014)

17. Kusner, M., Sun, Y., Kolkin, N., Weinberger, K.: From word embeddings to document distances. In: International Conference on Machine Learning, pp. 957–966 (2015)

18. Masood, R., Aker, A.: The fake news challenge: stance detection using traditional machine learning approaches. In: Proceedings of the 10th International Joint Conference on Knowledge Discovery, Knowledge Engineering and Knowledge Management (KMIS), pp. 128–135 (2018)

19. Mikolov, T., Le, Q.V., Sutskever, I.: Exploiting similarities among languages for machine translation. arXiv preprint arXiv:1309.4168 (2013)

20. Mohtarami, M., Baly, R., Glass, J., Nakov, P., Màrquez, L., Moschitti, A.: Automatic stance detection using end-to-end memory networks. arXiv preprint arXiv:1804.07581 (2018)

21. Pennington, J., Socher, R., Manning, C.: Glove: Global vectors for word representation. In: Proceedings of the 2014 Conference on Empirical Methods in Natural Language Processing (EMNLP), pp. 1532–1543 (2014)

22. Pomerleau, D., Rao, D.: Fake News Challenge (2017). http://www.fakenewschallenge.org/

23. Riedel, B., Augenstein, I., Spithourakis, G.P., Riedel, S.: A simple but tough-to-beat baseline for the fake news challenge stance detection task. arXiv preprint arXiv:1707.03264 (2017)

24. Rossiello, G., Basile, P., Semeraro, G.: Centroid-based text summarization through compositionality of word embeddings. In: Proceedings of the MultiLing 2017 Workshop on Summarization and Summary Evaluation Across Source Types and Genres, pp. 12–21 (2017)

25. Shu, K., Sliva, A., Wang, S., Tang, J., Liu, H.: Fake news detection on social media: a data mining perspective. ACM SIGKDD Explor. Newsl. **19**(1), 22–36 (2017). ACM

26. Staiano, J., Guerini, M.: Depechemood: a lexicon for emotion analysis from crowd-annotated news. arXiv preprint arXiv:1405.1605 (2014)

27. Thorne, J., Chen, M., Myrianthous, G., Pu, J., Wang, X., Vlachos, A.: Fake news stance detection using stacked ensemble of classifiers. In: Proceedings of the 2017 EMNLP Workshop: Natural Language Processing meets Journalism, pp. 80–83 (2017)

28. Tommasel, A., Rodriguez, J.M., Godoy, D.: Textual aggression detection through deep learning. In: Proceedings of the First Workshop on Trolling, Aggression and Cyberbullying (TRAC-2018), pp. 177–187 (2018)
29. Xu, B., Mohtarami, M., Glass, J.: Adversarial domain adaptation for stance detection. arXiv preprint arXiv:1902.02401 (2019)
30. Zhang, D., Yang, Z.: Word embedding perturbation for sentence classification. arXiv preprint arXiv:1804.08166 (2018)
31. Zhang, X., LeCun, Y.: Text understanding from scratch. arXiv preprint arXiv:1502.01710 (2015)

Prediction Uncertainty Estimation
for Hate Speech Classification

Kristian Miok[1]([⊠]), Dong Nguyen-Doan[1], Blaž Škrlj[2], Daniela Zaharie[1],
and Marko Robnik-Šikonja[3]

[1] Computer Science Department, West University of Timisoara,
Bulevardul Vasile Pârvan 4, 300223 Timişoara, Romania
{kristian.miok,dong.nguyen10,daniela.zaharie}@e-uvt.ro
[2] Jožef Stefan Institute and Jožef Stefan International Postgraduate School,
Jamova 39, 1000 Ljubljana, Slovenia
blaz.skrlj@ijs.si
[3] Faculty of Computer and Information Science, University of Ljubljana,
Večna pot 113, 1000 Ljubljana, Slovenia
marko.robnik@fri.uni-lj.si

Abstract. As a result of social network popularity, in recent years, hate speech phenomenon has significantly increased. Due to its harmful effect on minority groups as well as on large communities, there is a pressing need for hate speech detection and filtering. However, automatic approaches shall not jeopardize free speech, so they shall accompany their decisions with explanations and assessment of uncertainty. Thus, there is a need for predictive machine learning models that not only detect hate speech but also help users understand when texts cross the line and become unacceptable.

The reliability of predictions is usually not addressed in text classification. We fill this gap by proposing the adaptation of deep neural networks that can efficiently estimate prediction uncertainty. To reliably detect hate speech, we use Monte Carlo dropout regularization, which mimics Bayesian inference within neural networks. We evaluate our approach using different text embedding methods. We visualize the reliability of results with a novel technique that aids in understanding the classification reliability and errors.

Keywords: Prediction uncertainty estimation ·
Hate speech classification · Monte Carlo dropout method ·
Visualization of classification errors

1 Introduction

Hate speech represents written or oral communication that in any way discredits a person or a group based on characteristics such as race, color, ethnicity, gender, sexual orientation, nationality, or religion [35]. Hate speech targets disadvantaged social groups and harms them both directly and indirectly [33]. Social

© Springer Nature Switzerland AG 2019
C. Martín-Vide et al. (Eds.): SLSP 2019, LNAI 11816, pp. 286–298, 2019.
https://doi.org/10.1007/978-3-030-31372-2_24

networks like Twitter and Facebook, where hate speech frequently occurs, receive many critics for not doing enough to deal with it. As the connection between hate speech and the actual hate crimes is high [4], the importance of detecting and managing hate speech is not questionable. Early identification of users who promote such kind of communication can prevent an escalation from speech to action. However, automatic hate speech detection is difficult, especially when the text does not contain explicit hate speech keywords. Lexical detection methods tend to have low precision because, during classification, they do not take into account the contextual information those messages carry [11]. Recently, contextual word and sentence embedding methods capture semantic and syntactic relation among the words and improve prediction accuracy.

Recent works on combining probabilistic Bayesian inference and neural network methodology attracted much attention in the scientific community [23]. The main reason is the ability of probabilistic neural networks to quantify trustworthiness of predicted results. This information can be important, especially in tasks were decision making plays an important role [22]. The areas which can significantly benefit from prediction uncertainty estimation are text classification tasks which trigger specific actions. Hate speech detection is an example of a task where reliable results are needed to remove harmful contents and possibly ban malicious users without preventing the freedom of speech. In order to assess the uncertainty of the predicted values, the neural networks require a Bayesian framework. On the other hand, Srivastava et al. [32] proposed a regularization approach, called dropout, which has a considerable impact on the generalization ability of neural networks. The approach drops some randomly selected nodes from the neural network during the training process. Dropout increases the robustness of networks and prevents overfitting. Different variants of dropout improved classification results in various areas [1]. Gal and Ghahramani [14] exploited the interpretation of dropout as a Bayesian approximation and proposed a Monte Carlo dropout (MCD) approach to estimate the prediction uncertainty. In this paper, we analyze the applicability of Monte Carlo dropout in assessing the predictive uncertainty.

Our main goal is to accurately and reliably classify different forms of text as hate or non-hate speech, giving a probabilistic assessment of the prediction uncertainty in a comprehensible visual form. We also investigate the ability of deep neural network methods to provide good prediction accuracy on small textual data sets. The outline of the proposed methodology is presented in Fig. 1.

Our main contributions are:

- investigation of prediction uncertainty assessment to the area of text classification,
- implementation of hate speech detection with reliability output,
- evaluation of different contextual embedding approaches in the area of hate speech,
- a novel visualization of prediction uncertainty and errors of classification models.

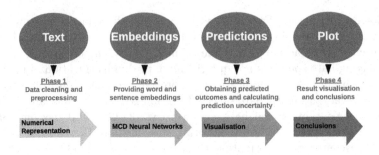

Fig. 1. The diagram of the proposed methodology.

The paper consists of six sections. In Sect. 2, we present related works on hate speech detection, prediction uncertainty assessment in text classification context, and visualization of uncertainty. In Sect. 3, we propose the methodology for uncertainty assessment using dropout within neural network models, as well as our novel visualization of prediction uncertainty. Section 4 presents the data sets and the experimental scenario. We discuss the obtained results in Sect. 5 and present conclusions and ideas for further work in Sect. 6.

2   Related Work

We shortly present the related work in three areas which constitute the core of our approach: hate speech detection, recurrent neural networks with Monte Carlo dropout for assessment of prediction uncertainty in text classification, and visualization of predictive uncertainty.

2.1   Hate Speech Detection

Techniques used for hate speech detection are mostly based on supervised learning. The most frequently used classifier is the Support Vector Machines (SVM) method [30]. Recently, deep neural networks, especially recurrent neural network language models [20], became very popular. Recent studies compare (deep) neural networks [9,12,28] with the classical machine learning methods.

Our experiments investigate embeddings and neural network architectures that can achieve superior predictive performance to SVM or logistic regression models. More specifically, our interest is to explore the performance of MCD neural networks applied to the hate speech detection task.

2.2   Prediction Uncertainty in Text Classification

Recurrent neural networks (RNNs) are a popular choice in text mining. The dropout technique was first introduced to RNNs in 2013 [34] but further research revealed negative impact of dropout in RNNs, especially within language modeling. For example, the dropout in RNNs employed on a handwriting recognition task, disrupted the ability of recurrent layers to effectively model sequences [25]. The dropout was successfully applied to language modeling by [36] who applied it only on fully connected layers. The then state-of-the-art results were explained with the fact that by using the dropout, much deeper neural networks can be constructed without danger of overfitting. Gal and Ghahramani [15] implemented the variational inference based dropout which can also regularize recurrent layers. Additionally, they provide a solution for dropout within word embeddings. The method mimics Bayesian inference by combining probabilistic parameter interpretation and deep RNNs. Authors introduce the idea of augmenting probabilistic RNN models with the prediction uncertainty estimation. Recent works further investigate how to estimate prediction uncertainty within different data frameworks using RNNs [37]. Some of the first investigation of probabilistic properties of SVM prediction is described in the work of Platt [26]. Also, investigation how Bayes by Backprop (BBB) method can be applied to RNNs was done by [13].

Our work combines the existing MCD methodology with the latest contextual embedding techniques and applies them to hate speech classification task. The aim is to obtain high quality predictions coupled with reliability scores as means to understand the circumstances of hate speech.

2.3 Prediction Uncertainty Visualization in Text Classification

Visualizations help humans in making decisions, e.g., select a driving route, evacuate before a hurricane strikes, or identify optimal methods for allocating business resources. One of the first attempts to obtain and visualize latent space of predicted outcomes was the work of Berger et al. [2]. Prediction values were also visualized in geo-spatial research on hurricane tracks [10,29]. Importance of visualization for prediction uncertainty estimation in the context of decision making was discussed in [17,18].

We are not aware of any work on prediction uncertainty visualization for text classification or hate speech detection. We present visualization of tweets in a two dimensional latent space that can reveal relationship between analyzed texts.

3 Deep Learning with Uncertainty Assessment

Deep learning received significant attention in both NLP and other machine learning applications. However, standard deep neural networks do not provide information on reliability of predictions. Bayesian neural network (BNN) methodology can overcome this issue by probabilistic interpretation of model parameters. Apart from prediction uncertainty estimation, BNNs offer robustness to overfitting and can be efficiently trained on small data sets [16]. However, neural networks that apply Bayesian inference can be computationally expensive, especially the ones with the complex, deep architectures. Our work is based on Monte Carlo Dropout (MCD) method proposed by [14]. The idea of this approach is to capture prediction uncertainty using the dropout as a regularization technique.

In contrast to classical RNNs, Long Short-term Memory (LSTM) neural networks introduce additional gates within the neural units. There are two sources of information for specific instance t that flows through all the gates: input values x_t and recurrent values that come from the previous instance h_{t-1}. Initial attempts to introduce dropout within the recurrent connections were not successful, reporting that dropout brakes the correlation among the input values. Gal and Ghahramani [15] solve this issue using predefined dropout mask which is the same at each time step. This opens the possibility to perform dropout during each forward pass through the LSTM network, estimating the whole distribution for each of the parameters. Parameters' posterior distributions that are approximated with such a network structure, $q(\omega)$, is used in constructing posterior predictive distribution of new instances y^*:

$$p(y^*|x^*, D) \approx \int p(y^*|f^\omega(x^*))\, q(\omega)d\omega, \tag{1}$$

where $p(y^*|f^\omega(x^*))$ denotes the likelihood function. In the regression tasks, this probability is summarized by reporting the means and standard deviations while for classification tasks the mean probability is calculated as:

$$\frac{1}{K}\sum_{k=1}^{K} p(y^*|x^*, \hat{\omega}_k) \tag{2}$$

where $\hat{\omega}_k \sim q(\omega)$. Thus, collecting information in K dropout passes throughout the network during the training phase is used in the testing phase to generate (sample) K predicted values for each of the test instance. The benefit of such results is not only to obtain more accurate prediction estimations but also the possibility to visualize the test instances within the generated outcome space.

3.1   Prediction Uncertainty Visualization

For each test instance, the neural network outputs a vector of probability estimates corresponding to the samples generated through Monte Carlo dropout. This creates an opportunity to visualize the variability of individual predictions. With the proposed visualization, we show the correctness and reliability of individual predictions, including false positive results that can be just as informative as correctly predicted ones. The creation of visualizations consists of the following five steps, elaborated below.

1. Projection of the vector of probability estimates into a two dimensional vector space.
2. Point coloring according to the mean probabilities computed by the network.
3. Determining point shapes based on correctness of individual predictions (four possible shapes).
4. Labeling points with respect to individual documents.
5. Kernel density estimation of the projected space—this step attempts to summarize the instance-level samples obtained by the MCD neural network.

As the MCD neural network produces hundreds of probability samples for each target instance, it is not feasible to directly visualize such a multi-dimensional space. To solve this, we leverage the recently introduced UMAP algorithm [19], which projects the input d dimensional data into a s-dimensional (in our case $s = 2$) representation by using computational insights from the manifold theory. The result of this step is a two dimensional matrix, where each of the two dimensions represents a latent dimension into which the input samples were projected, and each row represents a text document.

In the next step, we overlay the obtained representation with other relevant information, obtained during sampling. Individual points (documents) are assigned the mean probabilities of samples, thus representing the reliability of individual predictions. We discretize the $[0, 1]$ probability interval into four bins of equal size for readability purposes. Next, we shape individual points according to the correctness of predictions. We take into account four possible outcomes (TP - true positives, FP - false positives, TN - true negatives, FN - false negatives).

As the obtained two dimensional projection represents an approximation of the initial sample space, we compute the kernel density estimation in this subspace and thereby outline the main neural network's predictions. We use two dimensional Gaussian kernels for this task.

The obtained estimations are plotted alongside individual predictions and represent densities of the neural network's focus, which can be inspected from the point of view of correctness and reliability.

4 Experimental Setting

We first present the data sets used for the evaluation of the proposed approach, followed by the experimental scenario. The results are presented in Sect. 5.

4.1 Hate Speech Data Sets

We use three data sets related to the hate speech.

1 - HatEval data set is taken from the SemEval task "Multilingual detection of hate speech against immigrants and women in Twitter (hatEval)[1]". The competition was organized for two languages, Spanish and English; we only processed the English data set. The data set consists of 100 tweets labeled as 1 (hate speech) or 0 (not hate speech).

2 - YouToxic data set is a manually labeled text toxicity data, originally containing 1000 comments crawled from YouTube videos about the Ferguson unrest in 2014[2]. Apart from the main label describing if the comment is hate

---

[1] https://competitions.codalab.org/competitions/19935.
[2] https://zenodo.org/record/2586669#.XJiS8ChKi70.

speech, there are several other labels characterizing each comment, e.g., if it is a threat, provocative, racist, sexist, etc. (not used in our study). There are 138 comments labeled as a hate speech and 862 as non-hate speech. We produced a data set of 300 comments using all 138 hate speech comments and randomly sampled 162 non-hate speech comments.

3 - OffensiveTweets data set[3] originates in a study regarding hate speech detection and the problem of offensive language [11]. Our data set consists of 3000 tweets. We took 1430 tweets labeled as hate speech and randomly sampled 1670 tweets from the collection of remaining 23 353 tweets.

Data Preprocessing. Social media text use specific language and contain syntactic and grammar errors. Hence, in order to get correct and clean text data we applied different prepossessing techniques without removing text documents based on the length. The pipeline for cleaning the data was as follows:

- Noise removal: user-names, email address, multiple dots, and hyper-links are considered irrelevant and are removed.
- Common typos are corrected and typical contractions and hash-tags are expanded.
- Stop words are removed and the words are lemmatized.

4.2 Experimental Scenario

We use logistic regression (LR) and Support Vector Machines (SVM) from the scikit-learn library [5] as the baseline classification models. As a baseline RNN, the LSTM network from the Keras library was applied [8]. Both LSTM and MCD LSTM networks consist of an embedding layer, LSTM layer, and a fully connected layer within the Word2Vec and ELMo embeddings. The embedding layer was not used in TF-IDF and Universal Sentence encoding.

To tune the parameters of LR (i.e. *liblinear* and *lbfgs* for the solver functions and the number of component C from 0.01 to 100) and SVM (i.e. the *rbf* for the kernel function, the number of components C from 0.01 to 100 and the gamma γ values from 0.01 to 100), we utilized the random search approach [3] implemented in scikit-learn. In order to obtain best architectures for the LSTM and MCD LSTM models, various number of units, batch size, dropout rates and so on were fine-tuned.

5   Evaluation and Results

We first describe experiments comparing different word representations, followed by sentence embeddings, and finally the visualization of predictive uncertainty.

---

[3] https://github.com/t-davidson/hate-speech-and-offensive-language.

5.1   Word Embedding

In the first set of experiments, we represented the text with word embeddings (sparse TF-IDF [31] or dense word2vec [21], and ELMo [24]). We utilise the gensim library [27] for word2vec model, the scikit-learn for TFIDF, and the ELMo pretrained model from TensorFlow Hub[4]. We compared different classification models using these word embeddings. The results are presented in Table 1.

The architecture of LSTM and MCD LSTM neural networks contains an embedding layer, LSTM layer, and fully-connected layer (i.e. dense layer) for word2vec and ELMo word embeddings. In LSTM, the recurrent dropout is applied to the units for linear transformation of the recurrent state and the classical dropout is used for the units with the linear transformation of the inputs. The number of units, recurrent dropout, and dropout probabilities for LSTM layer were obtained by fine-tuning (i.e. we used 512, 0.2 and 0.5 for word2vec and TF-IDF, 1024, 0.5, and 0.2 for ELMo in the experiments with MCD LSTM architecture). The search ranges for hyper parameter tuning are described in Table 2.

Table 1. Comparison of classification accuracy (with standard deviation in brackets) for word embeddings, computed using 5-fold cross-validation. All the results are expressed in percentages and the best ones for each data set are in bold.

Model	HatEval			YouToxic			OffensiveTweets		
	TF-IDF	W2V	ELMo	TF-IDF	W2V	ELMo	TF-IDF	W2V	ELMo
LR	68.0 [2.4]	54.0 [13.6]	62.0 [6.8]	69.3 [3.0]	54.0 [3.0]	**76.6 [6.1]**	**77.2 [1.1]**	68.0 [2.4]	75.6 [1.2]
SVM	63.0 [5.1]	66.0 [3.7]	62.0 [12.9]	70.6 [4.2]	55.0 [3.4]	73.3 [5.5]	77.0 [0.7]	59.6 [1.5]	73.0 [1.9]
LSTM	69.0 [7.3]	67.0 [6.8]	66.0 [12.4]	66.6 [2.3]	59.3 [4.6]	74.3 [2.7]	73.4 [0.8]	75.0 [1.7]	74.7 [1.9]
MCD LSTM	67.0 [10.8]	**69.0 [6.6]**	67.0 [9.8]	66.0 [3.7]	59.3 [3.8]	75.3 [5.5]	71.1 [1.6]	72.0 [1.6]	75.2 [0.9]

Table 2. Hyper-parameters for LSTM and MCD LSTM models

Name	Parameter type	Values
Optimizers	Categorical	Adam, rmsprop
Batch size	Discrete	4 to 128, step=4
Activation function	Categorical	tanh, relu and linear
Number of epochs	Discrete	10 to 100, step=5
Number of units	Discrete	128, 256, 512, or 1024
Dropout rate	Float	0.1 to 0.8, step=0.05

The classification accuracy for HatEval data set is reported in the Table 1 (left). The difference between logistic regression and the two LSTM models indicates accuracy improvement once the recurrent layers are introduced. On the other hand, as the ELMo embedding already uses the LSTM layer to take into account semantic relationship among the words, no notable difference between logistic regression and LSTM models can be observed using this embedding.

---

[4] https://tfhub.dev/google/elmo/2.

Results for YouToxic and OffensiveTweets data sets are presented in Table 1 (middle) and (right), respectively. Similarly to the HatEval data set, there is a difference between the logistic regression and the two LSTM models using the word2vec embeddings. For all data sets, the results with ELMo embeddings are similar across the four classifiers.

5.2 Sentence Embedding

In the second set of experiments, we compared different classifiers using sentence embeddings [6] as the representation. Table 3 (left) displays results for HatEval. We can notice improvements in classification accuracy for all classifiers compared to the word embedding representation in Table 1. The best model for this small data set is MCD LSTM. For larger YouToxic and OffensiveTweets data sets, all the models perform comparably. Apart from the prediction accuracy the four models were compared using precision, recall and F1 score [7].

We use the Universal Sentence Encoder module[5] to encode the data. The architecture of LSTM and MCD LSTM contains a LSTM layer and dense layer. With MCD LSTM architecture in the experiments, the number of neurons, recurrent dropout and dropout value for LSTM is 1024, 0.75 and 0.5, respectively. The dense layer has the same number of units as LSTM layer, and the applied dropout rate is 0.5. The hyper-parameters used to tune the LSTM and MCD LSTM models are presented in the Table 2.

Table 3. Comparison of predictive models using sentence embeddings. We present average classification accuracy, precision, recall and F_1 score (and standard deviations), computed using 5-fold cross-validation. All the results are expressed in percentages and the best accuracies are in bold.

Model	HatEval				YouToxic				OffensiveTweets			
	Accuracy	Precision	Recall	F1	Accuracy	Precision	Recall	F1	Accuracy	Precision	Recall	F1
LR	66.0 [12.4]	67.3 [15.3]	65.2 [15.9]	65.2 [13.1]	77.3 [4.1]	74.3 [7.3]	77.3 [3.6]	75.7 [5.3]	80.8 [1.0]	79.6 [1.9]	84.9 [1.2]	82.2 [1.1]
SVM	67.0 [12.1]	68.2 [15.2]	65.0 [15.8]	65.8 [13.3]	77.3 [6.2]	72.6 [8.6]	80.7 [7.4]	76.3 [7.6]	80.7 [1.3]	78.6 [2.0]	86.7 [1.0]	82.4 [1.2]
LSTM	70.0 [8.4]	70.8 [11.0]	63.1 [17.5]	66.2 [14.4]	76.6 [8.6]	73.4 [11.2]	79.2 [8.0]	75.8 [8.6]	80.7 [1.6]	82.8 [2.1]	79.7 [2.3]	81.1 [1.5]
MCD LSTM	**74.0 [10.7]**	73.4 [12.7]	78.4 [13.6]	74.9 [10.0]	**78.7 [5.8]**	74.7 [9.2]	80.9 [6.5]	77.5 [7.4]	**81.0 [1.2]**	81.5 [1.8]	82.5 [2.7]	81.9 [1.3]

5.3 Visualizing Predictive Uncertainty

In Fig. 2 we present a new way of visualizing dependencies among the test tweets. The relations are result of applaing the MCD LSTM network to the HetEval data set. This allows further inspection of the results as well as interpretation of correct and incorrect predictions. To improve comprehensibility of predictions and errors, each point in the visualization is labeled with a unique identifier, making the point tractable to the original document, given in Table 4.

---

[5] https://tfhub.dev/google/universal-sentence-encoder-large/3.

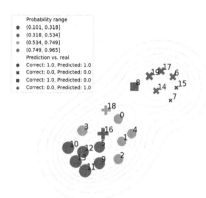

Fig. 2. Visualization of individual predictions along with their probability ranges. The numeric labels besides each point correspond to individual tweets, summarized in Table 4. The size of the points corresponds to the neural network's mean probability for a given prediction. True positives are marked with circles, true negatives with crosses, false positives with squares, and false negatives as pluses.

As Fig. 2 shows, the tweets are grouped into two clusters. According to the kernel density isometric lines, two centers are identified: the tweets assigned lower probability of being hate speech and the tweets with higher probability of being hate speech. Let us focus on the wrongly classified tweets and their positions in the graph (tweets 8, 16 and 18). While for tweets 8 and 18 the classifier wasn't certain and a mistake seems possible according to the plot, the tweet 16 was predicted to be hate speech with high probability. Analyzing the words that form this tweet, we notice that not only that most of them often do appear in the hate speech but also this combination of the words used together is very characteristic for the offensive language.

Table 4. Test documents (one fold) used for the visualization in Fig. 2.

Index (see Figure 2) — Corresponding text

0 — ok if you fu***** said leave block me. but dm me first i am gonna kick your ass. shut the f*** up you

1 — wow mo c*** got hard. want to pull your panties down, push you against these door and let you feel my c*** deep in your wet p****

2 — no i said hope. i hope you women learn your place! sitdowninthekitchen

3 — so this b**** blocked me but check her pinned tweet out, she is accusing men of rape on tumblr with ...

4 — this is not rape, or anything like it. this a crude pass. this is not 'rape culture.' it is w**** culture.

5 — rape is bad but alota h*** be lying bout that s*** women are crazy dawg & will f*** ya life up be careful

6 — lol. they did not applaud a rape threat .how long are you going to ride that s***?

7 — you had a great chance to show everyone that youre not a c*** on tv tonight and you f***** it up.

8 — lmao!!! stuff i say. some stuff be so damn funny.especially when it comes to illegal immigrants who are full fledged criminals coming to america ...

9 — the german government pays for 3 week vacation for refugees to go home -

10 — burundian refugees should go home!

11 — no benefits to any immigrants !! come here & make it on your own, or go back home !!

12 — illegal aliens waiting for first class msc container-shipping in antwerpen stop immigration

13 — sick barstewards! this is what happens when we put up the refugees welcome signs! they not only rape our wives or girlfriends ...

14 — i am not forcing anyone. i am just putting out my view. people believe it or not is their prerogative.even if a single person agree with me ...

15 — tajikistani charged with beating new york university professor to death another animal./

16 — child prostitution: feds break up mexico-to-queens sex trafficking ring via ...

17 — home office guilty of a serious breach of the duty of candour and cooperation regarding children entitled to enter uk. where did these children go? ...

18 — p.s why do you not pay unemployed people who do endless hours of voluntary work they do that to give something to the community

19 — seriously, amy and cindy are bffs, i know that for sure. hmm, mmm.

Our short demonstration shows the utility of the proposed visualization which can identify different types of errors and helps to explain weaknesses in the classifier or wrongly labeled data.

6   Conclusions

We present the first successful approach to assessment of prediction uncertainty in hate speech classification. Our approach uses LSTM model with Monte Carlo dropout and shows performance comparable to the best competing approaches using word embeddings and superior performance using sentence embeddings. We demonstrate that reliability of predictions and errors of the models can be comprehensively visualized. Further, our study shows that pretrained sentence embeddings outperform even state-of-the-art contextual word embeddings and can be recommended as a suitable representation for this task. The full Python code is publicly available[6].

As persons spreading hate speech might be banned, penalized, or monitored not to put their threats into actions, prediction uncertainty is an important component of decision making and can help humans observers avoid false positives and false negatives. Visualization of prediction uncertainty can provide better understanding of the textual context within which the hate speech appear. Plotting the tweets that are incorrectly classified and inspecting them can identify the words that trigger wrong classifications.

Prediction uncertainty estimation is rarely implemented for text classification and other NLP tasks, hence our future work will go in this direction. A recent emergence of cross-lingual embeddings possibly opens new opportunities to share data sets and models between languages. As evaluation in rare languages is difficult, the assessment of predictive reliability for such problems might be an auxiliary evaluation approach. In this context, we also plan to investigate convolutional neural networks with probabilistic interpretation.

Acknowledgments. The work was partially supported by the Slovenian Research Agency (ARRS) core research programme P6-0411. This project has also received funding from the European Union's Horizon 2020 research and innovation programme under grant agreement No 825153 (EMBEDDIA).

References

1. Baldi, P., Sadowski, P.J.: Understanding dropout. In: Advances in Neural Information Processing Systems, pp. 2814–2822 (2013)
2. Berger, W., Piringer, H., Filzmoser, P., Gröller, E.: Uncertainty-aware exploration of continuous parameter spaces using multivariate prediction. In: Computer Graphics Forum, pp. 911–920 (2011)
3. Bergstra, J., Bengio, Y.: Random search for hyper-parameter optimization. J. Mach. Learn. Res. **13**, 281–305 (2012)

---

[6] https://github.com/KristianMiok/Hate-Speech-Prediction-Uncertainty.

4. Bleich, E.: The rise of hate speech and hate crime laws in liberal democracies. J. Ethnic Migr. Stud. **37**(6), 917–934 (2011)
5. Buitinck, L., et al.: API design for machine learning software: experiences from the scikit-learn project. In: ECML PKDD Workshop: Languages for Data Mining and Machine Learning, pp. 108–122 (2013)
6. Cer, D., et al.: Universal sentence encoder. arXiv preprint arXiv:1803.11175 (2018)
7. Chinchor, N.: Muc-4 evaluation metrics. In: Proceedings of the Fourth Message Understanding Conference, p. 22–29 (1992)
8. Chollet, F., et al.: Keras (2015). https://keras.io
9. Corazza, M., et al.: Comparing different supervised approaches to hate speech detection. In: EVALITA 2018 (2018)
10. Cox, J., Lindell, M.: Visualizing uncertainty in predicted hurricane tracks. Int. J. Uncertain. Quantif. **3**(2), 143–156 (2013)
11. Davidson, T., Warmsley, D., Macy, M., Weber, I.: Automated hate speech detection and the problem of offensive language. In: Eleventh International AAAI Conference on Web and Social Media (2017)
12. Del Vigna12, F., Cimino23, A., Dell'Orletta, F., Petrocchi, M., Tesconi, M.: Hate me, hate me not: Hate speech detection on facebook (2017)
13. Fortunato, M., Blundell, C., Vinyals, O.: Bayesian recurrent neural networks. arXiv preprint arXiv:1704.02798 (2017)
14. Gal, Y., Ghahramani, Z.: Dropout as a Bayesian approximation: representing model uncertainty in deep learning. In: International Conference on Machine Learning, pp. 1050–1059 (2016)
15. Gal, Y., Ghahramani, Z.: A theoretically grounded application of dropout in recurrent neural networks. In: Advances in Neural Information Processing Systems, pp. 1019–1027 (2016)
16. Kucukelbir, A., Tran, D., Ranganath, R., Gelman, A., Blei, D.M.: Automatic differentiation variational inference. J. Mach. Learn.Res. **18**(1), 430–474 (2017)
17. Liu, L., et al.: Uncertainty visualization by representative sampling from prediction ensembles. IEEE Trans. Vis. Comput. Graph. **23**(9), 2165–2178 (2016)
18. Liu, L., Padilla, L., Creem-Regehr, S.H., House, D.H.: Visualizing uncertain tropical cyclone predictions using representative samples from ensembles of forecast tracks. IEEE Trans. Vis. Comput. Graph. **25**(1), 882–891 (2019)
19. McInnes, L., Healy, J., Saul, N., Grossberger, L.: UMAP: Uniform manifold approximation and projection. J. Open Source Softw. **3**(29), 861 (2018)
20. Mehdad, Y., Tetreault, J.: Do characters abuse more than words? In: Proceedings of the 17th Annual Meeting of the Special Interest Group on Discourse and Dialogue, pp. 299–303 (2016)
21. Mikolov, T., Chen, K., Corrado, G., Dean, J.: Efficient estimation of word representations in vector space. arXiv preprint arXiv:1301.3781 (2013)
22. Miok, K.: Estimation of prediction intervals in neural network-based regression models. In: 20th International Symposium on Symbolic and Numeric Algorithms for Scientific Computing (SYNASC), pp. 463–468, September 2018
23. Myshkov, P., Julier, S.: Posterior distribution analysis for Bayesian inference in neural networks. In: Workshop on Bayesian Deep Learning, NIPS (2016)
24. Peters, M.E., et al.: Deep contextualized word representations. arXiv preprint arXiv:1802.05365 (2018)
25. Pham, V., Bluche, T., Kermorvant, C., Louradour, J.: Dropout improves recurrent neural networks for handwriting recognition. In: 2014 14th International Conference on Frontiers in Handwriting Recognition, pp. 285–290. IEEE (2014)

26. Platt, J.C.: Probabilistic outputs for support vector machines and comparisons to regularized likelihood methods. In: Advances in large margin classifiers, pp. 61–74. MIT Press (1999)
27. Rehurek, R., Sojka, P.: Software framework for topic modelling with large corpora. In: Proceedings of the LREC 2010 Workshop on New Challenges for NLP Frameworks, pp. 45–50. ELRA, Valletta, Malta, May 2010
28. Rother, K., Allee, M., Rettberg, A.: Ulmfit at germeval-2018: a deep neural language model for the classification of hate speech in German tweets. In: 14th Conference on Natural Language Processing KONVENS 2018, p. 113 (2018)
29. Ruginski, I.T., et al.: Non-expert interpretations of hurricane forecast uncertainty visualizations. Spat. Cogn. Comput. **16**(2), 154–172 (2016)
30. Schmidt, A., Wiegand, M.: A survey on hate speech detection using natural language processing. In: Proceedings of the Fifth International Workshop on Natural Language Processing for Social Media, pp. 1–10 (2017)
31. Sparck Jones, K.: A statistical interpretation of term specificity and its application in retrieval. J. Doc. **28**(1), 11–21 (1972)
32. Srivastava, N., Hinton, G., Krizhevsky, A., Sutskever, I., Salakhutdinov, R.: Dropout: a simple way to prevent neural networks from overfitting. J. Mach. Learn. Res. **15**(1), 1929–1958 (2014)
33. Waldron, J.: The Harm in Hate Speech. Harvard University Press, Cambridge (2012)
34. Wang, S., Manning, C.: Fast dropout training. In: International Conference on Machine Learning, pp. 118–126 (2013)
35. Warner, W., Hirschberg, J.: Detecting hate speech on the world wide web. In: Proceedings of the Second Workshop on Language in Social Media, pp. 19–26. Association for Computational Linguistics (2012)
36. Zaremba, W., Sutskever, I., Vinyals, O.: Recurrent neural network regularization. arXiv preprint arXiv:1409.2329 (2014)
37. Zhu, L., Laptev, N.: Deep and confident prediction for time series at uber. In: 2017 IEEE International Conference on Data Mining Workshops (ICDMW), pp. 103–110. IEEE (2017)

Authorship Attribution in Russian in Real-World Forensics Scenario

Polina Panicheva[1,2(✉)] 🆔 and Tatiana Litvinova[2,3] 🆔

[1] National Research University Higher School of Economics,
16 Soyuza Pechatnikov Street, St. Petersburg 190121, Russia
ppanicheva@hse.ru
[2] RusProfiling Lab, Voronezh State Pedagogical University, 86 Lenina Street,
Voronezh 394043, Russia
centr_rus_yaz@mail.ru
[3] Plekhanov Russian University of Economics, Stremyanny Lane 36,
Moscow 117997, Russia

Abstract. Recent demands in authorship attribution, specifically, cross-topic authorship attribution with small numbers of training samples and very short texts, impose new challenges on corpora design, feature and algorithm development. In the current work we address these challenges by performing authorship attribution on a specifically designed dataset in Russian. We present a dataset of short written texts in Russian, where both authorship and topic are controlled. We propose a pairwise classification design closely resembling a real-world forensic task. Semantic coherence features are introduced to supplement well-established n-gram features in challenging cross-topic settings. Distance-based measures are compared with machine learning algorithms. The experiment results support the intuition that for very small datasets, distance-based measures perform better than machine learning techniques. Moreover, pairwise classification results show that in difficult cross-topic cases, content-independent features, i.e., part-of-speech n-grams and semantic coherence, are promising. The results are supported by feature significance analysis for the proposed dataset.

Keywords: Authorship identification · Plagiarism and spam filtering · Forensic authorship identification · Distributional semantics · Russian language

1 Introduction

Authorship Attribution (AA), the task of identifying unknown authorship of a document, has gained weight in recent years, especially in view of challenges posed by online security requirements [1, 20]. Traditionally, AA was aimed at identifying authors of large samples of fiction texts. However, the recent progress of the World Wide Web has influenced a drift in AA goals towards identifying authorship based on a very small number of extremely short text samples spanning different topics and genres.

These changes call for new methods to be developed for AA. First, small numbers of training examples render the traditional machine learning methods impractical. Second, traditional content-based features are ineffective in cross-domain scenarios.

C. Martín-Vide et al. (Eds.): SLSP 2019, LNAI 11816, pp. 299–310, 2019.
https://doi.org/10.1007/978-3-030-31372-2_25

In the current work, we set out to address these issues by performing AA experiments in a real-world forensic scenario with a dataset in Russian.

We present a corpus which is controlled over both authors and topics and includes short written texts in Russian. Apart from a traditional machine learning setting with dozens of training documents, we also perform experiments following a scenario similar to a real-world forensic task, with only two training documents by different authors and one test document, with various distributions of topics over the training and test documents; in the latter case, distance-based measures are effectively used. We apply well-established word, character and part-of-speech n-grams, and introduce semantic coherence features to handle the author-topic interference cases.

To our knowledge, this is the first report of a topic-author-controlled AA setting in Russian. It is also the first application of semantic coherence to AA. Moreover, this is the first attempt to date to perform AA in a setting similar to a real-world forensics scenario for Russian texts.

2 Related Work

2.1 Authorship Attribution

Recent challenges in authorship attribution include cross-domain attribution tasks (mostly cross-topic and cross-genre scenarios) and tasks where very few learning documents are available. In such cases, distance-based measures and clustering are preferred over traditional supervised learning approaches [6, 12, 29].

In fact, in order to tackle cross-topic AA, character n-grams were introduced, as opposed to well-established lexical features [26, 28]. The former are currently considered the best performing features in cross-topic AA [18, 24]. However, character n-grams were recently demonstrated to capture also topic-related information, depending on the type of character n-grams [21]. Some researchers propose techniques to remove content-dependent information and improve cross-topic AA results [27]. Some functional linguistic features were introduced to address the problem of content-dependency in cross-topic AA: namely, part-of-speech and syntactic n-grams, text complexity indices, etc. are applied to both English and Russian language data [14, 24, 29].

Comparing the AA results is a difficult issue, as the applied corpora differ a lot. For instance, for an English corpus controlled by topic and genre by 21 authors the AA results range from 0.19 to 0.65 [27]. Widely used in AA experiments is the *Reuters* corpus, with the most 10 or 50 productive authors and 50 documents by each in the training and test sets; this dataset has been classified as accurately as 0.71–0.85 with word n-grams and their embeddings. The *Guardian* corpus consists of at most 10 articles per each of 13 authors on different topics, and the AA accuracy ranges from 0.24 to 0.98, depending on the choice of the training and test topics [18]. All the above cited approaches, however, are rather far from typical forensic authorship attribution task (for details see Sect. 3.1).

2.2   RusIdioStyle Corpus

The mentioned benchmark corpora comprise of texts written mostly in English or other European languages. Unfortunately, works in AA in Russian is scarce. There have been attempts to perform AA in social media texts with large numbers of authors and large (over hundreds) collections of documents by each author based on social media texts [3], although it is well-known that they typically contain non-marked citations and other non-authorial elements. Litvinova et al. [15; see also 14] have used a corpus of hundreds of authors with two texts by each author from four different topics to identify stable authorship features across topics, and have identified the proportions of some functional, part-of-speech and LIWC categories as stable and therefore potentially useful for AA, although their contribution in AA has not been evaluated yet.

To date there has been no topic-balanced dataset reported for AA in Russian. We introduce a dataset we have compiled for the current study from *RusIdioStyle* corpus. The corpus has been gathered since September 2018 and includes both experimental texts (answers by different Russian speakers to a questionnaire) and text produced (both orally and in writing) in different natural communicative situations. The questionnaire consists of a range of questions, including everyday and controversial social topics, a number of picture descriptions, narratives, etc. In some tasks, these texts were to be composed in both written and oral manner. Currently the corpus includes texts by 125 authors. However, in the current experiments we only use written texts by five authors belonging to three topics (register) and produced in experimental conditions, as we aim at a highly topic-and-author-controlled dataset designed for experiments in AA which simulate the most typical forensic scenarios.

2.3   Semantic Coherence

Semantic coherence measurement stems from topic modelling coherence evaluation [16]. The idea behind topic coherence is that given a symmetric similarity measure between words, an overall score of word similarity can be measured in a list of words. Semantic coherence is measured as similarity between words in a context window in a text, whereas similarity is typically based on a distance metric between word meanings in a distributional semantic space. In recent years semantic coherence has been effectively applied to a range of cognitive and profiling tasks in NLP, including lexical error identification by learners of English [9], metaphor identification [23], thought disorder evaluation [4], psychosis onset prediction [2] and diagnosing schizophrenia [10].

Some approaches based on distributional word representations have been utilized in AA tasks with exceptional results. As early as 1997 [25] word-n-gram-based Latent Semantic Indexing was effectively applied to AA in biblical Hebrew texts. In a dissertation by Gritta [8], a number of target terms are represented in distributional semantic spaces; the representations are further applied to AA among three English authors. The term representations have shown highly significant differences between authors, despite the use of small corpora for modelling.

The abovementioned research has demonstrated that semantic coherence, and, more broadly, distributional word representations are capable of capturing important characteristics of individual authors in their texts. For these reasons, we find it necessary to investigate the performance of semantic coherence features in AA, and specifically, to find out whether semantic coherence can provide content-independent measures to advance cross-topic AA.

3   Experiment

The goal of the experiments is to perform cross-topic AA with a topic-author-controlled dataset in Russian and evaluate the widely used n-gram features and semantic coherence features. Apart from a traditional machine learning setting, we perform a specifically designed experiment replicating a common forensic task, with a single training document by each author and diverse topical relations between the training and test documents. For these purposes, we analyze feature distribution and classification performance in a specifically designed Russian language dataset.

3.1   Dataset

We have sampled a subcorpus of *RusIdioStyle* (Table 1). We aimed at constructing a dataset by various authors, which is controlled for both topic and authorship information. At the same time, we intended to replicate a typical forensic authorship attribution task [11, 19]:

- a very small but finite set of candidate authors (very often only two, see [7] as an example);
- a very small number of documents per author and topic;
- very short document length;
- training and test documents typically belong to different topics (see also [7], where it is stated that "forensic authorship identification usually has to deal with cross-register of mixed text type data").

In RusIdioStyle, different topics are addressed by different authors. We have sampled five authors, who have written texts on each of the three topics:

- Thematic Apperception Test: descriptions of seven drawings (**TAT**);
- description of the previous day in normal and reversed order (**Day**);
- essays on two of the following topics (**Essay**):
  - "My dream city";
  - "An annoying person";
  - "Family is the basis of the society";
  - "Surrogacy: is it good or bad?";
  - "Obesity: is it a problem?";
  - "Retirement age increase: is it good or bad?".

Only two of the six possible topics were chosen by every author, which further contributes to the topic diversification of the dataset. Number of documents and document length statistics of the sampled dataset are illustrated in Tables 1 and 2.

Table 1. Document numbers in the dataset

Topic	TAT	Day	Essay	All
№ of documents per author	7	2	2	11
№ of documents by all authors	35	10	10	55

Table 2. Document length properties in the dataset (in words)

Topic\AuthorID	153	154	155	157	158	Mean	Std
TAT	19.4	38.0	267.4	141.6	283.4	**150.0**	**126.9**
Day	78.5	194.5	200.0	234.0	294.5	**200.3**	**106.7**
Essay	53.5	76.5	177.5	499.0	302.0	**221.7**	**190.9**
Mean	**36.4**	**73.5**	**238.8**	**223.4**	**288.8**	172.2	
Std	**30.0**	**65.0**	**83.2**	**191.3**	**18.5**		**137.9**

Thus, we have compiled a dataset by five different authors stratified by topic (besides, texts belong to different registers). The dataset replicates a typical real-life AA dataset with texts differing in length, including very short texts; a small number of authors and texts by each author, and a variety of topics. The dataset was designed to allow for a cross-topic AA scenario, where training and test samples belong to different topics.

3.2 Features

The goal of the experiment is to evaluate the performance of a number of established n-gram features and the semantic coherence features. As a preprocessing step, we perform tokenization, morphological analysis and lemmatization with MyStem morphological analyzer [22].

N-grams. We apply n-grams of lemmas, parts-of-speech and characters, with n ranging from 1 to 3. These features are straightforward by design, but capable of achieving state-of-the-art performance in a number of AA tasks, including cross-topic AA [20, 21, 28, 29]. As a number of authors have shown [5, 27], cross-topic AA typically gains from the most frequent few hundred character and lemma features. We apply the most frequent 100 features for lemmatized **Word**, character (**Char**) and part of speech (**POS**) n-grams, with the resulting n-gram feature space containing 300 dimensions.

Semantic Coherence. Semantic coherence features are computed based on the cosine distance between the neighboring words in a distributional semantic model.

We have used the word2vec Continuous Skipgram model by Kutuzov[1] [13]. The model is trained with the Russian National Corpus and Wikipedia dump as of December 2017, a combined corpus of 600M tokens. The context window is set to ±2,

---

[1] The model is made available for free download by the *RusVectōrēs* project at https://rusvectores.org/ru/models/.

the resulting dimensionality is 300, and the word frequency cutoff is 40. Besides, every word is supplied with its part-of-speech tag, as defined by MyStem [22].

As our dataset contains very short texts, we apply the measure of semantic coherence between neighboring words in a sliding window ranging from 3 to 8. The semantic coherence in the sliding window is computed as follows (Eq. 1):

$$Coh(Win) = Mean\{\cos(w_i, w_j) \,|\, w_i, w_j \in Win, i > j\} \tag{1}$$

The result is the mean cosine distance between the words in a sliding window. The following statistics are computed from the semantic coherence measures in the consecutive sliding windows:

- mean, standard deviation;
- minimum, maximum;
- 90- and 10-percentile.

Six coherence features are computed for every sliding window size between 3 and 8; the final feature set consists of 6 * 6 = 36 semantic coherence features (**SemCoh**).

Experiment Settings. N-gram features are normalized with tf-idf and 'l2' normalization. All the features are further normalized by subtracting their mean and dividing by their standard deviation, so that Z-scores are obtained, which are suitable for Delta-based distance measurement [5]. Normalization is performed based on the whole dataset statistics.

First, we perform a **Group AA** experiment: we apply Support Vector Machine (**SVM**) and K-Nearest Neighbors (**KNN**) algorithms to the resulting feature sets. While **SVM** is a basic algorithm widely used for text classification problems, **KNN** can be seen as an extension of the Delta measure [5] to more than one nearest elements. The dataset is grouped by three topics, where we perform Leave-One-Group-Out cross-validation: the algorithm is trained with two topics and tested with the third topic. All the experiments have been performed with *scikit-learn* for *Python*[2] [17].

Second, we perform a **pairwise AA experiment**. It is organized similar to a typical real-world forensic task in it's most difficult scenario: there are only two documents by two authors with known authorship in the training set, and a third anonymous test document. The author of the test document is one of the authors of the two training documents. The author of the test document has to be chosen automatically between the two authors represented in the training set. As the training set in each experiment only includes two documents, classification algorithms are expected to fail. This task can be solved with distance-based measures [11], when the test document is compared to the training documents in a pairwise manner. We apply Cosine, Manhattan and Euclidean distance measures to the task, which correspond to the respective types of the Delta measure for AA [5]. The data for the task is obtained by a three-fold Cartesian product of the above dataset, and contains 55 * 10 * 44 = 24,200 document triples of the following structure:

$$(Dtest(Ai), Dtrain(Ai), Dtrain(Aj)), \qquad (2)$$

- where Dtest(Ai) and Dtrain(Ai) are different test and training documents by author i;
- Dtrain(Aj) is a training document by author j;
- i ! = j.

As the dataset size is small, we expect some random effects on feature performance when the dataset is further divided into training and test sets. To estimate the contribution of different features in the authorship classes, we also report feature importance measurement results based on the Z-scores of their distribution in the dataset.

4 Results and Discussion

4.1 Group Authorship Attribution Performance

We have performed AA experiments with **Word**, **Char** and **POS** n-grams (n = {1, 2, 3}), **SemCoh** features, and combined feature sets. We report a number of the highest results; different parameter values result in similar patterns with lower performance.

Table 3 illustrates the accuracy values for a number of experiment settings. The baseline is 0.2, as there are five equally distributed authors. The best result was obtained by the **Word + Char** n-grams and **KNN** with 5 nearest neighbors and distance-based weights (**N = 5, dist**). In most of the other settings, **Word + Char** features have the highest performance, whereas in **SVM** their combination with **SemCoh** is also among the highest results.

4.2 Pairwise Authorship Attribution Performance

We have performed distance-based AA in a pairwise distance-based setting, i.e. with two training documents, one by the author of a test document and the other written by a different author.

Cosine and Manhattan distance measures performed similarly whereas Euclidean distance resulted in lower performance; we report the accuracy results only for Manhattan distance for the sake of brevity (Table 4). The baseline for the task is 0.5, as it implies a choice between two authors equally represented in the training sets.

In the pairwise setting, there are four types of topical relation scenarios between the triples of documents. First, both training documents belong to the same topic as the test document (**Same**, 6,040 document triples). Second, both training documents belong to different topic(s) than the test document (**Diff**, 8,720 triples). Third, the training document by the same author as the test document belongs also to the same topic as the test document (**TrueSame**, 4,080 triples). Finally, the training document by a different author than the test document belongs to the same topic as the test document (**False-Same**, 5,360 triples). The results are reported for the full dataset (**All**) and for every kind of topical relations.

It is obvious that the **Same** relation is essentially a single-topic scenario, where the training and test documents belong to the same topic. The **TrueSame** scenario is relatively simple, because both authorship and topic information contribute to the

Table 3. AA performance (accuracy) with different feature sets (the highest two results in each setting are highlighted in bold).

Features	KNN (N = 5, dist)	KNN (N = 1, uni)	SVM (l2, C = 0.1)
Word	**0.564**	**0.564**	0.436
POS	0.436	0.309	0.255
Char	0.364	0.364	0.382
SemCoh	0.382	0.418	0.345
Word + SemCoh	0.473	0.473	0.4
POS + SemCoh	0.309	0.255	0.273
Char + SemCoh	0.473	0.382	0.4
Word + POS	0.436	0.473	0.4
Word + Char	**0.618**	**0.545**	**0.582**
POS + Char	0.455	0.327	0.309
Word + POS + SemCoh	0.436	0.4	0.345
Word + Char + SemCoh	0.473	0.455	**0.491**
POS + Char + SemCoh	0.364	0.345	0.345
Word + POS + Char	0.509	0.491	0.455
Word + POS + Char + SemCoh	0.418	0.418	0.455

Table 4. Accuracy of pairwise AA using Manhattan distance (the highest two results in each setting are highlighted in bold).

Features	All	Same	Diff	TrueSame	FalseSame
Word	**0.703**	**0.805**	**0.653**	**0.875**	0.537
POS	0.633	0.695	0.577	0.707	0.596
Char	0.575	0.625	0.536	0.613	0.553
SemCoh	0.575	0.627	0.533	0.6	0.567
Word + SemCoh	0.665	0.742	0.617	0.78	0.568
POS + SemCoh	0.629	0.692	0.569	0.701	**0.601**
Char + SemCoh	0.588	0.64	0.538	0.632	0.577
Word + POS	**0.696**	**0.776**	**0.633**	**0.829**	**0.605**
Word + Char	0.639	0.703	0.582	0.763	0.566
POS + Char	0.593	0.64	0.546	0.65	0.571
Word + POS + SemCoh	0.682	0.762	0.615	0.816	0.6
Word + Char + SemCoh	0.636	0.711	0.572	0.753	0.568
POS + Char + SemCoh	0.602	0.658	0.545	0.668	0.581
Word + POS + Char	0.642	0.703	0.584	0.752	0.584
Word + POS + Char + SemCoh	0.641	0.71	0.575	0.758	0.58

choice of the correct candidate. **Diff** scenario is complicated, because the test document belongs to a different topic than both training documents. Finally, **FalseSame** scenario is even more complicated, as the 'correct' training document by the same author as the test document belongs to a different topic, whereas the 'wrong' document by a different author belongs to the same topic as the test document: thus the topic information interferes with authorship, and it is only by ignoring topical features that the correct result can be obtained here.

In straightforward single-topic scenarios (**Same, TrueSame**) mostly content features result in the best performance: the best result is obtained with **Word** and **Char** features, and **POS** and **SemCoh** are in some cases useful as additional features. However, in the most complicated **FalseSame** scenario, **POS** and **SemCoh** features are promising, as their addition results in performance increase. In fact, it is different combinations of **Word, POS** and **SemCoh** features, which allow to reach the accuracy values of 0.6 in these difficult conditions. Despite the complications introduced in the task by topic and authorship information interference, the suggested features allow to overcome the baseline by 10%.

4.3   Feature Importance

We have identified the most important features for authorship classes based on the whole dataset. The most important features with their type (**char, word, POS** n-grams, or **SemCoh**) are presented in Table 5. The **word** features are represented by their English translation and the original Russian word.

There are 17 features significant for authorship classification ($p < 0.01$). They are distributed as follows: **SemCoh** – 6, **word** – 5, **char** – 4, and **POS** – 2 features.

Table 5. Feature importance (F-ANOVA) and feature count. Only significant features with $p < 0.01$ are shown (corrected with the Benjamini-Hochberg false discovery rate).

Feature name	Feature type	F	$\log_{10}P$	Feature count
VERB PRON	POS	18.38	−9	132
to start - начинать	word	11.47	−6	15
PRON	POS	10.31	−6	784
Std_win8	SemCoh	9.03	−5	−
ол	char	7.95	−5	329
she - она	word	7.51	−5	37
Min_win6	SemCoh	7.47	−5	−
Min_win8	SemCoh	7.41	−5	−
не	char	7.37	−5	531
you - ты	word	7.17	−4	33
._	char	6.75	−4	594
Std_win7	SemCoh	6.74	−4	−
Max_win6	SemCoh	6.69	−4	−
this - это	word	6.50	−4	63
Std_win6	SemCoh	6.33	−4	−
although - однако	word	6.13	−4	120
от	char	6.01	−4	414

It is obvious that based on significance in the corpus, the semantic coherence features play an important role in authorship characteristics. It is standard deviation, maximum and minimum of the semantic coherence values which are significant. All three of these measures for sliding window size = 6 belong to the significant features, some are also present for window size = 7, 8. These facts show that for the genre and mode of written short stories, semantic coherence for window size = 6 is probably most characteristic of individual style among other window sizes.

However, n-gram features are also important, and they confirm the established results based on English data: mostly stop words and some specific part-of-speech and character n-grams are reliable in AA based on our dataset.

5   Conclusions

We have performed experiments on cross-topic AA with a specifically designed topic-author-controlled dataset in Russian. The widely used **n-gram** features were applied, and **semantic coherence** features were introduced for the first time in AA. Traditional machine learning and distance-based algorithms were used. We have performed experiments in two scenarios: **Group AA,** a traditional setting with dozens of training and test documents, and **Pairwise AA,** a specifically designed scenario replicating a real-world forensic task. The features were evaluated for authorship significance based on the whole dataset.

The results demonstrate that we have effectively tackled the difficulties introduced by the real-world cross-topic dataset:

1. Distance-based measures allow to obtain high results on the small dataset: the **Nearest Neighbors** algorithm works well for both experiment settings, with training sample size ranging from 2 to 45 documents;
2. Content-independent **part-of-speech** and **semantic coherence** features are promising in difficult cross-topic scenarios, allowing to overcome the baseline in extremely difficult conditions of topic-authorship interference.

Certainly, the results obtained especially in the most difficult (**FalseSame**) scenario are very far from the results which could be applicable in real cases. However, one should bear in mind that this scenario is rare even in real life where investigators try to gain more texts from suspects.

Our future plans involve application of features types, including features related to discourse levels, and testing their stability in register-shift scenario, as well as estimating the volume of texts needed for high (above 0.9) level of accuracy of AA models.

Acknowledgment. Authors acknowledge support of this study by the Russian Science Foundation, grant № 18-78-10081 "Modelling of the idiolect of a modern Russian speaker in the context of the problem of authorship attribution".

References

1. Chaski, C.: The keyboard dilemma and authorship identification. In: Craiger, P., Shenoi, S. (eds.) DigitalForensics 2007. ITIFIP, vol. 242, pp. 133–146. Springer, New York (2007). https://doi.org/10.1007/978-0-387-73742-3_9

2. Corcoran, C.M., et al.: Prediction of psychosis across protocols and risk cohorts using automated language analysis. World Psychiatry 17(1), 67–75 (2018)

3. Dmitrin, Y., Botov, D., Klenin, J., Nikolaev, I.: Comparison of deep neural network architectures for authorship attribution of Russian social media texts. In: Computational Linguistics and Intellectual Technologies: Proceedings of the International Conference "Dialogue 2018" (Online articles). RSUH (2018)

4. Elvevåg, B., Foltz, P.W., Weinberger, D.R., Goldberg, T.E.: Quantifying incoherence in speech: an automated methodology and novel application to schizophrenia. Schizophr. Res. 93(1–3), 304–316 (2007)

5. Evert, S., et al.: Understanding and explaining Delta measures for authorship attribution. Digit. Sch. Hum. 32(2), ii4–ii16 (2017)

6. Gómez-Adorno, H., et al.: Hierarchical clustering analysis: the best-performing approach at PAN 2017 author clustering task. In: Bellot, P., et al. (eds.) CLEF 2018. LNCS, vol. 11018, pp. 216–223. Springer, Cham (2018). https://doi.org/10.1007/978-3-319-98932-7_20

7. Grant, T.: Txt 4n6: describing and measuring consistency and distinctiveness in the analysis of SMS text messages. J. Law Policy XXI(2), 467–494 (2013)

8. Gritta, M.: Distributional Semantics and Authorship Differences (MPhil Diss.). University of Cambridge (2015)

9. Herbelot, A., Kochmar, E.: 'Calling on the classical phone': a distributional model of adjective-noun errors in learners' English. In: Proceedings of COLING 2016, the 26th International Conference on Computational Linguistics: Technical Papers, pp. 976–986. COLING (2016)

10. Iter, D., Yoon, J., Jurafsky, D.: Automatic detection of incoherent speech for diagnosing schizophrenia. In: Proceedings of the Fifth Workshop on Computational Linguistics and Clinical Psychology: From Keyboard to Clinic, pp. 136–146. Association for Computational Linguistics (2018)

11. Juola, P.: The rowling protocol, Steven Bannon, and Rogue POTUS staff: a study in computational authorship attribution. Language and Law/Linguagem e Direito 5(2), 77–94 (2018)

12. Kestemont, M., et al.: Overview of the author identification task at PAN-2018: cross-domain authorship attribution and style change detection. In: Cappellato, L., et al. (eds.) Working Notes Papers of the CLEF 2018 Evaluation Labs, pp. 1–25. CEUR-WS.org (2018)

13. Kutuzov, A., Kuzmenko, E.: WebVectors: a toolkit for building web interfaces for vector semantic models. In: Ignatov, Dmitry I., et al. (eds.) AIST 2016. CCIS, vol. 661, pp. 155–161. Springer, Cham (2017). https://doi.org/10.1007/978-3-319-52920-2_15

14. Litvinova, T., Litvinova, O., Seredin, P.: Assessing the level of stability of idiolectal features across modes, topics and time of text production. In: 23rd Conference of Open Innovations Association: FRUCT 2018, pp. 223–230. IEEE (2018)

15. Litvinova, T., Seredin, P., Litvinova, O., Dankova, T., Zagorovskaya, O.: On the stability of some idiolectal features. In: Karpov, A., Jokisch, O., Potapova, R. (eds.) SPECOM 2018. LNCS (LNAI), vol. 11096, pp. 331–336. Springer, Cham (2018). https://doi.org/10.1007/978-3-319-99579-3_35

16. Newman, D., Lau, J.H., Grieser, K., Baldwin, T.: Automatic evaluation of topic coherence. In: The 2010 Annual Conference of the North American Chapter of the Association for Computational Linguistics: Human Language Technologies, pp. 100–108. Association for Computational Linguistics (2010)

17. Pedregosa, F., et al.: Scikit-learn: machine learning in Python. J. Mach. Learn. Res. **12**, 2825–2830 (2011)

18. Posadas-Durán, J.P., et al.: Application of the distributed document representation in the authorship attribution task for small corpora. Soft. Comput. **21**(3), 627–639 (2017)

19. Queralt, S.: The creation of Base Rate Knowledge of linguistic variables and the implementation of likelihood ratios to authorship attribution in forensic text comparison. Language and Law/Linguagem e Direito **5**(2), 59–76 (2018)

20. Rocha, A., et al.: Authorship attribution for social media forensics. IEEE Trans. Inf. Forensics Secur. **12**(1), 5–33 (2016)

21. Sapkota, U., Bethard, S., Montes, M., Solorio, T.: Not all character n-grams are created equal: a study in authorship attribution. In: Proceedings of the 2015 Conference of the North American Chapter of the Association for Computational Linguistics: Human Language Technologies, pp. 93–102. Association for Computational Linguistics (2015)

22. Segalovich, I.: A fast morphological algorithm with unknown word guessing induced by a dictionary for a web search engine. In: International Conference on Machine Learning; Models, Technologies and Applications, pp. 273–280. CSREA Press (2003)

23. Shutova, E., Kiela, D., Maillard, J.: Black holes and white rabbits: metaphor identification with visual features. In: Proceedings of the 2016 Conference of the North American Chapter of the Association for Computational Linguistics: Human Language Technologies, pp. 160–170. Association for Computational Linguistics (2016)

24. Sidorov, G., Velasquez, F., Stamatatos, E., Gelbukh, A., Chanona-Hernández, L.: Syntactic n-grams as machine learning features for natural language processing. Expert Syst. Appl. **41**(3), 853–860 (2014)

25. Soboroff, I.M., Nicholas, C.K., Kukla, J.M., Ebert, D.S.: Visualizing document authorship using n-grams and latent semantic indexing. In: Proceedings of the 1997 Workshop on New Paradigms in Information Visualization and Manipulation, pp. 43–48. ACM (1997)

26. Stamatatos, E., Fakotakis, N., Kokkinakis, G.: Computer-based authorship attribution without lexical measures. Comput. Humanit. **35**(2), 193–214 (2001)

27. Stamatatos, E.: Masking topic-related information to enhance authorship attribution. J. Assoc. Inf. Sci. Technol. **69**(3), 461–473 (2018)

28. Stamatatos, E.: On the robustness of authorship attribution based on character n-gram features. J. Law Policy **21**(2), 421–439 (2013)

29. Tschuggnall, M., et al.: Overview of the author identification task at PAN-2017: style breach detection and author clustering. In: Working Notes of CLEF 2017, CEUR Workshop Proceedings, vol. 1866. CEUR-WS.org (2017)

RaKUn: Rank-based Keyword Extraction via Unsupervised Learning and Meta Vertex Aggregation

Blaž Škrlj[1,2](✉), Andraž Repar[1,2], and Senja Pollak[2,3]

[1] Jožef Stefan International Postgraduate School, Ljubljana, Slovenia
blaz.skrlj@ijs.si, repar.andraz@gmail.com
[2] Jožef Stefan Institute, Ljubljana, Slovenia
senja.pollak@ijs.si
[3] Usher Institute, Medical School, University of Edinburgh, Edinburgh, UK

Abstract. Keyword extraction is used for summarizing the content of a document and supports efficient document retrieval, and is as such an indispensable part of modern text-based systems. We explore how load centrality, a graph-theoretic measure applied to graphs derived from a given text can be used to efficiently identify and rank keywords. Introducing meta vertices (aggregates of existing vertices) and systematic redundancy filters, the proposed method performs on par with state-of-the-art for the keyword extraction task on 14 diverse datasets. The proposed method is unsupervised, interpretable and can also be used for document visualization.

Keywords: Keyword extraction · Graph applications · Vertex ranking · Load centrality · Information retrieval

1 Introduction and Related Work

Keywords are terms (i.e. expressions) that best describe the subject of a document [2]. A good keyword effectively summarizes the content of the document and allows it to be efficiently retrieved when needed. Traditionally, keyword assignment was a manual task, but with the emergence of large amounts of textual data, automatic keyword extraction methods have become indispensable. Despite a considerable effort from the research community, state-of-the-art keyword extraction algorithms leave much to be desired and their performance is still lower than on many other core NLP tasks [13]. The first keyword extraction methods mostly followed a supervised approach [14,24,31]: they first extract keyword features and then train a classifier on a gold standard dataset. For example, KEA [31], a state of the art supervised keyword extraction algorithm is based on the Naive Bayes machine learning algorithm. While these methods offer quite good performance, they rely on an annotated gold standard dataset and require a (relatively) long training process. In contrast, unsupervised approaches need no training and can

© Springer Nature Switzerland AG 2019
C. Martín-Vide et al. (Eds.): SLSP 2019, LNAI 11816, pp. 311–323, 2019.
https://doi.org/10.1007/978-3-030-31372-2_26

be applied directly without relying on a gold standard document collection. They can be further divided into statistical and graph-based methods. The former, such as YAKE [6,7], KP-MINER [10] and RAKE [25], use statistical characteristics of the texts to capture keywords, while the latter, such as Topic Rank [3], TextRank [22], Topical PageRank [29] and Single Rank [30], build graphs to rank words based on their position in the graph. Among statistical approaches, the state-of-the-art keyword extraction algorithm is YAKE [6,7], which is also one of the best performing keyword extraction algorithms overall; it defines a set of five features capturing keyword characteristics which are heuristically combined to assign a single score to every keyword. On the other hand, among graph-based approaches, Topic Rank [3] can be considered state-of-the-art; candidate keywords are clustered into topics and used as vertices in the final graph, used for keyword extraction. Next, a graph-based ranking model is applied to assign a significance score to each topic and keywords are generated by selecting a candidate from each of the top-ranked topics. Network-based methodology has also been successfully applied to the task of topic extraction [28].

The method that we propose in this paper, RaKUn, is a graph-based keyword extraction method. We exploit some of the ideas from the area of graph aggregation-based learning, where, for example, graph convolutional neural networks and similar approaches were shown to yield high quality vertex representations by aggregating their neighborhoods' feature space [5]. This work implements some of the similar ideas (albeit not in a neural network setting), where redundant information is aggregated into meta vertices in a similar manner. Similar efforts were shown as useful for hierarchical subnetwork aggregation in sensor networks [8] and in biological use cases of simulation of large proteins [9].

The main contributions of this paper are as follows. The notion of load centrality was to our knowledge not yet sufficiently exploited for keyword extraction. We show that this fast measure offers competitive performance to other widely used centralities, such as for example the PageRank centrality (used in [22]). To our knowledge, this work is the first to introduce the notion of meta vertices with the aim of aggregating similar vertices, following similar ideas to the statistical method YAKE [7], which is considered a state-of-the-art for the keyword extraction. Next, as part of the proposed RaKUn algorithm we extend the extraction from unigrams also to bigram and threegram keywords based on load centrality scores computed for considered tokens. Last but not least, we demonstrate how arbitrary textual corpora can be transformed into weighted graphs whilst maintaining *global* sequential information, offering the opportunity to exploit potential context not naturally present in statistical methods.

The paper is structured as follows. We first present the text to graph transformation approach (Sect. 2), followed by the introduction of the RaKUn keyword extractor (Sect. 3). We continue with qualitative evaluation (Sect. 4) and quantitative evaluation (Sect. 5), before concluding the paper in Sect. 6.

2   Transforming Texts to Graphs

We first discuss how the texts are transformed to graphs, on which RaKUn operates. Next, we formally state the problem of keyword extraction and discuss its relation to graph centrality metrics.

2.1   Representing Text

In this work we consider *directed* graphs. Let $G = (V, E)$ represent a graph comprised of a set of vertices V and a set of edges ($E \subseteq V \times V$), which are ordered pairs. Further, each edge can have a real-valued weight assigned. Let $\mathcal{D}$ represent a document comprised of tokens $\{t_1, \ldots, t_n\}$. The order in which tokens in text appear is known, thus $\mathcal{D}$ is a totally ordered set. A potential way of constructing a graph from a document is by simply observing word co-occurrences. When two words co-occur, they are used as an edge. However, such approaches do not take into account the sequence nature of the words, meaning that the order is lost. We attempt to take this aspect into account as follows. The given corpus is traversed, and for each element t_i, its successor t_{i+1}, together with a given element, forms a directed edge $(t_i, t_{i+1}) \in E$. Finally, such edges are weighted according to the number of times they appear in a given corpus. Thus the graph, constructed after traversing a given corpus, consists of all local neighborhoods (order one), merged into a single joint structure. Global contextual information is potentially kept intact (via weights), even though it needs to be detected via network analysis as proposed next.

2.2   Improving Graph Quality by Meta Vertex Construction

A naïve approach to constructing a graph, as discussed in the previous section, commonly yields noisy graphs, rendering learning tasks harder. Therefore, we next discuss the selected approaches we employ in order to reduce both the computational complexity and the spatial complexity of constructing the graph, as well as increasing its quality (for the given down-stream task).

First, we consider the following heuristics which reduce the complexity of the graph that we construct for keyword extraction: Considered token length (while traversing the document $\mathcal{D}$, only tokens of length $\mu > \mu_{\min}$ are considered), and next, lemmatization (tokens can be lemmatized, offering spatial benefits and avoiding redundant vertices in the final graph). The two modifications yield a potentially "simpler" graph, which is more suitable and faster for mining.

Even if the optional lemmatization step is applied, one can still aim at further reducing the graph complexity by merging similar vertices. This step is called *meta vertex construction*. The motivation can be explained by the fact, that even similar lemmas can be mapped to the same keyword (e.g., mechanic and mechanical; normal and abnormal). This step also captures spelling errors (similar vertices that will not be handled by lemmatization), spelling differences (e.g., British vs. American English), non-standard writing (e.g., in Twitter data), mistakes in lemmatization or unavailable or omitted lemmatization step.

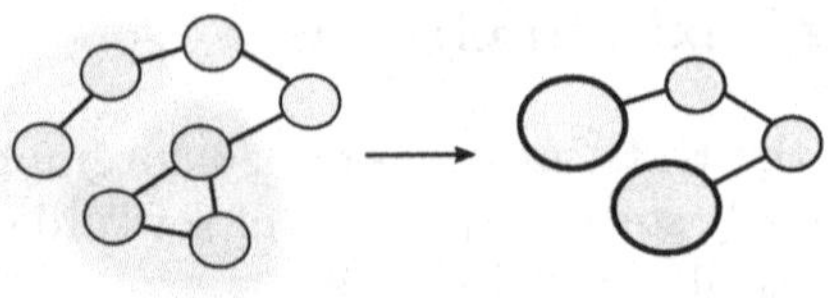

Fig. 1. Meta vertex construction. Sets of highlighted vertices are merged into a single vertex. The resulting graph has less vertices, as well as edges.

The meta-vertex construction step works as follows. Let V represent the set of vertices, as defined above. A meta vertex M is comprised of a set of vertices that are elements of V, i.e. $M \subseteq V$. Let M_i denote the i-th meta vertex. We construct a given M_i so that for each $u \in M_i$, u's initial edges (prior to merging it into a meta vertex) are rewired to the newly added M_i. Note that such edges connect to vertices which are not a part of M_i. Thus, both the number of vertices, as well as edges get reduced substantially. This feature is implemented via the following procedure:

1. Meta vertex candidate identification. Edit distance and word lengths distance are used to determine whether two words should be merged into a meta vertex (only if length distance threshold is met, the more expensive edit distance is computed).
2. The meta vertex creation. As common identifiers, we use the stemmed version of the original vertices and if there is more than one resulting stem, we select the vertex from the identified candidates that has the highest centrality value in the graph and its stemmed version is introduced as a novel vertex (meta vertex).
3. The edges of the words entailed in the meta vertex are next rewired to the meta vertex.
4. The two original words are removed from the graph.
5. The procedure is repeated for all candidate pairs.

A schematic representation of meta vertex construction is shown in Fig. 1. The yellow and blue groups of vertices both form a meta vertex, the resulting (right) graph is thus substantially reduced, both with respect to the number of vertices, as well as the number of edges.

3   Keyword Identification

Up to this point, we discussed how the graph used for keyword extraction is constructed. In this work, we exploit the notion of load centrality, a fast measure for estimating the importance of vertices in graphs. This metric can be defined as follows.

Load Centrality. The load centrality of a vertex falls under the family of centralities which are defined based on the number of shortest paths that pass through a given vertex v, i.e. $c(v) = \sum_{t \in V} \sum_{s \in V} \frac{\sigma(s,t|v)}{\sigma(s,t)}; t \neq s$, where $\sigma(s,t|v)$ represents the number of shortest paths that pass from vertex s to vertex t via v and $\sigma(s,t)$ the number of all shortest paths between s and t (see [4, 11]). The considered load centrality measure is subtly different from the better known betweenness centrality; specifically, it is assumed that each vertex sends a package to each other vertex to which it is connected, with routing based on a priority system: given an input of flow x arriving at vertex v with destination v', v divides x equally among all neighbors of minimum shortest path to the target. The total flow passing through a given v via this process is defined as v's load. Load centrality thus maps from the set of vertices V to real values. For detailed description and computational complexity analysis, see [4]. Intuitively, vertices of the graph with the highest load centrality represent key vertices in a given network. In this work, we assume such vertices are good descriptors of the input document (i.e. keywords). Thus, ranking the vertices yields a priority list of (potential) keywords.

Formulating the RaKUn Algorithm. We next discuss how the considered centrality is used as part of the whole keyword extraction algorithm RaKUn, summarized in Algorithm 1. The algorithm consists of three main steps described next. First, a graph is constructed from a given ordered set of tokens (e.g., a document) (lines 1 to 8). The resulting graph is commonly very sparse, as most of the words rarely co-occur. The result of this step is a smaller, denser graph, where both the number of vertices, as well as edges is lower. Once constructed, load centrality (line 10) is computed for each vertex. Note that at this point, should the top k vertices by centrality be considered, only single term keywords emerge. As it can be seen from line 11, to extend the selection to 2- and 3-grams, the following procedure is proposed:

2-gram keywords. Keywords comprised of two terms are constructed as follows. First, pairs of first order keywords (all tokens) are counted. If the support (= number of occurrences) is higher than f (line 11 in Algorithm 1), the token pair is considered as potential 2-gram keyword. The load centralities of the two tokens are averaged, i.e. $c_v = \frac{c_1 + c_2}{2}$, and the obtained keywords are considered for final selection along with the computed ranks.

3-gram keywords. For construction of 3-gram keywords, we follow a similar idea to that of bigrams. The obtained 2-gram keywords (previous step) are further explored as follows. For each candidate 2-gram keyword, we consider two extension scenarios: Extending the 2-gram from the left side. Here, the in-neighborhood of the left token is considered as a potential extension to a given keyword. Ranks of such candidates are computed by averaging the centrality scores in the same manner as done for the 2-gram case. Extending the 2-gram from the right side. The difference with the previous point is that all outgoing connections of the rightmost vertex are considered as potential

---

Algorithm 1. RaKUn algorithm.

Data: Document D, consisting of n tokens $t_1, \ldots, t_n$
Parameters : General: number of keywords k, minimal token length μ; Meta
          vertex parameters: edit distance threshold α, word length
          difference threshold l, Multi-word keywords parameters: path
          length p, 2-gram frequency threshold f
Result: A set of keywords $\mathcal{K}$

1 *corpusGraph* ← EmptyGraph;   ▷ Initialization.
2 **for** $t_i \in D$ **do**
3 | *edge* ← (t_i, t_{i+1});
4 | **if** *edge not in corpusGraph and len(t_i)* $\geq \mu$ **then**
5 | | add *edge* to *corpusGraph* ;   ▷ Graph construction.
6 | **end**
7 | updateEdgeWeight(*corpusGraph, edge*) ;   ▷ Weight update.
8 **end**
9 *corpusGraph* ← generateMetaVertices(*corpusGraph*, α, l);
10 tokenRanks ← loadCentrality(*corpusGraph*) ;   ▷ Initial token ranks.
11 scoredKeywords ← generateKeywords(p, f, tokenRanks) ; ▷ Keyword search.
12 $\mathcal{K}$ = scoredKeywords[:k];
13 **return** $\mathcal{K}$

---

extensions. The candidate keywords are ranked, as before, by averaging the
load centralities, i.e. $c_v = \frac{1}{3} \sum_{i=1}^{3} c_i$.

Having obtained a set of (keyword, score) pairs, we finally sort the set according to the scores (descendingly), and take top k keywords as the result. We next discuss the evaluation the proposed algorithm.

4   Qualitative Evaluation

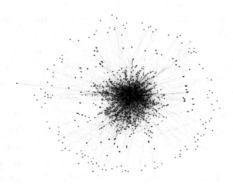

Fig. 2. Keyword visualization. Red dots represent keywords, other dots represent the remainder of the corpus graph. (Color figure online)

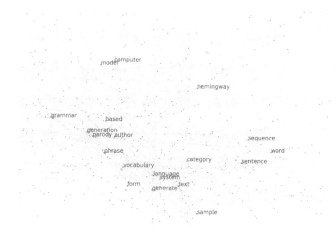

Fig. 3. Keyword visualization. A close-up view shows some examples of keywords and their location in the corpus graph. The keywords are mostly located in the central part of the graph.

RaKUn can be used also for visualization of keywords in a given document or document corpus. A visualization of extracted keywords is applied to an example from wiki20 [21] (for dataset description see Sect. 5.1), where we visualize both the global corpus graph, as well as a local (document) view where keywords are emphasized, see Figs. 2 and 3, respectively. It can be observed that the global graph's topology is far from uniform—even though we did not perform any tests of scale-freeness, we believe the constructed graphs are subject to distinct topologies, where keywords play prominent roles.

5   Quantitative Evaluation

This section discusses the experimental setting used to validate the proposed RaKUn approach against state-of-the-art baselines. We first describe the datasets, and continue with the presentation of the experimental setting and results.

5.1   Datasets

For RaKUn evaluation, we used 14 gold standard datasets from the list of [6,7], from which we selected datasets in English. Detailed dataset descriptions and statistics can be found in Table 1, while full statistics and files for download can be found online[1]. Most datasets are from the domain of computer science or contain multiple domains. They are very diverse in terms of the number of documents—ranging from *wiki20* with 20 documents to *Inspec* with 2,000

---

[1] https://github.com/LIAAD/KeywordExtractor-Datasets.

Table 1. Selection of keyword extraction datasets in English language

Dataset	Desc	No. docs	Avg. keywords	Avg. doc length
500N-KPCrowd-v1.1 [18]	Broadcast news transcriptions	500	48.92	408.33
Inspec [15]	Scientific journal papers from Computer Science collected between 1998 and 2002	2000	14.62	128.20
Nguyen2007 [23]	Scientific conference papers	209	11.33	5201.09
PubMed	Full-text papers collected from PubMed Central	500	15.24	3992.78
Schutz2008 [26]	Full-text papers collected from PubMed Central	1231	44.69	3901.31
SemEval2010 [17]	Scientific papers from the ACM Digital Library	243	16.47	8332.34
SemEval2017 [1]	500 paragraphs selected from 500 ScienceDirect journal articles, evenly distributed among the domains of Computer Science, Material Sciences and Physics	500	18.19	178.22
citeulike180 [19]	Full-text papers from the CiteULike.org	180	18.42	4796.08
fao30 [20]	Agricultural documents from two datasets based on Food and Agriculture Organization (FAO) of the UN	30	33.23	4777.70
fao780 [20]	Agricultural documents from two datasets based on Food and Agriculture Organization (FAO) of the UN	779	8.97	4971.79
kdd [12]	Abstracts from the ACM Conference on Knowledge Discovery and Data Mining (KDD) during 2004-2014	755	5.07	75.97
theses100	Full master and Ph.D. theses from the University of Waikato	100	7.67	4728.86
wiki20 [21]	Computer science technical research reports	20	36.50	6177.65
www [12]	Abstracts of WWW conference papers from 2004-2014	1330	5.80	84.08

documents, in terms of the average number of gold standard keywords per document—from 5.07 in *kdd* to 48.92 in *500N-KPCrowd-v1.1*—and in terms of the average length of the documents—from 75.97 in *kdd* to *SemEval2017* with 8332.34.

5.2 Experimental Setting

We adopted the same evaluation procedure as used for the series of results recently introduced by YAKE authors [6][2]. Five fold cross validation was used to determine the overall performance, for which we measured Precision, Recall and F1 score, with the latter being reported in Table 2.[3] Keywords were stemmed prior to evaluation.[4] As the number of keywords in the gold standard document

---

[2] We attempted to reproduce YAKE evaluation procedure based on their experimental setup description and also thank the authors for additional explanation regarding the evaluation. For comparison of results we refer to their online repository https://github.com/LIAAD/yake [7].

[3] The complete results and the code are available at https://github.com/SkBlaz/rakun.

[4] This being a standard procedure, as suggested by the authors of YAKE.

is not equal to the number of extracted keywords (in our experiments k=10), in the recall we divide the correctly extracted keywords by the number of keywords parameter k, if in the gold standard number of keywords is higher than k.

Selecting Default Configuration. First, we used a dedicated run for determining the default parameters. The cross validation was performed as follows. For each train-test dataset split, we kept the documents in the test fold intact, whilst performing a grid search on the train part to find the best parametrization. Finally, the selected configuration was used to extract keywords on the unseen test set. For each train-test split, we thus obtained the number of true and false positives, as well as true and false negatives, which were summed up and, after all folds were considered, used to obtain final F1 scores, which served for default parameter selection. The grid search was conducted over the following parameter range Num keywords: 10, Num tokens (the number of tokens a keyword can consist of): Count threshold (minimum support used to determine potential bigram candidates): Word length difference threshold (maximum difference in word length used to determine whether a given pair of words shall be aggregated): [0, 2, 4], Edit length difference (maximum edit distance allowed to consider a given pair of words for aggregation): [2, 3], Lemmatization: [yes, no].

Even if one can use the described grid-search fine-tuning procedure to select the best setting for individual datasets, we observed that in nearly all the cases the best settings were the same. We therefore selected it as the *default*, which can be used also on new unlabeled data. The default parameter setting was as follows. The number of tokens was set to 1, Count threshold was thus not needed (only unigrams), for meta vertex construction Word length difference threshold was set to 3 and Edit distance to 2. Words were initially lemmatized. Next, we report the results using these selected parameters (same across all datasets), by which we also test the general usefulness of the approach.

5.3   Results

The results are presented in Table 2, where we report on F1 with the default parameter setting of RaKUn, together with the results from related work, as reported in the github table of the YAKE [7][5]. We first observe that on the selection of datasets, the proposed RaKUn wins more than any other method. We also see that it performs notably better on some of the datasets, whereas on the remainder it performs worse than state-of-the-art approaches. Such results demonstrate that the proposed method finds keywords differently, indicating load centrality, combined with meta vertices, represents a promising research venue. The datasets, where the proposed method outperforms the current state-of-the-art results are: *500N-KPCrowd-v1.1*, *Schutz2008*, *fao30* and *wiki20*. In addition, RaKUn also achieves competitive results on *citeulike180*. A look at the

---

[5] https://github.com/LIAAD/yake/blob/master/docs/YAKEvsBaselines.jpg (accessed on: June 11, 2019).

gold standard keywords in these datasets reveals that they contain many single-word units which is why the default configuration (which returns unigrams only) was able to perform so well.

Table 2. Performance comparison with state-of-the-art approaches.

Dataset	RaKUn	YAKE	Single Rank	KEA	KP-MINER	Text Rank	Topic Rank	Topical PageRank
500N-KPCrowd-v1.1	**0.428**	0.173	0.157	0.159	0.093	0.111	0.172	0.158
Inspec	0.054	0.316	**0.378**	0.150	0.047	0.098	0.289	0.361
Nguyen2007	0.096	0.256	0.158	0.221	**0.314**	0.167	0.173	0.148
PubMed	0.075	0.106	0.039	**0.216**	0.114	0.071	0.085	0.052
Schutz2008	**0.418**	0.196	0.086	0.182	0.230	0.118	0.258	0.123
SemEval2010	0.091	0.211	0.129	0.215	**0.261**	0.149	0.195	0.125
SemEval2017	0.112	0.329	**0.449**	0.201	0.071	0.125	0.332	0.443
citeulike180	0.250	0.256	0.066	**0.317**	0.240	0.112	0.156	0.072
fao30	**0.233**	0.184	0.066	0.139	0.183	0.077	0.154	0.107
fao780	0.094	**0.187**	0.085	0.114	0.174	0.083	0.137	0.108
kdd	0.046	**0.156**	0.085	0.063	0.036	0.050	0.055	0.089
theses100	0.069	0.111	0.060	0.104	**0.158**	0.058	0.114	0.083
wiki20	**0.190**	0.162	0.038	0.134	0.156	0.074	0.106	0.059
www	0.060	**0.172**	0.097	0.072	0.037	0.059	0.067	0.101
#Wins	4	3	2	2	3	0	0	0

Four of these five datasets (*500N-KPCrowd-v1.1, Schutz2008, fao30, wiki20*) are also the ones with the highest average number of keywords per document with at least 33.23 keywords per document, while the fifth dataset (*citeulike180*) also has a relatively large value (18.42). Similarly, four of the five well-performing datasets (*Schutz2008, fao30, citeulike180, wiki20*) include long documents (more than 3,900 words), with the exception being *500N-KPCrowd-v1.1*. For details, see Table 1. We observe that the proposed RaKUn outperforms the majority of other competitive graph-based methods. For example, the most similar variants Topical PageRank and TextRank do not perform as well on the majority of the considered datasets. Furthermore, RaKUn also outperforms KEA, a supervised keyword learner (e.g., very high difference in performance on 500N-KPCrowd-v1.1 and Schutz2008 datasets), indicating unsupervised learning from the graph's structure offers a more robust keyword extraction method than learning a classifier directly.

6   Conclusions and Further Work

In this work we proposed RaKUn, a novel unsupervised keyword extraction algorithm which exploits the efficient computation of load centrality, combined with the introduction of meta vertices, which notably reduce corpus graph sizes. The method is fast, and performs well compared to state-of-the-art such as YAKE and graph-based keyword extractors. In further work, we will test the method on other languages. We also believe additional semantic background knowledge information could be used to prune the graph's structure even further, and potentially introduce keywords that are inherently not even present in the text (cf. [27]). The proposed method does not attempt to exploit meso-scale graph structure, such as convex skeletons or communities, which are known to play prominent roles in real-world networks and could allow for vertex aggregation based on additional graph properties. We believe the proposed method could also be extended using the Ricci-Oliver [16] flows on weighted graphs.

Acknowledgements. The work was supported by the Slovenian Research Agency through a young researcher grant [BŠ], core research programme (P2-0103), and projects Semantic Data Mining for Linked Open Data (N2-0078) and Terminology and knowledge frames across languages (J6-9372). This work was supported also by the EU Horizon 2020 research and innovation programme, Grant No. 825153, EMBED-DIA (Cross-Lingual Embeddings for Less-Represented Languages in European News Media). The results of this publication reflect only the authors' views and the EC is not responsible for any use that may be made of the information it contains. We also thank the authors of YAKE for their clarifications.

References

1. Augenstein, I., Das, M., Riedel, S., Vikraman, L., McCallum, A.: Semeval 2017 task 10: Scienceie - extracting keyphrases and relations from scientific publications. CoRR abs/1704.02853 (2017)
2. Beliga, S., Meštrović, A., Martinčić-Ipšić, S.: An overview of graph-based keyword extraction methods and approaches. J. Inf. Organ. Sci. **39**(1), 1–20 (2015)
3. Bougouin, A., Boudin, F., Daille, B.: Topicrank: graph-based topic ranking for keyphrase extraction. In: International Joint Conference on Natural Language Processing (IJCNLP), pp. 543–551 (2013)
4. Brandes, U.: On variants of shortest-path betweenness centrality and their generic computation. Soc. Netw. **30**(2), 136–145 (2008)
5. Cai, H., Zheng, V.W., Chang, K.C.C.: A comprehensive survey of graph embedding: Problems, techniques, and applications. IEEE Trans. Knowl. Data Eng. **30**(9), 1616–1637 (2018)
6. Campos, R., Mangaravite, V., Pasquali, A., Jorge, A.M., Nunes, C., Jatowt, A.: A text feature based automatic keyword extraction method for single documents. In: Pasi, G., Piwowarski, B., Azzopardi, L., Hanbury, A. (eds.) Advances in Information Retrieval, pp. 684–691. Springer International Publishing, Cham (2018)
7. Campos, R., Mangaravite, V., Pasquali, A., Jorge, A.M., Nunes, C., Jatowt, A.: YAKE! collection-independent automatic keyword extractor. In: Pasi, G., Piwowarski, B., Azzopardi, L., Hanbury, A. (eds.) ECIR 2018. LNCS, vol. 10772, pp. 806–810. Springer, Cham (2018). https://doi.org/10.1007/978-3-319-76941-7_80
8. Chan, H., Perrig, A., Song, D.: Secure hierarchical in-network aggregation in sensor networks. In: Proceedings of the 13th ACM Conference On Computer And Communications Security, pp. 278–287. ACM (2006)
9. Doruker, P., Jernigan, R.L., Bahar, I.: Dynamics of large proteins through hierarchical levels of coarse-grained structures. J. comput. chem. **23**(1), 119–127 (2002)

10. El-Beltagy, S.R., Rafea, A.: Kp-miner: a keyphrase extraction system for english and arabic documents. Inf. SysT. **34**(1), 132–144 (2009)
11. Goh, K.I., Kahng, B., Kim, D.: Universal behavior of load distribution in scale-free networks. Phys. Rev. Lett. **87**, 278701 (2001)
12. Gollapalli, S.D., Caragea, C.: Extracting keyphrases from research papers using citation networks. In: Twenty-Eighth AAAI Conference on Artificial Intelligence (2014)
13. Hasan, K.S., Ng, V.: Automatic keyphrase extraction: a survey of the state of the art. In: Proceedings of the 52nd Annual Meeting of the Association for Computational Linguistics (Volume 1: Long Papers), vol. 1, pp. 1262–1273 (2014)
14. Hulth, A.: Improved automatic keyword extraction given more linguistic knowledge. In: Proceedings of the 2003 Conference on Empirical Methods in Natural Language Processing, pp. 216–223. Association for Computational Linguistics (2003)
15. Hulth, A.: Improved automatic keyword extraction given more linguistic knowledge. In: Proceedings of the 2003 Conference on Empirical Methods in Natural Language Processing, EMNLP 2003, pp. 216–223 (2003)
16. Jin, M., Kim, J., Gu, X.D.: Discrete surface ricci flow: theory and applications. In: Martin, R., Sabin, M., Winkler, J. (eds.) Mathematics of Surfaces 2007. LNCS, vol. 4647, pp. 209–232. Springer, Heidelberg (2007). https://doi.org/10.1007/978-3-540-73843-5_13
17. Kim, S.N., Medelyan, O., Kan, M.Y., Baldwin, T.: Semeval-2010 task 5: automatic keyphrase extraction from scientific articles. In: Proceedings of the 5th International Workshop on Semantic Evaluation, pp. 21–26. SemEval 2010 (2010)
18. Marujo, L., Viveiros, M., da Silva Neto, J.P.: Keyphrase cloud generation of broadcast news. CoRR abs/1306.4606 (2013)
19. Medelyan, O., Frank, E., Witten, I.H.: Human-competitive tagging using automatic keyphrase extraction. In: Proceedings of the 2009 Conference on Empirical Methods in Natural Language Processing, EMNLP 2009, vol. 3, pp. 1318–1327 (2009)
20. Medelyan, O., Witten, I.H.: Domain-independent automatic keyphrase indexing with small training sets. J. Am. Soc. Inf. Sci. Technol. **59**(7), 1026–1040 (2008)
21. Medelyan, O., Witten, I.H., Milne, D.: Topic indexing with wikipedia. In: Proceedings of the AAAI WikiAI Workshop, vol. 1, pp. 19–24 (2008)
22. Mihalcea, R., Tarau, P.: Textrank: Bringing order into text. In: Proceedings of the 2004 Conference On Empirical Methods in Natural Language Processing (2004)
23. Nguyen, T.D., Kan, M.-Y.: Keyphrase extraction in scientific publications. In: Goh, D.H.-L., Cao, T.H., Sølvberg, I.T., Rasmussen, E. (eds.) ICADL 2007. LNCS, vol. 4822, pp. 317–326. Springer, Heidelberg (2007). https://doi.org/10.1007/978-3-540-77094-7_41
24. Nguyen, T.D., Luong, M.T.: Wingnus: keyphrase extraction utilizing document logical structure. In: Proceedings of the 5th international workshop on semantic evaluation, pp. 166–169. Association for Computational Linguistics (2010)
25. Rose, S., Engel, D., Cramer, N., Cowley, W.: Automatic keyword extraction from individual documents. In: Text mining: Applications and Theory, pp. 1–20 (2010)
26. Schutz, A.T., et al.: Keyphrase extraction from single documents in the open domain exploiting linguistic and statistical methods. Master's thesis, National University of Ireland (2008)
27. Škrlj, B., Kralj, J., Lavrač, N., Pollak, S.: Towards robust text classification with semantics-aware recurrent neural architecture. Mach. Learn. Knowl. Extr. **1**(2), 575–589 (2019)

28. Spitz, A., Gertz, M.: Entity-centric topic extraction and exploration: a network-based approach. In: Pasi, G., Piwowarski, B., Azzopardi, L., Hanbury, A. (eds.) ECIR 2018. LNCS, vol. 10772, pp. 3–15. Springer, Cham (2018). https://doi.org/10.1007/978-3-319-76941-7_1

29. Sterckx, L., Demeester, T., Deleu, J., Develder, C.: Topical word importance for fast keyphrase extraction. In: Proceedings of the 24th International Conference on World Wide Web, pp. 121–122. ACM (2015)

30. Wan, X., Xiao, J.: Single document keyphrase extraction using neighborhood knowledge. In: AAAI, vol. 8, pp. 855–860 (2008)

31. Witten, I.H., Paynter, G.W., Frank, E., Gutwin, C., Nevill-Manning, C.G.: KEA: Practical automated keyphrase extraction. In: Design and Usability of Digital Libraries: Case Studies in the Asia Pacific, pp. 129–152. IGI Global (2005)

Author Index

Printed in the United States
By Bookmasters